Technological Advancements and the Impact of Actor-Network Theory

Arthur Tatnall
Victoria University, Australia

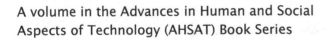

A volume in the Advances in Human and Social
Aspects of Technology (AHSAT) Book Series

Managing Director:	Lindsay Johnston
Production Editor:	Jennifer Yoder
Development Editor:	Erin O'Dea
Acquisitions Editor:	Kayla Wolfe
Typesetter:	Kaitlyn Kulp
Cover Design:	Jason Mull

Published in the United States of America by
Information Science Reference (an imprint of IGI Global)
701 E. Chocolate Avenue
Hershey PA 17033
Tel: 717-533-8845
Fax: 717-533-8661
E-mail: cust@igi-global.com
Web site: http://www.igi-global.com

                    Library of Congress Cataloging-in-Publication Data

Technological advancements and the impact of actor-network theory / Arthur Tatnall, editor.
     pages cm
  Includes bibliographical references and index.
    ISBN 978-1-4666-6126-4 (hardcover) -- ISBN 978-1-4666-6127-1 (ebook) -- ISBN 978-1-4666-6129-5 (print & perpetual access) 1. Technological innovations--Social aspects. 2. Information technology--Social aspects. 3. Actor-network theory. 4. Human-computer interaction. I. Tatnall, Arthur.
  HM846.T4355 2014
  303.01--dc23
                            2014013039

This book is published in the IGI Global book series Advances in Human and Social Aspects of Technology (AHSAT) (ISSN: 2328-1316; eISSN: 2328-1324)

British Cataloguing in Publication Data
A Cataloguing in Publication record for this book is available from the British Library.

For electronic access to this publication, please contact: eresources@igi-global.com.

Advances in Human and Social Aspects of Technology (AHSAT) Book Series

Ashish Dwivedi
The University of Hull, UK

ISSN: 2328-1316
EISSN: 2328-1324

MISSION

In recent years, the societal impact of technology has been noted as we become increasingly more connected and are presented with more digital tools and devices. With the popularity of digital devices such as cell phones and tablets, it is crucial to consider the implications of our digital dependence and the presence of technology in our everyday lives.

The **Advances in Human and Social Aspects of Technology (AHSAT) Book Series** seeks to explore the ways in which society and human beings have been affected by technology and how the technological revolution has changed the way we conduct our lives as well as our behavior. The AHSAT book series aims to publish the most cutting-edge research on human behavior and interaction with technology and the ways in which the digital age is changing society.

COVERAGE

- Activism & ICTs
- Computer-Mediated Communication
- Cultural Influence of ICTs
- Cyber Behavior
- End-User Computing
- Gender & Technology
- Human-Computer Interaction
- Information Ethics
- Public Access to ICTs
- Technoself

IGI Global is currently accepting manuscripts for publication within this series. To submit a proposal for a volume in this series, please contact our Acquisition Editors at Acquisitions@igi-global.com or visit: http://www.igi-global.com/publish/.

Titles in this Series

For a list of additional titles in this series, please visit: www.igi-global.com

Evolving Issues Surrounding Technoethics and Society in the Digital Age
Rocci Luppicini (University of Ottawa, Canada)
Information Science Reference • copyright 2014 • 349pp • H/C (ISBN: 9781466661226) • US $215.00 (our price)

Technological Advancements and the Impact of Actor-Network Theory
Arthur Tatnall (Victoria University, Australia)
Information Science Reference • copyright 2014 • 320pp • H/C (ISBN: 9781466661264) • US $195.00 (our price)

Gender Considerations and Influence in the Digital Media and Gaming Industry
Julie Prescott (University of Bolton, UK) and Julie Elizabeth McGurren (Codemasters, UK)
Information Science Reference • copyright 2014 • 313pp • H/C (ISBN: 9781466661424) • US $195.00 (our price)

Political Campaigning in the Information Age
Ashu M. G. Solo (Maverick Technologies America Inc., USA)
Information Science Reference • copyright 2014 • 359pp • H/C (ISBN: 9781466660625) • US $210.00 (our price)

Handbook of Research on Political Activism in the Information Age
Ashu M. G. Solo (Maverick Technologies America Inc., USA)
Information Science Reference • copyright 2014 • 498pp • H/C (ISBN: 9781466660663) • US $275.00 (our price)

Interdisciplinary Applications of Agent-Based Social Simulation and Modeling
Diana Francisca Adamatti (Universidade Federal do Rio Grande, Brasil) Graçaliz Pereira Dimuro (Universidade Federal do Rio Grande, Brasil) and Helder Coelho (Universidade de Lisboa, Portugal)
Information Science Reference • copyright 2014 • 376pp • H/C (ISBN: 9781466659544) • US $225.00 (our price)

Examining Paratextual Theory and its Applications in Digital Culture
Nadine Desrochers (Université de Montréal, Canada) and Daniel Apollon (University of Bergen, Norway)
Information Science Reference • copyright 2014 • 419pp • H/C (ISBN: 9781466660021) • US $215.00 (our price)

Global Issues and Ethical Considerations in Human Enhancement Technologies
Steven John Thompson (Johns Hopkins University, USA)
Medical Information Science Reference • copyright 2014 • 322pp • H/C (ISBN: 9781466660106) • US $215.00 (our price)

www.igi-global.com

701 E. Chocolate Ave., Hershey, PA 17033
Order online at www.igi-global.com or call 717-533-8845 x100
To place a standing order for titles released in this series, contact: cust@igi-global.com
Mon-Fri 8:00 am - 5:00 pm (est) or fax 24 hours a day 717-533-8661

Table of Contents

Detailed Table of Contents

Chapter 1

The term "Learning Difficulties" (sometimes also referred to as Special Needs) is used in reference to students who have significant difficulties in the acquisition of literacy and numeracy skills and need extra assistance with schooling. This is a large heterogeneous group. Another more specific term is "Learning Disabilities" that refers to the small sub-group of students who exhibit severe and unexplained problems. This chapter presents a report on an investigation, framed by the use of actor-network theory, of how the use of Information and Communications Technologies can aid in improving the education of students with Learning Disabilities. The study involved case studies and participant observation of the use of ICT in two outer Melbourne suburban Special Schools and an investigation of the impact of Education Department policies on these school environments. Research at the two Special Schools revealed that use of Information and Communications Technologies can have a very beneficial impact on these students by improving their self-esteem and facilitating their acquisition of useful life skills.

Chapter 2

Australia has designed, developed, and now implemented its national e-health solution known as the Personally Controlled Electronic Healthcare Record (PCEHR). This is a unique system as it subscribes to a shared governance model between patients and providers. To date, though, as with other national e-health solutions, there is poor uptake and much concern regarding the success of this multi-million dollar project. The authors contend that while these implementations and adoptions of e-health solutions are necessary, it is essential that an appropriate lens of analysis should be used in order to maximise and sustain the benefits of Information Systems/Information Technology (IS/IT) in healthcare delivery. Hence, in this chapter, the authors proffer Actor-Network Theory (ANT) as an appropriate lens to evaluate these various e-health solutions and illustrate in the context of the Personally Controlled Electronic Health Record (PCEHR), the chosen e-health solution for Australia.

    Maryam Sharifzadeh, Yasouj University, Iran
    Gholam Hossein Zamani, Shiraz University, Iran
    Ezatollah Karami, Shiraz University, Iran
    Davar Khalili, Shiraz University, Iran
    Arthur Tatnall, Victoria University, Australia

This chapter employed an interdisciplinary attempt to investigate agricultural climate information use, linking sociology of translation (actor-network theory) and actor analysis premises in a qualitative research design. The research method used case study approaches and purposively selected a sample consisting of wheat growers of the Fars province of Iran, who are known as contact farmers. Concepts from Actor-Network Theory (ANT) have been found to provide a useful perspective on the description and analysis of the cases. The data were analyzed using a combination of an Actor-Network Theory (ANT) framework and the Dynamic Actor-Network Analysis (DANA) model. The findings revealed socio political (farmers' awareness, motivation, and trust) and information processing factors (accuracy of information, access to information, and correspondence of information to farmers' condition) as the key elements in facilitating climate information use in farming practices.

    Rennie Naidoo, University of Pretoria, South Africa
    Awie Leonard, University of Pretoria, South Africa

This chapter extends existing metaphors used to conceptualise the unique features of contemporary IT artifacts. Some of these artifacts are innately complex, and current conceptualisations dominated by a "black box" metaphor seem to be too limited to further advance theory and offer practical design prescriptions. Using empirical material drawn from a longitudinal case study of an Internet-based self-service technology implementation, this chapter analyses various aspects of an artifact's fluidity. Post-actor network theory concepts are used to analyse the artifact's varying identities, its vague boundaries, its unexpected usage patterns, and its resourceful designers. The successes and failures of the artifact, its complex and elusive relations, and the unintended ways user practices emerged, are also analysed. This chapter contributes by extending orthodox metaphors that overemphasise a stable and enduring IT artifact—metaphors that conceal the increasingly unpredictable and transitory nature of IT artifacts—with the distinctive characteristics of fluidity. Several prescriptions for the design and management of fluid IT artifacts are offered.

The WHO has labelled diabetes the silent epidemic. This is because the instances of diabetes worldwide continue to grow exponentially. In fact, by 2030 it is expected that there will be a 54% global increase. Thus, it behooves all to focus on solutions that can result in superior management of this disease. Hence, this chapter presents findings from a longitudinal exploratory case study that examined the application of a pervasive technology solution, a mobile phone to provide superior diabetes self-care. Notably, the benefits of a pervasive technology solution for supporting superior self-care in the context of chronic disease are made especially apparent when viewed through the rich lens of Actor-Network Theory (ANT), and thus, the chapter underscores the importance of using ANT in such contexts to facilitate a deeper understanding of all potential advantages.

Chapter 6

Organisational reliance on Information and Communication Technology (ICT) continues to increase. This is informed and triggered by the premise that ICT will help them to yield solutions that will fulfil or exceed their expectations, thereby making the organisation realise the required return on investment. In order to realise return on ICT investment, many organisations deploy ICT solutions through projects. However, not all ICT projects realise their goals and objectives, due to associated risks. Unfortunately, risks are never easy to identify or managed. This chapter explores and examines the risks factors in the deployment of ICT projects in organisations. Using the case study method, the research employs actor-network theory in the analysis of the data to understand the factors that manifest themselves into risks during the deployment of ICT projects in organisations. The study reveals that factors, such as knowledge base, performance contract, and communicative structure, are used to enable and support and at the same time to constrain the deployment of ICT projects in organisations.

Public sector institutions continue to significantly invest in Information and Communication Technology (ICT) as a solution for many of their service provision challenges, for example, greater efficiency and quality of services. However, what has come to light is that there is a lack of research on understanding the contributory value or "success" of technological innovations. This chapter introduces a socio-technical view of public service innovation. The aim of this research is to extend on the notion of bureaucracy, which is traditionally focused on the politics of office environments. This socio-technical view extends this traditional view to include the politics of service networks, particularly within IT-enabled public service innovation. The chapter focuses on how service innovation is exploited to align specific interests through the process of translation and shifts the focus from value co-creation to value co-enactment. In essence, this chapter explains how public service technological innovations act as an agent of bureaucracy that alters the relational dynamics of power, risk, responsibility, and accountability. For demonstrative purposes, this chapter describes a case study that examines IT-enabled service innovation with an academic service environment.

This chapter focuses on two specific e-health solutions, the PCEHR in Australia and the German EHC. National e-health solutions are being developed by most if not all OECD countries, but few studies compare and contrast these solutions to uncover the true benefits and critical success criteria. The chapter provides an assessment of these two solutions, the possibility for any lessons learnt with regard to designing and implementing successful and appropriate e-health solutions, as well as understanding the major barriers and facilitators that must be addressed. Finally, ANT is used to provide a rich lens to investigate the key issues in these respective e-health solutions.

Software is intended to enable and support organisations, in order for them to function effectively and efficiently, towards achieving their strategic objectives. Hence, its deployment is critically vital to the focal actors (managers and sponsors). Software deployment involves two primary components, technology and non-technology actors. Both actors offer vital contributions to software deployment from different perspectives. Unfortunately, there has been more focus on the technological actors over the years. This could be attributed to why the same types of challenges, such as disjoint between development and implementation, persist. This chapter holistically examines the roles of non-technology actors in the deployment of software in organisations. The Actor-Network Theory (ANT) provides a lens in the analysis of the empirical data.

Approaches to innovation adoption often fail to explain why similar technologies in a single environment can have very different adoption outcomes. In this chapter, the single environment of education management systems in one country (Australia) are used to show how outcomes of similar technologies can be very different. An Actor-Network approach is used to explain how some technologies succeeded and others failed. Understandings reached in this case illuminate the power of the approach that includes listening to the technological actors in addition to the human. The chapter identifies actors and interactions and shows the connection between those interactions and the final outcomes of the innovations.

Chapter 11

Feminist methodologies and Actor-Network Theory (ANT) have often been considered opposing theoretical and intellectual traditions. This chapter imagines a conversation between these seemingly divergent fields and considers the theoretical and methodological challenges that ANT and particular branches of feminist thought raise for the other. This chapter examines an empirical project that calls for an engagement with both ANT and feminist methodologies. Through the lens of this empirical project, four methodological questions are considered, which an alliance between ANT and feminist methodologies would raise for any research project: 1) Where do we start our analysis? 2) Which actors should we follow? 3) What can we see when we begin to follow the actors? 4) What about politics? The potential places where ANT and feminist methodologies can meet and mutually shape research on scientific practice and technological innovation are explored. In doing so, this chapter moves towards envisioning new intersections between feminist methodologies and ANT.

Chapter 12

Healthcare delivery continues to be challenged in all OECD countries. To address these challenges, most are turning their attention to e-health as the panacea. Indeed, it is true that in today's global and networked world, e-health should be the answer for ensuring pertinent information, relevant data, and germane knowledge anywhere anytime so that clinicians can deliver superior healthcare. Sadly, healthcare has yet to realize the full potential of e-health, which is in stark contrast to other e-business initiatives such as e-government and e-education, e-finance, or e-commerce. This chapter asserts that it is only by embracing a rich theoretical lens of analysis that the full potential of e-health can be harnessed, and thus, it proffers Actor-Network Theory (ANT) as such a lens.

    Mohini Singh, RMIT University, Australia
    Jayan Kurian, RMIT University, Vietnam

This chapter analyses elements of social networking sites to establish how a combination of heterogeneous elements of technology, media, language, users, data, and information are networked together to provide this new communication media. Social networking sites are also referred to as social media sites, which can be explained using the Actor-Network Theory. Social networking sites have clearly achieved widespread adoption as a new means of communication in a very short time around the globe. An analysis of literature on social networking sites is included in this chapter to reflect the new social networking language and style, the content shared via this media, the mode of use, and the language used for communication, which is a combination of a number of technological and social entities. This chapter explains how the Actor-Network Theory (ANT) can be used to explain social networking and includes some issues for research on this topic.

    Fernando Abreu Gonçalves, CEG-IST, Portugal
    José Figueiredo, CEG-IST/DEG, Portugal & Technical University of Lisbon, Portugal

Most references to innovation relate to the development of new products. In this chapter, the authors do not address innovation in these terms; instead, they address them as changes in practices an engineer creatively adopts during engineering design projects. They adopt Actor-Network Theory as a way to understand these change processes (translations). The authors design a perturbation index inspired in Earned Value management to measure translation effort, having in mind the management of scope. Then they assess changes of regime in resource allocation of tasks and conclude some changes can lead to innovative results. That means we gain a wider view about scope, and scope management, being able to observe and change good practices, something crucial in engineering design projects where requirements and goals drift.

    Arthur Adamopoulos, RMIT University, Australia
    Martin Dick, RMIT University, Australia
    Bill Davey, RMIT University, Australia

Often actor-network theory studies find that technology has been translated through its relationships with human actors. This chapter reports on a study of online investing that found that the human actors were translated to more active and involved investors due to the changes, over time, in the online services that are available: the non-human actors. The Internet is a constantly evolving technological actor. New tools have the potential to change interactions with users. In this study, it became evident that new services had a noticeable effect on the behaviour of investors. Not only did investors report changes in their behaviour when they moved from offline to online investing, but they also reported changes in their investing strategies over time as new services became available. This study showed a new and interesting confirmation of the value of allowing non-human actors to be heard.

Preface

Actor-network theory (ANT) plays an important role in modern society, where technology is a natural extension of individuals and businesses. Understanding the interrelated nature of social, technical, professional, and personal networks is crucial to successfully navigating the modern information society landscape. *Technological Advancements and the Impact of Actor-Network Theory* applies concepts of socio-technical research to examine the implications and effects of human-non-human interactions and relationships. The chapters in this book will provide readers with a more thorough understanding of ANT, including fundamental principles, case studies, and emerging theories in the discipline, contributing fresh ideas to the body of knowledge on socio-technical phenomena and technological innovation.

ORGANIZATION OF THE BOOK

The first chapter, "The Impact of ICT in Educating Students with Learning Disabilities in Australian Schools: An ANT Approach" presents a report on an investigation, framed by the use of actor-network theory, of how the use of Information and Communications Technologies can aid in improving the education of students with Learning Disabilities. The study involved case studies and participant observation of the use of ICT in two outer Melbourne suburban Special Schools and an investigation of the impact of Education Department policies on these school environments.

Chapter 2 covers Australia's national e-health solution known as the Personally Controlled Electronic Healthcare Record (PCEHR). The authors contend that while implementations and adoptions of e-health solutions are necessary, it is essential that an appropriate lens of analysis should be used in order to maximize and sustain the benefits of Information Systems/Information Technology (IS/IT) in healthcare delivery. As such, the authors of this chapter proffer Actor-Network Theory (ANT) as an appropriate lens to evaluate these various e-health solutions and illustrate, in the context of the Personally Controlled Electronic Health Record (PCEHR), the chosen e-health solution for Australia.

"Climate Information Use: An Actor-Network Theory Perspective" employs an interdisciplinary attempt to investigate agricultural climate information use, linking sociology of translation (actor-network theory) and actor analysis premises in a qualitative research design. The research method uses case study approaches and purposively selected a sample consisting of wheat growers of the Fars province of Iran, who are known as contact farmers. The data is analyzed using a combination of an Actor-Network Theory (ANT) framework and the Dynamic Actor-Network Analysis (DANA) model. The findings reveal socio political (farmers' awareness, motivation, and trust) and information processing factors (accuracy of information, access to information, and correspondence of information to farmers' condition) as the key elements in facilitating climate information use in farming practices.

Chapter 4 extends existing metaphors used to conceptualize the unique features of contemporary IT artifacts. Some of these artifacts are innately complex, and current conceptualizations dominated by a "black box" metaphor seem to be too limited to further advance theory and offer practical design prescriptions. Using empirical material drawn from a longitudinal case study of an Internet-based self-service technology implementation, this chapter analyzes various aspects of an artifact's fluidity. Post-actor-network theory concepts are used to analyze the artifact's varying identities, its vague boundaries, its unexpected usage patterns, and its resourceful designers. The successes and failures of the artifact, its complex and elusive relations, and the unintended ways user practices emerged, are also analyzed. This chapter contributes by extending orthodox metaphors that overemphasize a stable and enduring IT artifact—metaphors that conceal the increasingly unpredictable and transitory nature of IT artifacts—with the distinctive characteristics of fluidity. Several prescriptions for the design and management of fluid IT artifacts are offered.

Chapter 5 presents findings from a longitudinal exploratory case study that examined the application of a pervasive technology solution: a mobile phone to provide superior diabetes self-care. Notably, the benefits of a pervasive technology solution for supporting superior self-care in the context of chronic disease are made especially apparent when viewed through the rich lens of Actor-Network Theory (ANT), and thus, the chapter underscores the importance of using ANT in such contexts to facilitate a deeper understanding of all potential advantages.

"Information and Communications Technology Projects and the Associated Risks" explores and examines the risk factors in the deployment of ICT projects in organizations. Using the case study method, the research employs actor-network theory in the analysis of the data to understand the factors that manifest themselves into risks during the deployment of ICT projects in organizations. The study reveals that factors, such as knowledge base, performance contract, and communicative structure, are used to enable and support and at the same time to constrain the deployment of ICT projects in organizations.

Chapter 7 introduces a socio-technical view of public service innovation. The aim of this research is to extend on the notion of bureaucracy, which is traditionally focused on the politics of office environments. This socio-technical view extends the traditional view to include the politics of service networks, particularly within IT-enabled public service innovation. The chapter focuses on how service innovation is exploited to align specific interests through the process of translation and shifts the focus from value co-creation to value co-enactment. In essence, this chapter explains how public service technological innovations act as an agent of bureaucracy that alters the relational dynamics of power, risk, responsibility, and accountability. For demonstrative purposes, this chapter describes a case study that examines IT-enabled service innovation with an academic service environment.

Chapter 8 focuses on two specific e-health solutions: the PCEHR in Australia and the German EHC. National e-health solutions are being developed by most if not all OECD countries, but few studies compare and contrast these solutions to uncover the true benefits and critical success criteria. The chapter provides an assessment of these two solutions, the possibility for any lessons learnt with regard to designing and implementing successful and appropriate e-health solutions, as well as understanding the major barriers and facilitators that must be addressed. Finally, ANT is used to provide a rich lens to investigate the key issues in these respective e-health solutions.

Chapter 9, "Human Interactions in Software Deployment: A Case of a South African Telecommunication" holistically examines the roles of non-technology actors in the deployment of software in organizations. The Actor-Network Theory (ANT) provides a lens in the analysis of the empirical data.

In the next chapter, "School Management Software in Australia and the Issue of Technological Adoption," the single environment of education management systems in one country (Australia) is used to show how outcomes of similar technologies can be very different. An actor-network approach is used to explain how some technologies succeeded and others failed. Understandings reached in this case illuminate the power of the approach that includes listening to the technological actors in addition to the human. This chapter identifies actors and interactions and shows the connection between those interactions and the final outcomes of the innovations.

Chapter 11 imagines a conversation between the seemingly divergent fields of feminist methodologies and actor-network theory, and considers the theoretical and methodological challenges that ANT and particular branches of feminist thought raise for the other. This chapter examines an empirical project that calls for an engagement with both ANT and feminist methodologies. Through the lens of this empirical project, four methodological questions are considered, which an alliance between ANT and feminist methodologies would raise for any research project: 1) Where do we start our analysis? 2) Which actors should we follow? 3) What can we see when we begin to follow the actors? 4) What about politics? The potential places where ANT and feminist methodologies can meet and mutually shape research on scientific practice and technological innovation are explored. In doing so, this chapter moves towards envisioning new intersections between feminist methodologies and ANT.

Chapter 12 considers the challenges that OECD countries face in the delivery of healthcare. To address these challenges, most are turning their attention to e-health as the panacea. Indeed, it is true that in today's global and networked world, e-health should be the answer for ensuring pertinent information, relevant data, and germane knowledge anywhere anytime so that clinicians can deliver superior healthcare. This chapter asserts that it is only by embracing a rich theoretical lens of analysis that the full potential of e-health can be harnessed, and thus, it proffers Actor-Network Theory (ANT) as such a lens.

The next chapter, "Evolving Digital Communication: An Actor-Network Analysis of Social Networking Sites," analyses elements of social networking sites to establish how a combination of heterogeneous elements of technology, media, language, users, data, and information are networked together to provide this new communication media. An analysis of literature on social networking sites is included in this chapter to reflect the new social networking language and style, the content shared via this media, the mode of use, and the language used for communication, which is a combination of a number of technological and social entities. This chapter examines how the Actor-Network Theory (ANT) can be used to explain social networking and includes some issues for research on this topic.

In Chapter 14, the authors do not address innovation in terms of development of new products; instead, they address them as changes in practices an engineer creatively adopts during engineering design projects. They adopt Actor-Network Theory as a way to understand these change processes (translations). The authors design a perturbation index inspired in Earned Value management to measure translation effort, having in mind the management of scope. Then they assess changes of regime in resource allocation of tasks and conclude with some changes that can lead to innovative results.

Chapter 15 reports on a study of online investing that found that the human actors were translated to more active and involved investors due to the changes, over time, in the online services that are available: the non-human actors. The Internet is a constantly evolving technological actor. New tools have the potential to change interactions with users. In this study, it became evident that new services had a noticeable effect on the behavior of investors. Not only did investors report changes in their behavior when they moved from offline to online investing, but they also reported changes in their investing strategies over time as new services became available. This study showed a new and interesting confirmation of the value of allowing non-human actors to be heard.

Introduction

APPLICATIONS OF ACTOR-NETWORK THEORY

Applications of Actor-Network Theory: Socio-Technical Research

Socio-technical research deals with issues involving both people and machines and with the interactions between various human and non-human entities (Tatnall 2011). Many situations need to deal with both people and things and the interactions and relationships between them. Various approaches to socio-technical research attempt to handle these situations in a way that gives appropriate credit to both the social and the technical, but they do this is different ways. In this article I will begin by briefly examining three approaches to socio-technical research, then concentrate on the use of Actor-Network Theory (ANT). I will provide a few examples from the founders of ANT then attempt to categorise the ANT articles from the *International Journal of Actor-Network Theory and Technological Innovation* (IJANTTI) over the last few years.

There are a number of approaches to socio-technical research but in this article I will consider only three: Activity Theory (Kaptelinin and Nardi 2006; Hashim and Jones 2007), Structuration Theory (Giddens 1984; Turner 1986) and Actor-Network Theory (Latour 1986; Law 1986; Callon 1999).

Activity Theory

Activity Theory is based upon the 1920s work of Russian developmental psychologist Aleksei Leontiev (Verenikina 2001). Hashim and Jones (2007) describe Activity Theory as a theoretical framework that can be used in the analysis and understanding of human interaction through use of their artefacts and see it as the integration of technology as tools which mediate social action. Kaptelinin and Nardi (2006:31) describe Activity Theory as an approach: "... *that aims to understand individual human beings, as well as the social entities they compose, in their natural everyday life circumstances, through an analysis of the genesis, structure and processes of their activities*". In Activity Theory, an activity is seen as the basic unit of analysis and one which is used to understand individual actions (Hashim and Jones 2007). Activity theory considers an entire system and not just a single actor and accounts for the environment, history, culture, motivation and complexity of the person and artefact (Kuutti 1996). Morf and Weber (2000:81) put it this way: "*Activity theory is a conceptual framework based on the idea that activity is primary, that doing precedes thinking, that goals, images, cognitive models, intentions, and abstract notions like 'definition' and 'determinant' grow out of people doing things"*. Kaptelinin and Nardi (2006) note that in Activity Theory, apart from their activities no properties exist of either the subject or object. An article from IJANTTI: 'Mediated Action and Network of Actors: From Ladders, Stairs and Lifts to Escalators (and Travelators)' by Samuel Ekundayo and Antonio Diaz Andrade (2011) from New Zealand offers a comparison of Activity Theory and Actor-Network Theory.

Structuration Theory

For Giddens (1984), Structuration Theory presents a focus on social processes that involve the interaction between actors and the structural properties of a system. Structuration Theory considers agents as a source of power. An agent can be a person, but also anything with the capability to exert power; anything that can influence others in a social context. Giddens sees 'power' as an actor's capability to encourage decisions in a way favourable to them. He sees power, not as a resource but as something used mainly as a medium for power. Human agency can be seen to represent the capacity to make a difference and influence the outcome in a given situation. Structures are rules and resources (Turner 1986), the means and medium through which power is exercised, and are instantiated in recurrent social practices (Iyamu 2013). Orlikowski (1992) notes that Structuration Theory investigates the creation and reproduction of social systems and focuses on social factors including both agents and structure without giving greater importance to one or the other. Structures do not exist independently but are enacted by human agents who continually produce and reproduce these structures. An IJANTTI article by Tiko Iyamu and Dewald Roode (2010): 'Structuration Theory and Actor-Network Theory for Analysis: Case Study of a Financial Institution in South Africa' provides a comparison between Structuration Theory and Actor Network Theory.

Actor-Network Theory

Actor-Network Theory (ANT), or the 'sociology of translations', originated from research in the social studies of science in the 1980s, its designers and main proponents being Bruno Latour, Michel Callon and John Law (Callon 1986b; Latour 1986; Law 1986). ANT was designed as an approach to socio-technical research that would treat the contributions of both human and non-human actors fairly and in the same way, and in an ANT framework, nothing is purely social and nothing is purely technical. ANT is concerned with studying the mechanics of power as this occurs through construction and maintenance of networks made up of both human and non-human actors (Al-Hajri and Tatnall 2011; Tatnall and Dakich 2011). In actor-network theory, "... *the extent of a network is defined by the presence of actors that are able to make their presence individually felt."* (Law 1987b:21). An actor is made up only of its interactions with these other actors, and a network can be hidden inside a 'black box' when its internal details are not under investigation (Davey and Tatnall 2013). ANT investigates the construction and maintenance of networks made up of both human and non-human actors and attempts impartiality towards all actors in consideration, whether human or non-human. It makes no distinction in approach between the social, the natural and the technological (Tatnall and Dakich 2011). ANT is based on three principles:

- Analytical impartiality is demanded of all actors, whether they be human or non-human
- Generalized symmetry explains conflicting viewpoints of different actors making use of the same vocabulary
- Elimination of all *a priori* distinctions between the technological and the social.

The rule which we must respect is not to change registers when we move from the technical to the social aspects of the problem studied. (Callon 1986b, p. 200)

Although there is value in both Activity Theory and Structuration Theory, the remainder of this article deals only with Actor-Network Theory.

Research Using ANT: Some Classical Examples

The work of Latour, Callon and Law provide some classical examples of how actor-network theory can be used productively to explore socio-technical situations. I will mention just a few.

In 'The Case of the Electric Vehicle Callon' (1986a) describes how, in 1973, EDF (Electricite de France) came up with a plan for the VEL (vehicule electrique). EDF prescribed both the precise details of this vehicle and also the social situation in which it would operate. Callon describes the 'ingredients' of the electric vehicle as: "... *electrons that jump effortlessly between electrodes; the consumers who reject the symbol of the motor car and who are ready to invest in public transport; the Ministry of the Quality of Life which imposes regulations about the level of acceptable noise pollution; Renault which accepts the fact that it will be turned into a manufacturer of car bodies; lead accumulators whose performance has been improved; and post-industrial society which is on its way.*" (Callon 1986a, p. 23). He argues that the 'co-operation' of every ingredients with every other is important, and that in the absence, or lack of co-operation of any one of these 'actors' the whole VEL would break down.

Another article by Callon (1986b): 'Some elements of a sociology of translation: domestication of the scallops and the fishermen of St Brieuc Bay' describes the attempts of three marine biologists to determine the reasons and find a solution to the decline of the number of scallops in St. Brieuc Bay, France. Callon describes how, at a 1972 conference in Brest, scientists and representatives of the local fishing community sought the means to increase production of scallops by controlling their cultivation. Discussions at the conference involved three points: firstly, in Japan scallops are intensively cultivated with their larvae being anchored to collectors to provide shelter from predators before later being harvested, secondly there is very little knowledge of the mechanisms by which the scallops develop and thirdly fishing had been so intensive that its consequences were apparent.

The article then introduces the concept of Innovation Translation with its four 'moments': Problematisation, Interessement, Enrolment and Mobilisation and uses these to describe what happened in St Brieuc Bay.

In 'Technology and Heterogeneous Engineering: The Case of Portuguese Expansion', Law (1987a) examines Portuguese exploration down the African coast on the way to India and how King Henry the Navigator acted to facilitate this. Law describes the actors as including: the boats, the crew, supplies, masters of the vessels, Cape Bojador (point of no return) and ocean currents. In describing the part played by the Portuguese King he suggests that: "*I call such activity heterogeneous engineering and suggest that the product can be seen as a network of juxtaposed components.*" (Law 1987a, p. 3).

In 'Engineering and Sociology in a Military Aircraft Project: A Network Analysis of Technological Change' (Law and Callon 1988) and 'The Life and Death of an Aircraft: A Network Analysis of Technical Change' (Law and Callon 1992), the authors describe the ill-fated British Aircraft Corporation TSR2 Cold War strike and reconnaissance military aircraft project in the UK in the early 1960s. The article considers two actor-networks, an external network made up by the Defence Ministry, Navy and Government bureaucracies, and an internal network comprising the contractors and the Air Ministry and argues that it is not just the technical features but also human factors that determine the outcome: "*... we must study not only the social but also the technical features of the engineer's work; in other words, we have to understand the content of engineering work because it is in this content that the technical and the social are simultaneously shaped.*" (Law and Callon 1988, p. 284). "*Our object, then, is to trace the interconnections built up by technologists as they propose projects and then seek the resources required to bring these projects to fruition.*" (Law and Callon 1988, p. 285).

Perhaps my favourite is: 'Aramis or the Love of Technology' by Bruno Latour (1996). The book tells the story of a revolutionary 'guided-transportation' system intended for the Petite Ceinture district and to become part of the Parisian public transportation system in the 1970s. Apart from its allusion to The Three Musketeers, 'Aramis' is an acronym for 'Agencement en Rames Automatisées de Modules Indépendants dans les Stations' (Arrangement in Automated Trains of Independent Modules in Stations). The idea was to produce a system that combined the flexibility of a car with the efficiency of a railway. Work on Aramis began in 1969 but was abandoned in 1987. The book takes for form of a detective story and tells the story of Aramis simultaneously from several different perspectives using a number of different 'voices'. A young engineer and the sociologist, Norbert, conducting the investigation carry the story line. The engineers and administrators who worked on Aramis speak through both interviews and documents. The 'author' interjects from time to time to provide a sociological commentary, and later Aramis speaks on its own behalf, bewailing its fate

Research Using ANT: Some Examples from IJANTTI

The *International Journal of Actor-Network Theory and Technological Innovation* (IJANTTI) has, since 2009, acted as a vehicle for the publication of articles both on approaches to technological innovation and also, more importantly in regards to this article, on actor-network theory (ANT). In this section I will attempt a categorisation in terms of research subject of the articles published from 2009 to 2013. I will, here, include only those dealing with ANT which I have categorised as from the fields of: Information Systems, Healthcare, Education, Business, eGovernment, Science and Engineering and other miscellaneous topics. It should be noted that this categorisation is, at best, somewhat subjective as many of the articles cover more than a single field, and some also look at the application of ANT itself. Much socio-technical research using ANT is cross-disciplinary and so covers multiple fields.

Information Systems (IS) / Information and Communication Technology (ICT and IT)

While there were few articles relating to both information systems and actor-network theory before the mid-1990s, in recent times the number has increased dramatically. In relation to IJAN-TTI, the category of information systems (information technology) is the largest, comprising 25 articles. There is, however, considerable overlap between this category and that of business, given that many of the applications of information systems are in business.

Many of the articles in this category deal with general aspects of the field of information systems while some look at specific examples. Relating to overall aspects of this field, 'Have you taken your Guys on the Journey? – An ANT Account of IS Project Evaluation' (Cecez-Kecmanovic and Nagm 2009) deals with ANT and IS project evaluation research. An article by Amany Elbanna: 'Actor-Network Theory in ICT Research' also deals with this topic. 'Having a Say: Voices for all the Actors in ANT Research?' (Kennan, Cecez-Kecmanovic and Underwood 2010) deals with the issues associated with giving non-human actors a voice of their own in ANT based research. A number of articles, such as 'Linux Kernel Developers Embracing Authors Embracing Licenses' (Linden and Saunders 2009) look at specific topics with this field.

Several business-related articles use ANT to consider enterprise IT issues. These include: 'Theoretical Analysis of Strategic Implementation of Enterprise Architecture' (Iyamu 2010), 'Institutionalisation of the Enterprise Architecture: The Actor-Network Perspective' (Iyamu) and 'Information Infrastructure: an Actor-Network Perspective' (Cordella 2010). As mentioned earlier, one article compares Structuration Theory and Actor Network Theory and this is in the context of information systems. In 'The Use of Structuration Theory and Actor-Network Theory for Analysis: Case Study of a Financial Institution in South Africa' Tiko Iyamu and Dewald Roode (2010) note that many organisations are increasingly dependent on their IT strategy both to increase their competitiveness or just to survive. They point out that despite this, little is known about the non-technical factors, such as people, and their impact on the development and implementation of IT strategy. Making use of both Structuration Theory and Actor-Network Theory they analyse how non-technical factors influence IT strategy. IS Project management is another significant with articles including: 'Assessment of Risk on Information Technology Projects through Moments of Translation' (Iyamu) and 'Critical Success factors in the Management of Projects Using Innovative Approaches' (Bali and Wickramasinghe 2010).

Other IS articles include: 'Exploring Multi-organizational Interaction Issues' (Kasimin and Ibrahim), 'An Actor-Network Analysis of a Case of Development and Implementation of IT Strategy' (Iyamu and Tatnall), 'Future Research on Cloud Computing Adoption by Small and Medium-Sized Enterprises: A Critical Analysis of Relevant Theories' (Saedi and A.Iahad), 'On Actors, Networks, Hybrids, Black Boxes and Contesting Programming Languages' (Tatnall), 'Social Network Services: the Science of Building and Maintaining Online Communities, a perspective from Actor-Network Theory' (Zammar), 'A Socio-Technical Account of an Internet-Based self-Service Technology Implementation: Why Call-Centres Sometimes 'Prevail' in a Multi-Channel Context?' (Naidoo), 'Desituating Context in Ubiquitous Computing: Exploring

Strategies for the Use of Remote Diagnostic Systems for Maintenance Work' (Holmstrom, Jonsson, *Kalle Lyytinen and Agneta Nilsson),* 'Empirically examined the Disjoint in Software Deployment: A Case of Telecommunication' (Iyamu), Information Systems and Actor-Network Theory Analysis (Iyamu and Sekgweleo), 'Negotiating the Socio-Material in and about Information Systems: An Approach to Native Methods' (Cole), 'Making Information Systems Material through Blackboxing: Allies, Translation and Due Process' (Underwood and Tabak), 'Emerging Standardization' (Cordella), Innovation in communication: an actor network analysis of social websites (Singh, Dwivedi, Hackney and Peszynski), 'Observing the 'fluid' continuity of an IT artefact' (Naidoo and Leonard) and 'Enhancing Understanding of Cross-Cultural ERP Implementation Impact with A FVM Perspective Enriched By ANT' (Muhammad and Wickramasinghe).

Business and Government

'Actor-Network Theory for Service Innovation' by Lorna Uden and Janet Francis (2009) from the UK looks at the growth of the service sector as the dominant industrial world economy, and how there is a need for new conceptual understandings and theoretical underpinnings to systematically describe its nature and behaviour. They point out that the development and adoption of service innovation requires the integration of multiple elements, including people, technologies and networks across organisations. The article describes how ANT can help in an understanding of the relationships among the actors and how these actors have their needs shaped by the network formation. On a similar note, in an article titled: 'Service Science: An Actor-Network Theory Approach', Noel Carrol, Ita Richardson, and Eoin Whelan from Ireland note that service networks that implement technology to execute processes to deliver a specific service become increasingly complex, and how ANT can be used to examine their complexity. Given their, they describe 'service' as comprising socio-technical factors which exchange various resources and competencies, and use service networks to transfer resources and competencies.

Spanning both information systems and business is an article by Fabian Muniesa, from France. Through a case study on the Arizona Stock Exchange 'Is a Stock Exchange a Computer Solution? Explicitness, algorithms and the Arizona Stock Exchange', the article looks at how computerisation and automation challenged the definition of the stock exchange in the context of North-American financial markets in the nineteen-nineties. Muniesa notes that for a market to be formulated in algorithmic terms implies in part the possibility of being 'executed by a machine'. Also relating to financial markets, 'Actor-Network Theory and the Online Investor' by Arthur Adamopoulos, Martin Dick and Bill Davey (2012) from Australia offers an understanding of the way in which technology is used, rather than adopted and involves interactions between people and the Internet. They describe an interesting confirmation of the power of allowing non-human actors to be heard and how, in this case, as the non-human actor (on-line services) changed over the course of the investigation it became clear that the human actors were changing as a result. They found that the human actor had been translated to a more active and involved investor due to the change in the online services that are available.

'Neither Heroes nor Chaos: The Victory of VHS Against Betamax' by Diego Ponte and Pier Franco Camussone (2013) from Italy sets out to contribute to the ongoing debate on standardisation processes by addressing economic and social aspects that shape creation and diffusion of standards. Their topic is the contest of videotape formats VHS and BETAMAX in the 1970s and 1980s. They demonstrate how the evolution of a standard is a complex process where creation and diffusion are co-evolving dimensions that cannot easily be distinguished and how the coupling of economic and social aspects permits a better understanding of the factors and events that affect the evolution of a standard and the dominance of one standard among competing ones. In an article relating to both information systems and business: 'Competitive Intelligence in the Enterprise: Power Relationships', Relebohile Moloi from South Africa and Tiko Iyamu from Namibia examine factors that may lead to decisions in choosing Competitive Intelligence products in organisations. An article: 'Using Actor Network Theory to Facilitate a Superior Understanding of Knowledge Creation and Knowledge Transfer' (Wickramasinghe, Tatnall and Bali) argues that as current knowledge creation techniques tend to focus on either human or technology aspects of organisational development and less often on process-centric aspects of knowledge generation, it is important to view knowledge creation and all socio-technical organisational operations that result in knowledge generation through the richer lens of ANT.

Johanes Eka Priyatma from Indonesia and Zainal Abidin Mohamed from Malaysia point out that leadership has been identified as a factor critical to successful development of e-government projects. They further note that this is especially the case in developing countries. In their article: 'Opening the Black Box of Leadership in the Successful Development of Local E-Government Initiative in a Developing Country' they make use of the four moments of innovation translation as a framework to trace and monitor how leadership has been practiced effectively in an e-government project in a developing country. In another article, 'A Critical Review of the Ontological Assumptions of Actor-Network Theory for Representing E-Government Initiatives' Johanes Eka Priyatma suggests that the potential contribution of ANT for representing e-government initiatives flows from its ontological assumptions, but that these assumptions have never been critically reviewed using real e-government cases. The article then examines two e-government cases from Indonesia to achieve this.

Healthcare

A substantial number (12) of articles relate to ANT and healthcare. 'The S'ANT Imperative for Realizing the Vision of Healthcare Network Centric Operations' (Wickramasinghe and Bali 2009) argues that in in the information-intensive environment of healthcare a network-centric approach offers rapid sharing of information and the effective knowledge building required for development of coherent objectives and their attainment. The article suggests the use of rich analysis tools from Social Network Analysis combined with Actor-Network Theory (S'ANT). Along a similar line is 'The S'ANT Approach to Facilitate a Superior Chronic Disease Self-Management Model' (Wickramasinghe, Bali and Goldberg).

'Actor-Network-Theory in Medical e-Communication – The Role of Websites in Creating and Maintaining Healthcare Corporate Online Identity' (Bielenia-Grajewska 2011) deals with how websites create and maintain the online identity of medical care providers. The author notes that she has chosen an ANT approach as this makes it possible to study the role of living and nonliving entities in shaping the online identity of healthcare suppliers and to concentrate on the networks and systems within e-healthcare as well as the flows and interrelations constituting it.

ICT has been used in medical General Practice throughout Australia for some years, but although most General Practices make use of ICT for administrative purposes such as billing, prescribing and medical records, many individual General Practitioners themselves do not make full use of these ICT systems for clinical purposes. This issue is discussed in the article: 'Adoption of ICT in Rural Medical General Practices in Australia – an Actor-Network Study' (Deering, Tatnall and Burgess 2010) in which an ANT lens is used to examine the human and non-human actors that contribute to this situation.

'The (Re-)Socialization of Technical Objects in Patient Networks: The Case of the Cochlear Implant' (Spöhrer 2013) describes the processes of technical stabilization of the cochlear implant in certain stabilised scientific environments. The author notes that there is a problem here as these technical stabilisations can only be accomplished by rigorously excluding attributes of the social, but that as the cochlear implant is born out of the need to enable participation in 'normal' social life it is certainly a social actor attributed with certain social attachments.

Other healthcare articles include: 'Why Using Actor Network Theory (ANT) Can Help to Understand the Personally Controlled Electronic Health Record (PCEHR) in Australia' (Muhammad, Teoh and Wickramasinghe), 'A Tale of Two Cities: E-Health in Germany and Australia' (Zwicker, Seitz and Wickramasinghe), 'A Manifesto for E-Health Success – The Key Role for ANT' (Wickramasinghe, Bali and Tatnall), 'Understanding the Advantages of Mobile Solutions for Chronic Disease Management: the Role of ANT as a Rich Theoretical Lens' (Wickramasinghe, Tatnall and Goldberg), 'Using ANT to Uncover the Full Potential of an Intelligent Operational Planning and Support Tool (IOPST) for Acute HealthCare Contexts' (Muhammad, Moghimi, Taylor, Redly, Nguyen, Stein, Kent, Botti and Wickramasinghe), 'How Using ANT Can Assist to Understand Key Issues for Successful e-Health Solutions' (Muhammad, Zwicker and Wickramasinghe) and 'Future Research on Dimensions of E-Service Quality in Interactive Health Portals: The Relevancy of Actor Network Theory' (Foroutani, Iahad and Rahman).

Education

Education is another popular area for ANT research with 15 articles in this category. A number of these articles refer to university education and some to schools. Beginning with university education, focussing on assessment of one teacher-training course for the learning and skills sector in the UK, 'Higher education in further education in England: an actor-network ethnography' (Tummons 2009) offers a way to conceptualise the responses of those Further Education colleges where the course is actually run, to the systems and procedures established by the university which provides the course.

'Knowledge in Networks – Knowing in Transactions?' (Rimpiläinen 2011) investigates a large, publicly-funded interdisciplinary project undertaken at a university in Scotland. The specific focus of the article is what might be termed the epistemology of actor-network theory, and the paper goes on to consider the implications for epistemology of one of the more significant elements of ANT: the principle of symmetry. In 'Distance, climate, demographics and the development of online courses in Newfoundland and Labrador' (Reid 2013) the author notes that one of ANT's assertions is that physical factors can be actors within a network of other factors and determine development and use of technology. The article documents the impact of climate, distance and demographics on the adoption of online courses at a university in Canada.

An interesting article: 'Between Blackboxing and Unfolding: Professional Learning Networks of Pastors' (Reite 2013) points out that pastors can be thought of as an example of a value-oriented profession being both a keeper of traditions and an innovator facing the challenges of globalisation and secularisation. The analysis of pastor networks is therefore an interesting case of professional learning in a changing society, and the author presents an ethnographic study of five pastors from the Church of Norway doing their everyday work. The article argues that professional learning is a process of moving between 'black boxing' and 'unfolding'.

There are also several examples of relating ANT to school education. '"What's Your Problem?" ANT Reflections on a Research Project Studying Girls Enrolment in Information Technology Subjects in Post-compulsory Education' (Rowan and Bigum 2009) notes that despite more than 30 years of gender reform in schools, the number of girls enrolled in information technology subjects in the post-compulsory years of education has remained under 25%. The authors discuss the difference between the researchers' perception of the problem under consideration, and the participants' perception of the same issue using ANT to highlight the gaps, tensions and contradictions within the data and to ask key questions about the extent to which the enrolment of girls in IT is indeed 'a problem'.

In 'A Petri Net Model for Analysing e-Learning and Learning Difficulties', Tas Adam (2011) examines the issue of using an actor-network framework to develop a model for e-Learning for students with Learning Difficulties. In the article Adam points out that Petri Nets are tools for the modelling and analysis of the behaviour of systems and can reveal important information about the structure and dynamic behaviour of the modelled system. He argues that Petri Net concepts (when used qualitatively) are not fundamentally different from those of ANT. A related article: 'School Children with Learning Disabilities: An Actor-Network Analysis of the Use of ICT to Enhance Self-Esteem and Improve Learning Outcomes' (Adam and Tatnall 2012) presents a report on an investigation into the use of ICT to aid in the education of students with Learning Disabilities.

Other education related articles include: '(Un)locating learning: agents of change in case-based learning' (Tscholl, Patel and Carmichael), 'Deconstructing Professionalism: an Actor-Network critique of professional standards for teachers in the UK lifelong learning sector' (Tummons), 'Performativity in practice: An Actor-Network Account of Professional Teaching Standards' (Mulcahy), Aspects of e-Learning in a University (Manning, Wong and Tatnall), 'Complexifying the 'visualised' curriculum with actor-network theory' (Vincentis), 'Two Computer Systems in Victorian Schools and the Actors and Networks Involved in their Implementation and Use' (Davey and Tatnall), 'Performing Actor-Network Theory in the Post-Secondary Classroom' (A. Quinlan, E. Quinlan and Nelson) and 'Using ANT to Guide Technological Adoption: the Case of School Management Software' (Davey and Tatnall).

Science and Engineering

This is not an area with a large number of ANT articles, despite two of Latour's early ANT works relating to science. 'Science in Action' (Latour 1987) describes how, as well as technical factors, social context is important to an understanding of how science works. Latour argues that science and technology need to be studied 'in the making', or while in action to see why scientific discoveries could have moved in different directions. Similarly, 'The Pasteurization of France' (Latour 1988) suggests that in this case what matters is the role of each actor and how they play a role in power relationships, rather than whether these actors are microbes, scientists or bureaucrats.

Having said that the number of articles on Science and Engineering is small in number, they make very interesting reading. I have classified six articles as being relevant to this field. The first, 'Opening the Indonesian Bio-Fuel Box: How Scientists Modulate the Social' by Yuti Ariani and Sonny Yuliar uses the notion of translation to study present bio-fuel developments in Indonesia. The authors identify the actors as including: scientists, businessmen, policy makers, farmers, the land itself, bio-fuel, oil and politicians. Their article describes how scientists seek to establish their role through a heterogeneous network modulating the social, the politics and the trajectory of bio-fuel development in Indonesia (Ariani and Yuliar 2009). Fernando Gonçalves and José Figueiredo from Portugal have contributed three important articles relating to engineering and the process of engineering design. In: 'How to Recognize an Immutable Mobile When You Find One: Translations on Innovation and Design' (Gonçalves and Figueiredo 2010) they explore some ways to build practical methodological approaches into engineering design, discussing in particular Obligatory Passage Points and Immutable Mobiles. In two more articles they discuss 'Negotiating Meaning – an ANT Approach to the Building of Innovations' and 'Engineering Innovative Practice in Managing Design Projects'.

Other Topics

A number of articles do not fit neatly into any of the above categories and are included here. Florian Neisser from Germany has contributed an article titled: 'Fostering Knowledge Transfer for Space Technology Utilization in Disaster Management – An Actor-Network Perspective' (Neisser 2013). He points out that, due to the human and property costs, it is important that efforts to mitigate potential, and respond to actual disasters, be coordinated and that Disaster Management is an issue of global importance. The article describes a study of the design and use of a web-portal to handle the complex and constantly changing interrelations between human and non-human actors and provide timely and accurate information as well as clear and suitable communication technology to guarantee coordinated efforts. 'The Iranian Wheat Growers' Climate Information Use: An Actor-Network Theory Perspective' by Maryam Sharifzadeh et al. from Iran (2012) relates a study of agricultural climate information use, linking actor-network theory and actor analysis premises in a qualitative research design. The study's findings showed that socio-political factors: farmers' awareness, motivation and trust, along with information processing factors including accuracy of information, access to information, and correspondence of information to farmers' conditions were the key elements in facilitating climate information use in farming practice.

In 'Actor-Network Theory in Intercultural Communication – Translation through the Prism of Innovation, Technology, Networks and Semiotics' Magdalena Bielenia-Grajewska discusses the role of participants in one type of intercultural exchange, namely in translation (Bielenia-Grajewska 2009). 'From Intermediary to Mediator and Vice Versa: On Agency and Intentionality of a Mundane Sociotechnical System' (Antonio Diaz Andrade) examines the use of electronic mail systems, especially the automatically generated 'Out of Office' message, is examined in this article to emphasise the distinction between agency and intentionality. In 'Networks, agents and models: objections and explorations', Muniesa and Tchalakov uses the form of a dialog to argue that critical problems of computational modelling of network topologies can be well considered from an ANT point of view.

In 'Imagining a Feminist Actor-Network Theory' Andrea Quinlan (2012) points out that Feminism and Actor-Network Theory have often been considered opposing theoretical and intellectual traditions, but goes on to imagine a meeting between these seemingly divergent fields and consider the theoretical and methodological challenges that ANT and feminism raise for one another. 'Murphy's Law in Action: The Formation of the Film Production Network of Paul Lazarus' Barbarosa (1982): An Actor-Network-Theory Case Study', by Markus Spöhrer explores the possibilities of ANT as a methodological approach to Production Studies. Using a detailed production log written by producer Paul Lazarus III, the coming-into-being of the film Barbarosa (1982) is described.

Technological Innovation Using Other Investigative Frameworks

In investigating technological innovation several other articles made use of approaches, principally Innovation Diffusion (Rogers 2003), the Technology Acceptance Model (Davis 1986), hermeneutics/phenomenology and variations on these, rather than innovation translation. These articles were contributed by: Tatnall, Alawneh and Hattab, Kripanont and Tatnall, Bingley and Burgess, Tatnall, Lukaitis. Although these were interesting and worthwhile articles they will not be considered here as they do not relate directly to ANT.

Re-Analysis Using ANT

Several articles show how it is possible to do some 'after the fact' re-conceptualisation and re-analysis of the data from other approaches to technological innovation to make use of an innovation translation approach. Of course, if an ANT approach is to be used in any research project then ideally it would be used from the beginning of the study so that the data collection and interviews reflected the sort of questions and the approach appropriate to this framework. Nevertheless the following articles, all originally making use of an approach to technological innovation based on TAM, the Technology Acceptance Model (Davis 1986) show that re-interpretation is possible and even worthwhile.

'A Socio-Technical Study of the Adoption of Internet Technology in Banking, Re-Interpreted as an Innovation Using Innovation Translation' by Salim Al-Hajri and Arthur Tatnall took a study using TAM that investigated Internet banking in Oman and Australia and re-interpreted it using innovation translation. Another article: 'Knowledge Conversion Processes in Thai Public Organisations Seen as an Innovation: The Re-Analysis of a TAM Study Using Innovation Translation' by Puripat Charnkit and Arthur Tatnall took a similar line to re-interpret a TAM study relating to knowledge management in Thailand. Also relating to Thailand, 'Technological Innovation and the Adoption and Use of ICT in Thai Universities' by Arthur Tatnall took an article that originally used by TAM and Structured Equation Modelling and performed a similar re-interpretation.

CONCLUSION

In this article I have demonstrated the wide and multi-disciplinary nature of research framed by actor-network theory, and how the articles from IJANTTI cover a broad scope. The classification of ANT research applications presented here is, as previously mentioned, at best rather subjective as many articles span more than a single topic and are really cross-disciplinary. The largest number of IJANTTI articles came from the fields of Information Systems, Healthcare and Education, but a significant number also came from a variety of other areas. There are, however, still major areas that have not, as yet, attracted many IJANTTI articles. One of these is history of technology and events, despite the early classic articles by the founders of ANT relating to this area. Other possible opportunities for further ANT research publications include more articles on science and engineering, and articles on gardening, entertainment and many other areas. Since its initial conception 30 years ago, ANT has gone on to offer an important framework for much socio-technical research.

Arthur Tatnall
Victoria University, Australia

REFERENCES

Adam, T. (2011). A Petri Net Model for Analysing E-Learning and Learning Difficulties. *International Journal of Actor-Network Theory and Technological Innovation, 3*(4), 11–21. doi:10.4018/jantti.2011100102

Adam, T., & Tatnall, A. (2012). School Children with Learning Disabilities: An Actor-Network Analysis of the Use of ICT to Enhance Self-Esteem and Improve Learning Outcomes. *International Journal of Actor-Network Theory and Technological Innovation, 4*(2), 10–24. doi:10.4018/jantti.2012040102

Adamopoulos, A., Dick, M., & Davey, W. (2012). Actor-Network Theory and the Online Investor. *International Journal of Actor-Network Theory and Technological Innovation, 4*(2). doi:10.4018/jantti.2012040103

Al-Hajri, S., & Tatnall, A. (2011). *A Study of Adoption of Internet Banking in Oman in the Early 2000s, Re-Interpreted Using Innovation Translation Informed by Actor-Network Theory.* Paper presented at the International Information Systems Conference (iiSC-2011). Muscat, Oman.

Ariani, Y., & Yuliar, S. (2009). Opening the Indonesian Bio-Fuel Box: How Scientists Modulate the Social. *International Journal of Actor-Network Theory and Technological Innovation, 1*(2).

Bali, R. K., & Wickramasinghe, N. (2010). Critical Success factors in the Management of Projects Using Innovative Approaches. *International Journal of Electronic Healthcare, 2*(3), 33–39.

Bielenia-Grajewska, M. (2009). Actor-Network Theory in Intercultural Communication – Translation through the Prism of Innovation, Technology, Networks and Semiotics. *International Journal of Actor-Network Theory and Technological Innovation, 1*(4), 53–69. doi:10.4018/jantti.2009062304

Bielenia-Grajewska, M. (2011). Actor-Network-Theory in Medical e-Communication – The Role of Websites in Creating and Maintaining Healthcare Corporate Online Identity. *International Journal of Actor-Network Theory and Technological Innovation, 3*(1), 39–53. doi:10.4018/jantti.2011010104

Callon, M. (1986a). The Sociology of an Actor-Network: The Case of the Electric Vehicle. In *Mapping the Dynamics of Science and Technology*. Macmillan Press.

Callon, M. (1986b). Some Elements of a Sociology of Translation: Domestication of the Scallops and the Fishermen of St Brieuc Bay. In *Power, Action & Belief. A New Sociology of Knowledge?* London: Routledge & Kegan Paul.

Callon, M. (1999). Actor-Network Theory - The Market Test. In *Actor Network Theory and After*. Oxford, UK: Blackwell Publishers.

Cecez-Kecmanovic, D., & Nagm, F. (2009). Have you Taken your Guys on the Journey? An ANT Account of IS Project Evaluation. *International Journal of Actor-Network Theory and Technological Innovation, 1*(1), 1–22. doi:10.4018/jantti.2009010101

Cordella, A. (2010). Information Infrastructure: an Actor-Network Perspective. *International Journal of Actor-Network Theory and Technological Innovation, 2*(1), 27–53. doi:10.4018/jantti.2010071602

Davey, W., & Tatnall, A. (2013). Social Technologies in Education: An Actor-Network Analysis. In *Open and Social Technologies for Networked Learning*. Heidelberg, Germany: Springer. doi:10.1007/978-3-642-37285-8_17

Davis, F. (1986). *A Technology Acceptance Model for Empirically Testing New End-User Information Systems: Theory and Results*. Boston, MA: MIT.

Deering, P., Tatnall, A., & Burgess, S. (2010). Adoption of ICT in Rural Medical General Practices in Australia - An Actor-Network Study. *International Journal of Actor-Network Theory and Technological Innovation, 2*(1), 54–69. doi:10.4018/jantti.2010071603

Ekundayo, S., & Diaz Andrade, A. (2011). Mediated Action and Network of Actors: From Ladders, Stairs and Lifts to Escalators (and Travelators). *International Journal of Actor-Network Theory and Technological Innovation, 3*(3), 21–34. doi:10.4018/jantti.2011070102

Giddens, A. (1984). *The constitution of society: Outline of the theory of structuration*. Cambridge, MA: Polity Press.

Gonçalves, F., & Figueiredo, J. (2010). How to Recognize an Immutable Mobile When You Find One: Translations on Innovation and Design. *International Journal of Actor-Network Theory and Technological Innovation, 2*(2). doi:10.4018/jantti.2010040103

Hashim, N. H., & Jones, M. L. (2007). Activity theory: A framework for qualitative analysis. In *Proceedings of 4th International Qualitative Research Convention (QRC)*. Malaysia: QRC.

Iyamu, T. (2010). Theoretical Analysis of Strategic Implementation of Enterprise Architecture. *International Journal of Actor-Network Theory and Technological Innovation, 2*(3), 17–32. doi:10.4018/jantti.2010070102

Iyamu, T. (2013). *Enterprise Architecture: From Concept to Practice*. Melbourne, Australia: Heidelberg Press.

Iyamu, T., & Roode, D. (2010). The Use of Structuration Theory and Actor Network Theory for Analysis Case Study of a Financial Institution in South Africa. *International Journal of Actor-Network Theory and Technological Innovation, 2*(1), 1–26. doi:10.4018/jantti.2010071601

Kaptelinin, V., & Nardi, B. A. (2006). *Acting with Technology: Activity Theory and Interaction Design*. Cambridge, MA: MIT Press.

Kennan, M. A., Cecez-Kecmanovic, D., & Underwood, J. (2010). Having a Say: Voices for all the Actors in ANT Research? *International Journal of Actor-Network Theory and Technological Innovation, 2*(2), 1–16. doi:10.4018/jantti.2010040101

Kuutti, K. (1996). Activity Theory as a Potential Framework for Human-Computer Interaction Research. In *Context and Consciousness: Activity Theory and Human-Computer Interaction*. Cambridge, MA: MIT Press.

Latour, B. (1986). The Powers of Association. In *Power, Action and Belief: A New Sociology of Knowledge?* London: Routledge & Kegan Paul.

Latour, B. (1987). *Science in Action: How to Follow Engineers and Scientists Through Society*. Milton Keynes, UK: Open University Press.

Latour, B. (1988). *The Pasteurization of France*. Cambridge, MA: Harvard University Press.

Latour, B. (1996). *Aramis or the Love of Technology*. Cambridge, MA: Harvard University Press.

Law, J. (1986). The Heterogeneity of Texts. In *Mapping the Dynamics of Science and Technology*. Macmillan Press.

Law, J. (1987a). On the social explanation of technical change: The case of the Portuguese maritime expansion. *Technology and Culture, 28*(2), 227–252. doi:10.2307/3105566

Law, J. (1987b). Technology and Heterogeneous Engineering: The Case of Portuguese Expansion. In *The Social Construction of Technological Systems: New Directions in the Sociology and History of Technology*. Cambridge, MA: MIT Press.

Law, J., & Callon, M. (1988). Engineering and Sociology in a Military Aircraft Project: A Network Analysis of Technological Change. *Social Problems, 35*(3), 284–297. doi:10.2307/800623

Law, J., & Callon, M. (1992). The life and death of an aircraft: A network analysis of technical change. In *Shaping Technology/Building Society: Studies in Sociological Change*. Cambridge, MA: MIT Press.

Linden, L., & Saunders, C. (2009). Linux Kernel Developers Embracing Authors Embracing Licenses. *International Journal of Actor-Network Theory and Technological Innovation*, *1*(3), 15–35. doi:10.4018/jantti.2009070102

Morf, M. E., & Weber, W. G. (2000). I/O Psychology and the Bridging Potential of A. N. Leont'ev's Activity Theory. *Canadian Psychology*, *41*, 81–93. doi:10.1037/h0088234

Neisser, F. (2013). Fostering Knowledge Transfer for Space Technology Utilization in Disaster Management – An Actor-Network Perspective. *International Journal of Actor-Network Theory and Technological Innovation*, *5*(1). doi:10.4018/jantti.2013010101

Orlikowski, W. J. (1992). The duality of technology: Rethinking the concept of technology in organisations. *Management of Technology*, *3*(3), 398–427.

Ponte, D., & Camussone, P. F. (2013). Neither Heroes nor Chaos: The Victory of VHS Against Betamax. *International Journal of Actor-Network Theory and Technological Innovation*, *5*(1). doi:10.4018/jantti.2013010103

Quinlan, A. (2012). Imagining a Feminist Actor-Network Theory. *International Journal of Actor-Network Theory and Technological Innovation*, *4*(2). doi:10.4018/jantti.2012040101

Reid, S. (2013). Distance, climate, demographics and the development of online courses in Newfoundland and Labrador. *International Journal of Actor-Network Theory and Technological Innovation*, *5*(2). doi:10.4018/jantti.2013040102

Reite, I. C. (2013). Between Blackboxing and Unfolding: Professional Learning Networks of Pastors. *International Journal of Actor-Network Theory and Technological Innovation*, *5*(4). doi:10.4018/ijantti.2013100104

Rimpiläinen, S. (2011). Knowledge in Networks – Knowing in Transactions? *International Journal of Actor-Network Theory and Technological Innovation*, *3*(2), 46–56. doi:10.4018/jantti.2011040104

Rogers, E. M. (2003). *Diffusion of Innovations*. New York: The Free Press.

Rowan, L., & Bigum, C. (2009). What's your problem? ANT reflections on a research project studying girls enrolment in information technology subjects in postcompulsory education. *International Journal of Actor-Network Theory and Technological Innovation*, *1*(4), 1–20. doi:10.4018/jantti.2009062301

Sharifzadeh, M., Zamani, G. H., Karami, E., Khalili, D., & Tatnall, A. (2012). The Iranian wheat growers' climate information use: An Actor-Network Theory perspective. *International Journal of Actor-Network Theory and Technological Innovation*, *4*(4), 1–22. doi:10.4018/jantti.2012100101

Spöhrer, M. (2013). The (re-)socialization of technical objects in patient networks: The case of the cochlear implant. *International Journal of Actor-Network Theory and Technological Innovation*, *5*(3). doi:10.4018/jantti.2013070103

Tatnall, A. (2011). *Information Systems Research, Technological Innovation and Actor-Network Theory*. Melbourne, Australia: Heidelberg Press.

Tatnall, A., & Dakich, E. (2011). *Informing Parents with the Victorian Education Ultranet*. Novi Sad, Serbia. *Informing Science*.

Tummons, J. (2009). Higher Education in Further Education in England: An Actor-Network Ethnography. *International Journal of Actor-Network Theory and Technological Innovation*, *1*(3), 55–69. doi:10.4018/jantti.2009070104

Turner, J. H. (1986). The Theory of Structuration. *American Journal of Sociology*, *91*(4), 969–977. doi:10.1086/228358

Uden, L., & Francis, J. (2009). Actor-Network Theory for service innovation. *International Journal of Actor-Network Theory and Technological Innovation*, *1*(1), 23–44. doi:10.4018/jantti.2009010102

Verenikina, I. (2001). Cultural-historical psychology and activity theory in everyday practice. In information systems and activity theory: Volume 2 theory and practice. Wollongong, Australia: University of Wollongong Press.

Wickramasinghe, N., & Bali, R. (2009). The S'ANT Imperative for realizing the vision of healthcare network centric operations. *International Journal of Actor-Network Theory and Technological Innovation*, *1*(1), 45–58. doi:10.4018/jantti.2009010103

Chapter 1
The Impact of ICT in Educating Students with Learning Disabilities in Australian Schools:
An ANT Approach

Tas Adam
Independent Researcher, Australia

Arthur Tatnall
Victoria University, Australia

ABSTRACT

The term "Learning Difficulties" (sometimes also referred to as Special Needs) is used in reference to students who have significant difficulties in the acquisition of literacy and numeracy skills and need extra assistance with schooling. This is a large heterogeneous group. Another more specific term is "Learning Disabilities" that refers to the small sub-group of students who exhibit severe and unexplained problems. This chapter presents a report on an investigation, framed by the use of actor-network theory, of how the use of Information and Communications Technologies can aid in improving the education of students with Learning Disabilities. The study involved case studies and participant observation of the use of ICT in two outer Melbourne suburban Special Schools and an investigation of the impact of Education Department policies on these school environments. Research at the two Special Schools revealed that use of Information and Communications Technologies can have a very beneficial impact on these students by improving their self-esteem and facilitating their acquisition of useful life skills.

DOI: 10.4018/978-1-4666-6126-4.ch001

SPECIAL NEEDS SCHOOLS

In Australia, and around the world, there are a significant number of primary and secondary school students with Learning Disabilities, otherwise also known as Special Needs. These students require some assistance and support in their learning. There are over 100 Special Needs Schools in Victoria and over 1,200 in Australia (Australian Schools Directory 2011). An analysis revealed over 1,000 special schools in various categories, including Learning Difficulties/disabilities of general nature, intellectual and physical – ranging from mild to physical and severe Learning Difficulties. Table 1 below provides the various categories that can be found in the directory of Special Needs Schools.

Although policies have existed for several decades to integrate students with Learning Difficulties into the mainstream classroom, this has not always provided the best learning environment for all these students (Johnson, Gersten and Carmine 1998) and this contention is supported by more recent work from Shaw, Grimes and Bulman (2005). The literature provides many examples where the demands of the students in this category could not be catered for in an adequate manner in mainstream classrooms (for example Klinger (1998) and Zigmond (1995)). In the USA many students with Learning Disabilities do attend mainstream classrooms, but most of the research on Learning Difficulties which has discussed this environment produces findings that clearly indicate that the needs of these students are not being adequately met (Mac-Millan and Hendrick 1993). In Europe and the UK similar learning platforms and standards exist for Special Needs students, and these are referred to there as Special Needs Education (SNE). However, in the UK, there has been strong interest to establish an understanding and recognition of the rights and privileges of Special Needs students (Riddell and Watson 2003; Shakespeare 1999).

The advent of low price, high power Information and Communications Technologies (ICT) and use of the Internet have played a major role in enhancing and shaping the knowledge, skills and self-esteem of these students.

SCHOOL CHILDREN WITH LEARNING DISABILITIES

Assistive technology has introduced awareness for both educators and students and for the past decade there has been a growing effort in the design and development of ICT-based platforms to enhance the learning outcomes of

Table 1. Summary of Special Needs categories (Australian Schools Directory 2011)

General Disabilities	Intellectual Disabilities/Autism
Hearing Impaired	Learning Difficulties
English Learning	Moderate to High Needs
Distance Education	Multiple Disabilities
Autistic	Physical Disabilities
Emotional Behaviour	Speech / Language Disorders
High Needs	Vision Impaired
Intellectual Disabilities	Young Mothers

these students (Adam and Tatnall 2007; Adam and Tatnall 2008b; Adam and Tatnall 2012).

This chapter describes a study involving case studies and participant observation in two outer suburban special schools in Melbourne, Australia and also looked at the role and impact of Victorian Education Department policies on these school environments (Adam 2010). It identified several different categories of students with Special Needs, ranging from those with severe physical disabilities to mainstream students who have a need to maintain some continuity with their studies while temporarily residing in a hospital.

The background and account of Learning Disabilities (LD) has been well documented with respect to the nature of particular problems such as specific language and reading disorders and the behavioural correlates of brain injury. The structure of Special Education over the years has also been well documented and a compelling case has been made for the need for a category like LD (Kavale and Forness 1995). It should be noted that given the sensitive nature of this problem and the wide range of special needs, the researcher must be very careful in applying the definition, as inferences could be drawn that could lead to negative results. In some publications the term 'Learning Disabilities' has been used by organisations such as the Australian Department of Education, Employment Training and Youth Affairs (DEETYA 1999), and the definition below has been used to classify funding categories for Special Needs students:

a student, who has been assessed by a person with a relevant qualification, as having intellectual, sensory, physical, social/emotional or multiple impairments to a degree that satisfies the criteria for enrolment in special education services provided by the government of the state or territory in which the student is locate" (DEETYA 1999, p2).

One of the problems in working in this area is the terminology, with the terms: Learning Difficulties, Children at Risk, Special Needs and Learning Disabilities all being used in different countries and different contexts to describe these children. In this article the term Learning Difficulties (LD) will be used to cover all of these other terms (Commonwealth of Australia 1998), but most of the article refers to the more specific Learning Disabilities. The definition for the term Learning Difficulties that is used in Australia is similar to that used in the USA, and a useful definition of LD comes from the Learning Disabilities Association of Canada (LDAC) who defines LD as:

... a number of disorders which may affect the acquisition, organization, retention, understanding or use of verbal or nonverbal information. These disorders affect learning in individuals who otherwise demonstrate at least average abilities essential for thinking and/or reasoning. As such, Learning Disabilities are distinct from global intellectual deficiency (Learning Disabilities Association of Canada 2002).

Given the complexity of definitions, one way to represent these differences is with the following Venn diagram where the term Learning Difficulties is used to refer to a large group of children who need extra assistance with schooling, and Learning Disabilities to refer to students who constitute a small sub-group that exhibit severe and unexplained problems. The terminology of Learning Difficulties and Learning Disabilities is further reflected in a survey (Rivalland 2000, p69) where principals referred to these categories by the following percentages listed in Table 2.

The term 'Learning Disabilities (LD)' is widely accepted in the psychological field in USA but is more likely to be used by teachers with special education training and by

Table 2. Reference to Learning Difficulties (Rivalland 2000)

Factor	%
Learning Difficulties	47%
Children at Risk	37%
Special Needs	17%
Learning Disabilities	10%

school psychologists. In Australia it is usually qualified to specific learning needs (Chan and Dally 2000).

Although policies have existed for some time in many countries to integrate students with Learning Disabilities into the mainstream classroom, this has not always provided the best learning environment for these students (Shaw, Grimes et al. 2005); hence the need for some Special Schools. Bulgren (1998) and Agran (1997), support the view that some students with LD required an alternative approach to their learning and numerous integration or remedial programs have proved inefficient in relation to the overall learning of this group of students. The literature also shows that in some selected fields, for example in mathematics and social studies, specialist instruction has been applied to this group of individuals with little success.

RESEARCH METHOD

This research involved case studies at two special schools in suburban Melbourne. The case studies consisted of interviews with the School Principal, teachers and parents. (Due to the sensitive nature of this area school pupils were not interviewed.)

The case study data research described in this chapter was collected over several years by one of the authors who attended and observed specific classes at these two Special Schools on a regular basis. One school was visited in

2003 and 2004 while the other was approached in 2005 and 2006 (Adam and Tatnall 2012).

TWO SPECIAL SCHOOLS IN MELBOURNE

Schools of all types form part of the broad community in any city, and schools themselves form a series of communities within this broad community. Educational institutions that cater for children with Special Needs (Learning Disabilities) have a lot in common with each other, but less in common with mainstream schools, even when these are located quite close by. These special schools need to form community groupings, and particularly groupings consisting of other special schools and educational institutions offering support to students with Special Needs. The problem is that these educational institutions are often geographically well separated, but this is where the Internet offers a good means of providing the infrastructure to support virtual communities of special schools.

This study involved case studies with participant observation in two outer suburban special schools in Melbourne. One of these schools provides a basic ICT infrastructure for its staff and students, whilst the other provides a more enhanced 'state of the art' ICT environment that is integrated into the school's curriculum. These two schools provided two distinct geographical groups of students and were observed over a significant time period.

Special School-S1

The vision of this school community encompasses a commitment to achieving excellence in education for students with additional learning needs through a curriculum which integrates learning technologies with best practice in teaching and learning. The values embraced by the school community are: Respect, Personal Best, Happiness, Cooperation, and Honesty. These values are imbedded in the Student Code of Conduct and the Staff, Principal and School Council Codes of Practice (Adam and Tatnall 2010).

Located on two campuses in the northern suburbs of Melbourne, School-S1 is a day specialist school which caters for students with mild to moderate intellectual disability between the ages of 5 and 18 years. Students come from a wide geographical area and diverse socio-economic backgrounds. The school's junior annexe (on another site) provides three classrooms for students aged between 5 and 9 years of age. The school itself is situated in attractive, well maintained grounds with excellent facilities which include: a Technology Centre, fully equipped gymnasium, Healthy Living Centre, Art and Craft room, well-resourced library, modern playground equipment, four school excursion buses, computer networking across the school, multi-purpose room and shaded outdoor playing areas.

The school, with 71 teaching and ancillary staff has an enrolment of approximately 250 students. Enrolment is dependent on eligibility criteria as determined by the Department of Education and Training. A significant number of teachers at the school have post-graduate qualifications in special education. School-S1 supports integration into and from mainstream schools. Support services available to parent/carers and students include social workers, guidance officers, speech therapists and visiting teacher services.

The implementation of the Early Years Literacy and Numeracy Program at Junior and Middle school levels has increased opportunities for improved student skills in literacy and numeracy. The Secondary School has a focus on the enhancement of student engagement through Middle Years strategies based on improved Literacy and Numeracy, and the introduction of a 'Thinking Curriculum'. The Transition Centre caters for students sixteen to eighteen years of age with a focus on the development of dual pathways to cater for the diverse needs of these students. The school has a high level of commitment to a curriculum which integrates learning technologies with best practice in teaching and learning in order to enhance educational outcomes for its students. The Technology Centre has facilitated important opportunities for the school's students and its staff and also for staff from neighbouring schools (Adam and Tatnall 2008a).

In 2006 School-S1 introduced a number of social and networked learning activities and practices, with software such as Lumil, WordPressMU, ccHost, Urdit, Gregarius and Scuttle. This paper reports on some of the activities undertaken, technologies used and the progress made during this period. For the purposes of this research the individual items of social software used at School-S1 have been appraised separately, yet in practice these tools and technologies are complimentary and have been used concurrently. In fact, much of the power of social software is its interoperability. By using these tools the school expects its students to create and publish content and respond to the content creation of others. Created content can be aggregated to show progress and richness and depth of learning. Students can respond to the work of others, provide feedback and learn through their interactions with others online. Not only are the students learning, but they are also learning how to be independent learners (Adam and Tatnall 2008b).

Collaborative Web 2.0 technologies and practices strongly support effective knowledge management practices. By using open web-based standards, such as RSS and XML and open API web services, complimentary software can share data in rich and unique ways. In future years, interoperability may well be the key criteria for introducing new technologies and systems as teachers and administrators become more familiar with working in networked environments. In introducing the social software it was attempted to integrate with existing practices, using web-based tools and technologies to construct richer tasks for the students. For example, using web-based photo sharing, students and teachers tagged photos which were then used in student digital portfolios. By using this approach not only did they introduce the required skills and practices but the school also reduced the workload for teachers who had previously organised the resources for use in the earlier student digital portfolios (Adam and Tatnall 2008a; Adam and Tatnall 2008b).

By looking at their existing curriculum they identified opportunities to use web collaboration. For example, they found that they could use the social media sharing website ccHost within their loop-based music creation topic. Having traditionally used the audio samples that came with the software, they discovered they could easily integrate online networked learning to increase learning outcomes. Now students can find audio samples based on tags, use these samples in their composition and then share their composition online, highlighting the samples that they used. Other students could then make derivative works by taking samples from the composition of others. The school has attempted to provide all of the social networking web services on their school Intranet to ensure that the students' privacy and security can be carefully monitored.

Special School-S2

School-S2 is a purpose built 'Specialist School' providing a range of educational programs for students with special learning needs including global development delay, autism spectrum disorder, physical, social and emotional disabilities. It is located in Melbourne's north-west and is one of over 100 such Government funded Specialist School Facilities in Victoria (Adam and Tatnall 2010).

In addition to delivering the key learning areas as part of its curriculum the school provides a broad range of programs that are designed to further enhance the independence of its students. Some of the programs include Augmentative Communication, Work Education, Outdoor Education, Bike and Road Safety Education, Music Therapy, Swimming and Hydrotherapy Programs, Riding for the Disabled, Home Crafts, Recreation and Leisure, and Health and Human Relations Programs. The school has around 90 students, 8 full-time and 3 part-time teaching staff, 15 part-time school support officers and 2 part-time administrative staff. The students are quite diverse in their special needs, both physical and intellectual. In the early 2000s the school had limited ICT resources for their learning needs with several Acorn machines, but very few PCs. The majority of the software was based on the DOS operating system platform where graphics and sound quality were limited. The library had a PC that was used to allow the students' access to the Internet, and PCs were used predominantly to reinforce language and numeracy skills. By 2010 the Acorn computers had been replaced by PCs.

The list of computer skills as was evidenced by a student survey indicated that some key programs were used to assist the students in their learning. The teaching staff were quite happy with these programs as they felt these were adequate for their students' needs. An examination of the school's technology policy and curriculum showed that the use of ICT was an integral part of the classroom teaching and learning. The school relied heavily on ICT policies and support from the Victorian Education Department, both for network access and software supplies. The administration systems were also provided and supported by the Department.

A research project over several years investigated the infrastructure to set up links between different classes at the local level. Given the limited support and availability for video-conferencing by the Education Department's resources, the attempt was welcomed by relevant school staff. The students showed a tremendous level of enthusiasm and immediate engagement when they began to communicate via the webcams. The main issue was, as expected, the limited bandwidth from the Local Area Network. Another significant issue was the security constraints and filters that are imposed on the Education Department's VicOne Network. This matter was further investigated in the following year of the research, but it was discovered that the Education Department had found that video-conferencing was not very much in demand for Victorian schools, and hence its support was downgraded considerably.

ACTORS, NETWORKS, AND STUDENTS

A significant difficulty arises in framing research in a situation like this that involves both technological and human actors ranging from students and teachers to software and broadband connections. Special Schools are particularly complex socio-technical entities and research into their infrastructure, organisation and curriculum needs to take account of this complexity. When dealing with the related contributions of both human and non-human actors, actor-network theory (Callon 1986b; Law and Callon 1992; Latour 1996) provides a useful framework. ANT reacts against the idea that characteristics of humans and social organisations exist which distinguish actions from the inanimate behaviour of technological and natural objects, instead offering a socio-technical approach in which neither social nor technical positions are privileged.

Like any ANT study, the first step in this analysis was to identify the actors which were a part of the students' networks. The task was to identify those entities the students considered had reduced to a single entity and association. While some actors were easily identifiable, others were more difficult to identify. We were not only looking for physical entities but also patterned effects of associations which the students enrolled as actors. For example, the computer technology has multiple levels of grammar which may be enrolled at any level of a single actor.

The actors involved in the adoption of this technology to assist students with Special Needs were found to include: the students themselves, their parents, their teachers, school principals, school ICT specialist teachers, the School Council, the Web, microcomputers (PC and Macintosh), software, Education Department policies, learning technology policy, the school environment, classroom environments, learning approaches and paradigms, delivery methods of instruction, engagement methods, thinking processes, technology infrastructure-bandwidth, curriculum, Internet resources, digital libraries and other related Special Schools.

Viewed through an ANT framework, actors relating to Special Schools and to Learning Disabilities are seen to contest and negotiate with each other in an attempt to influence the final outcome in a direction to their own liking. The Education Department, for example, might want to ensure that all it schools offer a similar level of service to students and to ensure and verify their accountability. The parents of a student with LD, on the other hand, would want the best for their own child regardless of what was going on in other schools. The technology (both hardware and software) itself acts in the way it was designed, both intentionally and unintentionally, to act.

A major challenge to schools of this type is to get all these actors to form a common problematisation (Callon 1986b) of their task – to all see the problem in the same way. If this can be achieved then all the actors can work together to achieve a common goal and ANT offers some ideas on how this might be achieved. ANT considers associations and interactions between human and non-human actors but its advocates make no claim that this approach can do any more than shed a little light on how a given approach is taken or technology is adopted. Despite this, we believe that if a researcher understands how the factors involved in the adoption of a new technology interact in one situation then it is possible to affect the outcome of a similar situation by assisting favourable interactions and doing one's best to reduce unfavourable interactions.

The concept of an actor underlies ANT, where 'actor' is the term used to represent any physical entity (such as a Special Education teacher or some particular item of computer software) that has a significant effect on the phenomenon under investigation (Callon 1986b; Law 1992). An actor is an abstraction that enables the analysis of situations where heterogeneous entities are encountered (Law 1992). It is considered as any entity able to associate texts, humans, non-humans and money (Callon 1991):

Accordingly, it is any entity which more or less successfully defines and builds a world filled by other entities with histories, identities, and interrelationships of their own (Callon 1991).

The purely social or technological approaches of other frameworks are essentialist and deterministic in nature, whereas ANT is designed to be anti-essentialist and non-deterministic. The concept of an actor allows sociologists to write about the situatedness of innovation and technology without the need to use demarcations separating the social from the natural; or sociological conventions from technological ones. The abstraction frees the analysis from the boundaries of disciplines, thereby allowing the observer to resist the need to reduce complex phenomena to a few well-defined political, social or technological categories (Callon 1986a).

The main advice on method offered by Latour in conducting an ANT research project is to 'Follow the Actors' (Latour 1996), but this is just a beginning. When using ANT for an investigation like this, the research process that is followed is far from linear and not just simply a matter of collecting data, analysing the data then writing it up. The process could better be seen as an iterative one and after the actors have been identified and 'interviewed' and networks of interactions between actors have been examined, the process continues to look for new actors and for how the technology may have been translated in the process of adoption.

Another step taken in the analysis was to identify and describe different translation moments in which actors adopted or were given specific roles. In the first instance this meant looking at descriptions of the students' work and identifying specific roles and different actors, including the technology, and how these were assigned to each other. Role identification led us to consider competing enrolments that existed within the students' network. It also highlighted some of the differences in role assignment that existed between the students, the researcher and ICT. This analytical step highlighted the enrolments and counter enrolments that existed within the students' networks. For example, the level of LD differed in the class and this made the study more challenging as the researcher became an actor and was aligned with their networks.

Having identified different enrolments among the actors, a further step was to consider how these enrolments were stabilised, that is, what actions did particular actors undertake in order to ensure that other actors within the networks accepted or acted in accordance with their assigned role? It was also important to consider how other actors may have been disassociated or marginalised through the actions of particular actors. In other words, how did the actions of one actor attract another while marginalising a third actor? It quickly became apparent that not only was the researcher a part of the students' networks, but also that the researcher attempted to engineer the students as part of his network. Learning and teaching became a competition between, and reconciliation of, different networks where certain actors, at different times sought to stabilise their own networks, at times using enrolments unacceptable to the researcher

or to ICT. This led to an analysis of different programmes of action (Latour 1991) which were used to load each actor's definition of their network. Programmes of action are a series of (set of) interessements geared to enlisting an actor, which has displayed counter enrolments (Latour 1991).

CASE STUDY DATA COLLECTION AND ANALYSIS

It is impossible to describe the full detail of each student's actor-world and only some elements of the students' network are described. The main point is that each student's network is different as each brought together different resources. Such networks differ in origin largely because the students exhibited different skill levels and knowledge of ICT tools. The resources presented students with different problems and led them to develop networks of varying complexity. The students' networks also differed in enrolment, organisation, utility, significance and purpose but these are properties more aligned with the discussion of black-boxing and enrolment.

At School-S2 a specific project was set up by the school community, and the researcher was asked to take part in this through the assistance and co-operation of the Principal, as well as several key staff members. He became a team member and worked very closely with a class of senior students who were referred to as the Transition Group which showed varied ability and special needs or handicap. The project was called the 1-2-1 (one-to-one) project and it was a significant step to providing the state of the art 'hands-on' resources to the group. For example, each student was provided with a Laptop to use and carry in class or to take home. The research was conducted in a very supportive manner. It should be noted that funding was also provided by the school of Information Systems at Victoria University and this allowed an analysis of the ICT infrastructure and provided several PDAs (personal digital assistants – the forerunner of the smart phone) along with Dragon digital voice recording equipment. The school gave access to all available data at the time. The students were assessed in the main domains of knowledge and, in particular, data were provided to the research that showed an analysis of the key indicators for each class in ICT for teaching and learning. The 1-2-1 project was later extended to include many other groups in the middle of the school.

The original concept by now had gone beyond what anyone could have imagined with the introduction of Social Networking programs. The main observations highlighted the strong self-esteem and engagement by the groups of students. Other data were collected by the ICT coordinator to determine the background of teachers in ICT. In fact at this time, the Victorian Government introduced the 'ICT Potential Project' with the aim to diagnose the various ICT skills of teachers in schools and how these were applied to the curriculum. The researcher became an actor himself in this process whilst endeavouring to discover the impact and existence of other actors and actor-networks or ensembles. Using participant observation the researcher became an observer and participant at school meetings and regular class sessions that involved technology and computer programs through the Web. In addition to observations he interviewed these actors and collected relevant documents in order to identify further actors and their networks.

The research found that there are several actors whose work is particularly important to the achievement of a good outcome in the school. Firstly, the ICT system, social software and networked learning activities must be appropriate and also they must be induced to co-operate with the students and teachers. In particular, the way that a particular piece of software sets up its user interface and approaches the delivery of its content using appropriate language and metaphor is most important. Also the nature of computer itself is significant as the PC and Macintosh have different operating systems and cannot always run the same software. When School-S1 was using the Acorn computers then later using PCs running MS-DOS, the computers themselves put forward a quite different face to the students and acted in a quite different manner to the later Windows PCs and the Apple Macintosh. Close co-operation between these technological and human actors is essential to success and this co-operation can be facilitated by another important actor: the ICT specialist teacher who should have a good idea of the appropriate technology, what is required educationally and also of the capabilities of students and teachers to understand and use the software and hardware. It does not take much observation to see that the students were readily willing to work with ICT, enjoyed using it and gained substantial benefits from this use: they did not need much convincing. Hopefully the other teachers can also be brought to see the benefits of using ICT, even if they do not use it themselves, but this is not always the case and this is where another important actor comes in. The School Principal is crucial to the success of this program as without the Principal's support, many things would not be possible. One area where the Principal can be of considerable importance is in influencing, or perhaps coercing other teachers to support the program. Another is with the provision on funds for hardware and software purchases and the provision of time for the ICT co-ordinator to find out about new products and services. One thing to come out very clearly from our research is the importance of the School Principal, and without an actively supportive

Principal there is little chance that the project of using ICT with these children will succeed.

The way in which ICT was used in the early part of the research at Special School-S1 indicated willingness by staff to use technology to support teaching and learning in a limited way. However, this position was seen to be improved in later observations. The school adopted a forward looking policy in its use of ICT by appointing an ICT co-ordinator and extending its curriculum to incorporate VCAL studies (- the more practically oriented 'Victorian Certificate for Applied Learning'). In doing so, the school provided an extended curriculum to be on a par with other Special Schools and it now applies benchmarking through the Victorian Education Learning Standards (VELS). The way that ICT was used by teachers at Special School-S2 to provide support for students with Learning Disabilities also illustrates how a powerful and pervasive philosophy and ethos, tightly aligned networks within the school community and an alert and sensitive School Principal have ensured the school community's satisfaction with a dynamic model of support for ICT adoption via the 1-2-1 (one-to-one) ICT project. The model has developed and changed in response to need and the availability and type of resources have determined its overall shape. The model seemed to be based on curriculum innovation but also on technological innovation and is quite proactive; its focus is on addressing the current needs of students, parents and teachers. For example, the inclusion of social networking concepts like blogs that was introduced in 2007 enhanced the way ICT was used in the classroom with students engaging and producing work of higher standard and displaying stronger skills.

Making use of ANT concepts such as the 'moments' of innovation translation (Callon 1986b): problematisation, interessement, enrolment and mobilisation is also useful in the analysis here. Adoption of various aspects of ICT in Special Schools should definitely be seen in terms of technological innovation and it should be analysed as such. It is important that all the actors (both human and non-human) can be seen to adopt the same problematisation of ICT in Special Schools in order for interessement to produce full and complete enrolment of students, teachers, parents and the technology itself. In fact in both schools this was found to be the case with mobilisation also occurring in that both students and parents were keen to promote the use of ICT in this way to others.

The leadership of the key actors like the Principal, ICT Co-ordinator and several leading team teachers who worked in the 1-2-1 ICT project, all problematised their own vision of integrating ICT in the curriculum. The research observations in the period 2005 – 2007 concluded with a successful translation and mobilisation of the pilot project through its extension to other curriculum areas and levels in the school. The documents from the pilot and the social networking implementation testify to the way ICT was integrated successfully into the curriculum and school infrastructure and organisation with very positive learning outcomes.

CONCLUSION

The study showed that ICT does indeed assist students with Learning Disabilities and that there is no doubt that its use can have a beneficial impact on the education of children with Learning Difficulties. The study identified leadership, infrastructure, consistent policy from within the school and the Education Department as well as parental awareness and support as significant factors in this.

ICT can assist students both by improving their self-esteem by providing the means by which they can achieve something they consider worthwhile, and also by facilitating the acquisition of useful life skills. To achieve this result however, the actors involved, both human and non-human must be induced to work together to produce the desired result. This is where the benefits of using an ANT framework become very apparent, and in particular in demonstrating the importance of one particular actor. The role of this particular actor, the School Principal, is crucial. This focal actor has the power to provide appropriate funds when required, to coerce his colleagues into working towards a common goal, to encourage the school ICT specialist to make good use of the relevant software and make it easy to use, and to reassure the parents that the school is doing its job well. The use of actor-network theory as a research framework facilitates well the holistic analysis of schools such as these and how they operate.

The study also demonstrated that actor-network theory can be usefully applied to examining this. Particularly at Special School-S2, ICT was embedded into the curriculum and the beliefs of staff and the whole school community. ICT can be seen as a strong vehicle and enabler for good pedagogy as it reinforced students' communication skills and engagement and was a clear indicator and facilitator for school-to-work transition. The research found that the most effective manner that teachers can implement programs using computer-based technology is to integrate the technology fully into the curriculum. In doing this the Special Needs students potentially gain lifelong skills and also enhance their self-esteem and communication skills. Although the data collection period covered several years, and was completed four years ago, we do not believe that this has had any significant effect on what we can conclude from the study. Over this period things have

changed: computer hardware and software has improved and the schools have moved along, but a recent check has found that this is not to the extent that any of the reported findings would need to be reconsidered.

REFERENCES

Adam, T. (2010). *Determining an e-learning model for students with learning disabilities: An analysis of web-based technologies and curriculum.* (PhD Dissertation). Victoria University, Melbourne, Australia.

Adam, T., & Tatnall, A. (2007). Building a virtual knowledge community of schools for children with special needs. Prague, Czech Republic: Information Technologies for Education and Training (iTET), Charles University.

Adam, T., & Tatnall, A. (2008a). ICT and Inclusion: Students with Special Needs. In *The New 21st Century Workplace.* Heidelberg Press.

Adam, T., & Tatnall, A. (2008b). Using ICT to Improve the Education of Students with Learning Disabilities. In Learning to Live in the Knowledge Society. Springer.

Adam, T., & Tatnall, A. (2010). Use of ICT to Assist Students with Learning Difficulties: An Actor-Network Analysis. In Key Competencies in the Knowledge Society. Springer.

Adam, T., & Tatnall, A. (2012). School Children with Learning Disabilities: An Actor-Network Analysis of the Use of ICT to Enhance Self-Esteem and Improve Learning Outcomes. *International Journal of Actor-Network Theory and Technological Innovation, 4*(2), 10–24. doi:10.4018/jantti.2012040102

Agran, M. (1997). Teaching Self-Instructional Skills to Persons with Mental Retardation: A Descriptive and Experimental Analysis. *Education and Training of the Mentally Retarded, 21*, 273–281.

Australian Schools Directory. (2011). *Australian Schools Directory: The Only Guide to all Australian Primary and Secondary Schools.* Retrieved from http://www.australianschoolsdirectory.com.au/

Bulgren, J. (1998). Effectiveness of a concept teaching routine in enhancing the performance of LD students in secondary-level mainstream classes. *Learning Disability Quarterly*, 11.

Callon, M. (1986a). The Sociology of an Actor-Network: The Case of the Electric Vehicle. In *Mapping the Dynamics of Science and Technology*. Macmillan Press.

Callon, M. (1986b). Some Elements of a Sociology of Translation: Domestication of the Scallops and the Fishermen of St Brieuc Bay. In *Power, Action & Belief. A New Sociology of Knowledge?* Routledge & Kegan Paul.

Callon, M. (1991). Techno-Economic Networks and Irreversibility. In *A Sociology of Monsters: Essays on Power, Technology and Domination*. London: Routledge.

Chan, L., & Dally, K. (2000). Review of the Literature. In *Mapping the Territor. Primary Students with Learning Difficulties: Literature and Numeracy*. Canberra, Australia: Department of Education, Training and Youth Affairs.

Commonwealth of Australia. (1998). Literacy for All: The Challenge for Australian Schools. Commonwealth Literacy Policies for Australian Schools. *Australian Schooling Monograph Series No. 1/1998*. Retrieved from http://www.dest.gov.au/archive/schools/literacy&numeracy/publications/lit4all.htm

DEETYA. (1999). *Students with Learning Difficulties. 2010*. Retrieved from http://www.dest.gov.au/archive/sectors/school_education/publications.../swd_rtf.htm

Johnson, G., Gersten, R., & Carmine, D. (1998). Effects of Instructional Design Variables on Vocabulary acquisition of LD students: A Study of computer-assisted Instruction. *Journal of Learning Disabilities, 20*(4). PMID:3553400

Kavale, K. A., & Forness, S. R. (1995). *The Nature of Learning Disabilities: Critical Elements of Diagnosis and Classification*. Lawrence Erlbaum Associates, Inc.

Klinger, J. K. (1998). Outcomes for Students With and Without Learning Disabilities in Inclusive Classrooms. *Learning Disabilities Research & Practice, 13*(3), 153–161.

Latour, B. (1991). Technology is society made durable. In *A Sociology of Monsters: Essays on Power, Technology and Domination*. London: Routledge.

Latour, B. (1996). *Aramis or the Love of Technology*. Cambridge, MA: Harvard University Press.

Law, J. (1992). Notes on the Theory of the Actor-Network: Ordering, Strategy and Heterogeneity. *Systems Practice, 5*(4), 379–393. doi:10.1007/BF01059830

Law, J., & Callon, M. (1992). The Life and Death of an Aircraft: A Network Analysis of Technical Change. In *Shaping Technology/Building Society: Studies in Sociological Change*. Cambridge, MA: MIT Press.

Learning Disabilities Association of Canada. (2002). *Official Definition of Learning Disabilities*. Retrieved May 2007, from http://www.ldac-taac.ca/Defined/defined_new-e.asp

MacMillan, D. L., & Hendrick, I. G. (1993). Evolution and Legacies. In *Integrating General and Special Education*. Academic Press.

Riddell, S., & Watson, N. (2003). *Disability, Culture and Identity*. Harlow, UK: Pearson Prentice Hall.

Rivalland, J. (2000). Definitions & identification: Who are the children with learning difficulties? *Australian Journal of Learning Difficulties*, 5(2), 12–16. doi:10.1080/19404150009546621

Shakespeare, T. (1999). Disability, Genetics and Global Justice. *Social Policy and Society*, 4(1), 87–95. doi:10.1017/S1474746404002210

Shaw, S., Grimes, D., & Bulman, J. (2005). Educating Slow Learners: Are Charter Schools the Last, Best Hope for Their Educational Success?. *The Charter Schools Resource Journal, 1*(1).

Zigmond, N., Jenkins, J., Fuchs, L. S., Deno, S., Fuchs, D., & Baker, J. N. et al. (1995). Special education in restructured schools: Findings from three multi-year studies. *Phi Delta Kappan, 76*, 531–540.

KEY TERMS AND DEFINITIONS

Assistive Technology: In Special Education there has been awareness for the past decade of the importance of the design and development of ICT-based platforms to enhance the learning outcomes of students with Learning Disabilities.

School Children with Learning Disabilities (LD), Sometimes Known as Learning Difficulties or Special Needs: These include a number of disorders which may affect the acquisition, organization, retention, understanding or use of verbal or nonverbal information and may affect learning in individuals who otherwise demonstrate at least average abilities essential for thinking and/or reasoning.

School Communities: Schools of all types form part of the broad community in any city, and schools themselves form a series of communities within this broad community.

Schools and Special Schools: Educational institutions that cater for children with Special Needs (Learning Disabilities) have a lot in common with each other, but less in common with mainstream schools, even when these are located quite close by.

Special Education: A system of education designed to cater for the needs of students with Learning Disabilities.

Special Schools: Students with Learning Disabilities require some assistance and support in their learning, and one way this is offered is in Special (Needs) Schools. These special schools form community groupings, and particularly groupings consisting of other special schools and educational institutions offering support to students with Special Needs.

Chapter 2
How an Actor Network Theory (ANT) Analysis Can Help Us to Understand the Personally Controlled Electronic Health Record (PCEHR) in Australia

Imran Muhammad
RMIT University, Australia

Nilmini Wickramasinghe
Epworth Healthcare, Australia & RMIT University, Australia

ABSTRACT

Australia has designed, developed, and now implemented its national e-health solution known as the Personally Controlled Electronic Healthcare Record (PCEHR). This is a unique system as it subscribes to a shared governance model between patients and providers. To date, though, as with other national e-health solutions, there is poor uptake and much concern regarding the success of this multi-million dollar project. The authors contend that while these implementations and adoptions of e-health solutions are necessary, it is essential that an appropriate lens of analysis should be used in order to maximise and sustain the benefits of Information Systems/Information Technology (IS/IT) in healthcare delivery. Hence, in this chapter, the authors proffer Actor-Network Theory (ANT) as an appropriate lens to evaluate these various e-health solutions and illustrate in the context of the Personally Controlled Electronic Health Record (PCEHR), the chosen e-health solution for Australia.

INTRODUCTION

Globally, governments are increasingly investing in health information technology particularly in digitalising health records as well as other e-health solutions. This is in response to immense pressures of changing patient demographics, health, financial implications, work force shortages, advancements in medical technologies and their impact on healthcare demand and delivery as well as a move towards a system where interaction

DOI: 10.4018/978-1-4666-6126-4.ch002

between healthcare providers and consumers can achieve maximum output with limited human and financial resources (Wickramasinghe and Schaffer 2010).

It is well established that healthcare is an information rich industry (ibid). The underlying assumption in support of the introduction of IT(information technology) in healthcare service delivery is that by improving the ways of accessing and sharing information across healthcare systems and moving away from pen, paper and human memory towards a new environment, where key stakeholders (for example: service providers, consumers, government agencies and healthcare managers) can reliably and securely share information electronically, will significantly improve health outcomes and quality of care (Mort et al. 2007), help with cost savings, improve patient involvement and produce useable secondary data for further research and training (Car et al., 2008). However, such a transformation is not a straightforward proposition and is sometimes faced with many known and unknown hurdles such as (technological, organisational, financial and people issues) because of the complex and multifaceted environment of healthcare service delivery where different human and non-human actors interact with each other in multiple ways (Ammenwerth et al., 2006;Catwell and Sheikh, 2009; Cresswell et al. 2010; Lorenzi et al., 2009; DesRoches et al. 2008; André et al. 2008).

Further, given the inherent complexities of healthcare operations, it has been argued that these kinds of interventions are challenging and need to be evaluated with theoretically informed techniques (Wickramasinghe and Schaffer, 2010). One approach identified in the literature, to facilitate correctly and accurately capturing the complexities and levels of interventions in healthcare operations, is to use a Socio-Technical Systems (STS) perspective (Wickramasinghe, Bali, & Lehaney,

2009; Yusof et al., 2007; Aarts et al., 2004). A Socio-Technical system is described as a system where technical dimensions and social dimensions of a system are interrelated (Cresswell et al. 2010). To determine the functionality of a system, it is important to understand a better fit between technical sub-systems and social sub-systems in an organisation (Mitchell & Nault 2008). This emphasis then is on not only studying the impact of the technology on organisations and their work processes but also the impact of social and people issues on technology and processes (Cresswell et al. 2010). For this reason, it is also important to understand the inter-relationship and interactions of the two between each other (Coiera, 2004).

To provide an even richer and more accurate picture of key healthcare operations as well as the impact of technology on these scenarios several scholars have argued that Actor-Network Theory (ANT) should be used together with an STS perspective (Wickramasinghe, Bali, & Lehaney, 2009; Yusof et al., 2007; Aarts et al., 2004; Cresswell et al. 2010). Hence, this paper reflects on the use of Actor-Network Theory to evaluate the Personally Controlled Electronic Health Record (PCEHR) in the Australian context in an attempt to demonstrate the merits of such an approach.

THE PERSONALLY CONTROLLED ELECTRONIC HEALTH RECORD (PCEHR)

Before discussing the PCEHR and its benefits, it is important to first understand that there are many different terminologies and vocabularies used interchangeably for clinical communication and electronic record handling and storage. In general all these terms typically make up the myriad of e-health solutions currently discussed in most countries. The ambiguity in the use and significance of the terms used can become an obstacle in the progress of ehealth adoption. If the definition of the term used for the system is not clear, this can complicate the contractual matters along with policy expectations and directives and expected features of product. It further can instigate confusion in developing polices and regulations or legislation. For this reason, we look at publications in the past 10 years in order to understand the nomenclature and

terminologies used for a better understanding of the meanings of the term (refer to Table 1).

The terminology adopted in Australia for electronic record keeping and its e-health solution is known as the Personally Controlled Electronic Health Record (PCEHR) which sits between individually-controlled health records and healthcare provider health records (NEHTA and DoHA 2011; Figure 1). Thus, the PCEHR has a shared use and mixed governance model (NEHTA and DoHA 2011; Figure1).

Table 1. Nomenclature and terminologies

Term	Description	Reference
Electronic Medical Record (EMR)	Any electronic record maintained by a private practitioner or family health teams locally within their clinics computer systems is referred to as an EMR. Service providers can access patient's visit history, health record, previous diagnosis, plans of care and lab reports and their placement orders with in their clinical practice with-out any integration with other providers or hospital systems and with very limited scope.	(Nagle 2007; Fisher,1999)
Electronic Patient Record (EPR)	The EPR is an electronic record maintained and managed by healthcare organisations at large. These healthcare organisations like hospitals can share this record with in their jurisdiction if shared infrastructure is available along with same vendor solutions. The ownership of the record is held by the organisation and access is typically restricted to the authorised users and user groups within the organisation's healthcare circle. Patients cannot access this record electronically and their GPs can request information.	(Mohd & Mohamad 2005; Nagle 2007).
Electronic Health Record (EHR)	The Electronic Health Record is a comprehensive health record that consists of information collected from different service providers and point-of-service systems such as pharmacies and diagnostic centres for all healthcare encounters an individual may have with them. Within the larger concept of EHR, health information is proprietary to the person whom it relates to and other service providers within or outside the jurisdiction can access this information only with proper access rights.	(Nagle 2007; Neves et al. 2008)
Personal Health Record (PHR)	The PHR can be defined by its attributes including, but not limited to, the nature of information and its scope in the PHR, the source of information, the party who owns the record, the functionary of the system, data reposition location and type, approaches towards privacy and security, and definition of access roles. It can be differentiated from other systems by its maintenance, ownership and access rules. Depending on the type of PHR users can draw information from different sources including healthcare providers, medical devices, wellness promoter systems or websites, research institutions and government websites.	(Halamka et al. 2007; Krohn 2007; Thompson & Brailer 2004; Detmer et al. 2008; Jones et al. 2010).

Figure 1. The position of the PCEHR in the e-health solution spectrum

Specifically, the PCEHR is a person-centric secure repository of electronic health and medical records of individual's medical history that would act as a hub for linking hospital, medical and pharmaceutical systems using a patient unique identifier (NHHRC, 2009:134). One of its key features is that it captures information from different systems and presents this information in a single view to consumers and authorised service providers for better decision making about health and service delivery (NEHTA & DoHA 2011). This is a hybrid health information system that integrates web based personal health records with a clinical electronic health record system and allows shared access to both consumers and healthcare providers based on a shared responsibilities and mixed governance model. (Leslie 2011). The proposed PCEHR will have the key features as presented in Table 2.

Unlike most other eHealth solutions to date, the proposed PCEHR in Australia has a unique point of difference. Specifically, this PCEHR requires the patient (citizen) to authorise any additional information to be included in his/her record. This includes giving the access rights to providers with the exception of an emergency situation where a provider can access records other than those labelled as "no access" (NEHTA and DoHA, 2011). These records would be identified by Individual Health Identifiers (IHI) assigned by Medicare Australia, in which, IHIs will be created automatically but can be activated only at the request of the individual. More importantly, this system will operate on the principal of an "opt-in" model. That is an entirely voluntary basis, in which a user can "opt-in" or "opt-out" any information they wish to be included and can even withdraw from the system at any time. In addition, individuals would also have the final authority to allow and deny access of any service provider or organization (NEHTA & DoHA, 2011).

Further, this system will also include a reporting service, which can be used for the analysis of the information from multiple sources for better decision-making. Moreover, as this system matures many more types of reports would evolve and may help with research and training.

The benefits of ICT use in the healthcare sector may be enormous, however the uptake and diffusion of ICT in healthcare service and delivery is still a major challenge for healthcare service providers, governments and developers (Showell 2011; Leslie 2011; Hordern et al. 2011; Boonstra & Broekhuis 2010; Liu et al. 2011; Trudel 2010; Bernstein et al. 2007; HFMA 2006; Tang et al. 2006). There are very high stakes involved in health information system adoption and implementation for service providers and governments as well as system developers in the shape of financial,

Table 2. PCEHR features (NEHTA and DoHA 2011)

- It will provide patients health information including medication, medical history referrals, lab test results, prescriptions, discharge summaries, allergies and immunisations.
- Rigorous governance and privacy procedures.
- Fast and reliable access of patient's health information for both consumers and service providers remotely via fast Internet connection to enhance decision making for diagnoses and treatments and preventive actions.
- Provide high quality data for policy development as-well-as research and planning to government bodies.
- Provide all activity history of any actions preformed on the PCEHR by any service provider and/or consumer.
- Individuals can make enquires and complaints about the management of their information.
- Easy access to health literacy information via a direct link.
- Can collect information from consumer devices such as blood pressure monitors or blood glucose monitors.

clinical and organisational risks (Westbrook & Braithwaite 2010; Karsh et al. 2010). This makes it imperative to identify the challenges that the healthcare service sector is facing in the adoption of e-health solutions, especially from the PECHR perspective. The following section examines the key constraints to the PCEHR development, adoption, implementation and diffusion of the new eHealth solution. In so doing, it is possible to appreciate the complex nature of healthcare operations as well as the importance of people, process and technology issues.

CONSTRAINTS TO THE DEVELOPMENT, ADOPTION, IMPLEMENTATION AND DIFFUSION OF PCEHR TECHNOLOGY

A large number of health information systems have been implemented around the globe with mixed results. Despite the claims that eHealth solutions can play a significant role in efficiency and effectiveness of healthcare service delivery, literature still provides evidence of failed clinical systems and lack of adoption by users (Protti & Smit 2006; Basch 2005; DesRoches et al. 2008).

Challenges and barriers to develop, adopt and implement eHealth solutions have been extensively debated. Researchers have divided these barriers into different categories ranging from environmental, social, technical and organisational (André et al. 2008) and note that these categories play crucial roles in the decision-making processes of technology adoption (Huang & Palvia 2001). In this study, we believe that the barriers identified by André et al. (2008) while important and a necessary condition may not be all-inclusive or sufficient. The primary reason for this be-

ing that in a healthcare service context where organisations are now required to work as a networked framework, health information technology implementation and adoption becomes an even more complex and challenging endeavour because of the different business processes, available infrastructure, compatibility issues, decision centres, authorization mechanisms and hierarchies, enterprise systems and data semantics (Avgerou 2008; Liu et al. 2011; Trudel 2010).

In the Australian context, social issues are considered a significant topic in healthcare IT transformation. This includes topics relating to individual privacy, legal, health information security and ethics issues. Individual privacy is considered as the private property of a person to whom this information is relating, so information privacy is defined as the ability of an individual to control the access to personal information (Culnan & Armstrong 1997). The breach of privacy is a common concern among Australian consumers and privacy advocates despite the draft's (Personally Controlled Electronic health Record Act 2011) emphasis on the security and privacy of an electronic health record of individuals as well as any information that is protected by law. Further, the placement of these requirements and standards as well as the fact that the language is vague only serves to add more confusion and raise concerns among both healthcare providers and consumers (Hoffinan & Podguski 2008). It is noted that standardisation is important to set security and access rules of the system (Hoffinan & Podguski 2008) and this has been identified as a policy issue. In addition, there is a need for direct involvement of citizens in consultation about the development and implementation of the PCEHR system not just interest groups and citizen privacy and information security groups that have been emphasised as policy issues and to date this has proved challenging and contentious (Showell 2011).

Secondly, financial issues have been identified as a big hurdle in the adoption and implementation of health information technology especially an electronic health record (Aaronson et al. 2001; Aarts et al. 2004; Aarts & Koppel 2009; Abbott 2005; André et al. 2008; Bernstein et al. 2007; Boonstra & Broekhuis 2010; Culnan & Armstrong 1997; HFMA 2006; Kennedy 2011; Liu et al. 2011; Thweatt & Kleiner 2007; Trudel 2010; Ashish 2009; Bahensky et al. 2008; Bates 2005; Bath 2008; Weimar 2009; Kaplan & Harris-Salamone 2009). These issues range from start-up costs to software upgrades and staff training. It is observed that for sole practitioners; the capital cost is very high and not considered as a good investment for returns (Ashish 2009). Lack of incentives, budget over runs and high time costs are other financial concerns (DePhillips 2007; Cohn et al. 2009; Boonstra & Broekhuis 2010; Liu et al. 2011; Trudel 2010). This in turn makes the embracement of ICTs in healthcare problematic.

Thirdly, organizational issues have also been extensively reported in the literature. For example, poor governance, organisational culture and poor management of the change process harm the flow of transformation (Hoffinan 2009; Greenhalgh & Stones 2010; Kennedy 2011; Bernstein et al. 2007). Technological issues can also exacerbate the resistance to the adoption of health information technology and complicate the diffusion of PCEHR technology. Furthermore, the lack of infrastructure, standards and protocols which in turn results in fragmented healthcare information systems could further complicate a very complex situation for coordination (Davidson & Heslinga 2006; Hoffinan &

Podguski 2008; Kralewski et al. 2010; Vitacca et al. 2009; HFMA 2006; Kennedy 2011; Trudel 2010).

Fourthly, technology issues such as the lack of interpretability between different healthcare delivery and management systems can hinder the penetration of this technology and its sustainability (André et al. 2008; Kennedy 2011; Liu et al. 2011). Pre-implementation and post-implementation vendor support is another key concern for organisations (Kralewski et al. 2010; Cohn et al. 2009; Kennedy 2011; Liu et al. 2011; Trudel 2010; Tang et al. 2006). Lack of technical resources and experience with information technology implementation within healthcare settings is another problem faced by many (Torda et al. 2010; Trudel 2010; Liu et al. 2011; Kennedy 2011; André et al. 2008; Bath 2008; DePhillips 2007; Davidson & Heslinga 2006; McReavy et al. 2009). The accuracy of data obtained through the information system and its ability of sorting, querying and validating data in some cases is very poor and is considered as a big barrier for HIT adoption (Rosenbloom et al. 2006; Rosebaugh 2004; Kimaro & Nhampossa 2007).

Lastly, people issues are yet another big concern ranging from user acceptance (Frame et al. 2008; Agarwal & Prasad 1997), perceived ease of use (Al-Azmi et al. 2009), lack of knowledge about the system (Bath 2008; Elrod & Androwich 2009; Kaplan & Harris-Salamone 2009; André et al. 2008; Liu et al. 2011), lack of training , lack of stake holder consultation (Showell 2011), lack of willingness to assimilate the technology into daily routines and processes (Cash 2008; Ross et al. 2010; Davidson & Heslinga 2006; Kaplan & Harris-Salamone 2009), conflict between system and user embedded values (Cash 2008; Kaplan & Harris-Salamone 2009), complex and complicated user interfaces (Yusof et al. 2007), conflict between physician activities and training schedules (André et al. 2008; Yusof et al. 2007; Kaplan & Harris-Salamone 2009).

Given the complex nature of the healthcare system and the challenges and barriers described relating to the adoption and implementation of ICTs into healthcare contexts, the importance of conceptualising and framing the critical factors for evaluating the proposed PCEHR system cannot be over emphasised. Hence, the next section presents our conceptual model that attempts to capture all the key considerations as discussed above for further analysis.

CONCEPTUAL MODEL

Based on the literature discussed in the previous sections some key factors important for the successful implementation of a PCEHR have been identified and served to develop the proposed conceptual modal as shown in Figure 2. This conceptual model will serve to capture all the important aspects of the barriers and facilitators for the prediction of the successful adoption and implementation of the PCEHR. The proposed model identifies the network of different actors interconnected to each other. These actors include social, people, technology, finance, and organization. Based on the literature review, their interac-

Figure 2.

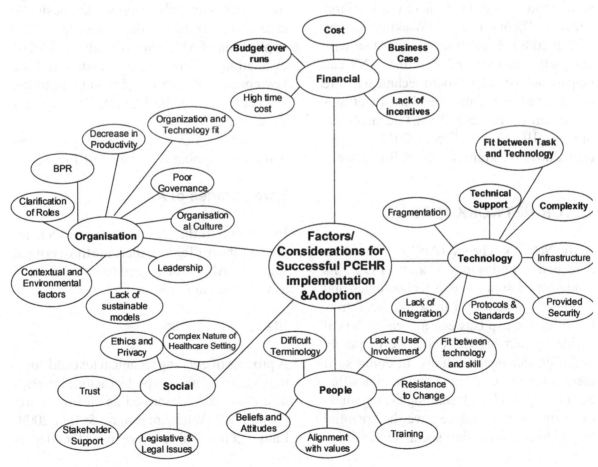

tions presented a very complex picture of relationships. To study this complex network of interactions of humans with technology in organisations and certain individual levels a Socio-Technical System (STS) perspective is proposed(Cresswell et al. 2010).

A Socio-Technical System is described as a system where technical dimensions and social dimensions are interrelated (Cresswell et al. 2010). To determine the functionality of a system, it is important to understand that a better fit between technical sub-systems and social sub-systems in an organisation is presented (Mitchell & Nault 2008). As seen in the conceptual model, it appears that this study is complex and consists of hybrid entities of both human and non-human elements such as organisation, people and social dimensions. Thus, to better understand this study an appropriate lens of analysis to use is Actor-network Theory (ANT) (Wickramasinghe & Bali 2009). This choice is consistent with many other researchers' view that ANT can help to understand the socio-technical nature of information systems in health care settings (Walsham 1997; Tobler 2008; Greenhalgh & Stones 2010; Cresswell et al. 2010; Timpka et al. 2007; Wickramasinghe & Bali 2009).

ACTOR-NETWORK THEORY

Actor-Network Theory (ANT) is based on a recursive philosophy (Latour 1992). Its fundamental stand is that technologies and people are linked in a network. ANT tries to bridge the gap between a socio-technical divide by denying the existence of purely social or technical relations. In doing so it takes a very radical stand and assumes that each entity (such as technologies, organisation) are actors therefore have the potential to transform and mediate social relationships

(Cresswell et al. 2010). It also emphasises the concept of heterogeneous networks because of the non-similar nature of elements and their relationship in network makes these networks open and evolving systems (Hanseth 2007). Therefore Actor-networks are highly dynamic and inherently unstable in their nature; and a better understanding of how alignment between people, technology, their roles, routines, values, training and incentives as well as understanding of the role of technology that how it can facilitate or negatively impact the work processes and tasks in an organisation can stabilise these network to some extent (Greenhalgh & Stones 2010; Wickramasinghe et al. 2011). For this reason, ANT can be a material-semiotic approach and can provide an appropriate lens to study the ordering of scientific, technological, social, and organisational processes and events (Wickramasinghe et al. 2011). To realise the importance of the application of ANT into the study of PCEHR evaluation it is important to understand the key concepts of ANT and map them to the critical issues of the PCEHR development in Australia.

An initial assessment of these key concepts and their mapping is provided in Table 3.

Three Stages of ANT

In addition to the key concepts of ANT described above (Table 3) there are three critical stages of ANT which need to be considered as presented in Figure 3.

Stage 1: Inscription

A process of creating technical text and communication artifacts to protect actor's interests in a network is described as an inscription (Leila 2009; Wickramasinghe & Bali 2009; Latour 2005). This is a term used for all texts

Table 3. Key concepts of ANT and their mapping with the PCEHR

Key Constructs of ANT	Initial Mapping of the PCEHR with ANT
Actor/Actant: Actors are the web of participants in the network including all human and non-human entities. Because of the strong biased interpretation of the word actor towards human; a word actant is commonly used to refer both human and non-human actors. Examples are human, information systems, technical artifacts, work process and graphical representation (Wickramasinghe et al. 2011).	In PCEHR context Actor/Actant are different stakeholders in healthcare delivery settings such as Technology (Web 2.0, Databases, Graphical User Interfaces, IHI and different Computer hardware and Software) and People (service providers, healthcare funders, healthcare service recipients, healthcare organisations, suppliers and private health insurers as well as clinical administrative technologies, work process and health records in the form of paper or electronic.
Heterogeneous Network: Is a network of aligned interests formed by the actors. This is a network of materially heterogeneous actors that is achieved by a great deal of work that both shapes those various social and non-social elements, and "disciplines" them so that they work together, instead of "making off on their own" (Latour, 1996, Latour, 2005; Wickramasinghe et al. 2011).	The PCEHR technology here is clearly a network of different applications in this context. But it is important to understand that the heterogeneous network in ANT requires conceptualising the network as aligned interest including people, organisations, standards and protocols and their interaction with technology. The key here is a better alignment and representation of the interests so that the healthcare delivery can be improved.
Tokens/Quasi Objects: Are created through the successful interaction of actors/actants in a network and are passed between actors within the network. As the token is increasingly transmitted or passed through the network, it becomes increasingly punctualised and also increasingly reified. When the token is decreasingly transmitted, or when an actor fails to transmit the token (e.g., disconnection of patient portal from PCEHR Server), punctualisation and reification are decreased as well (Wickramasinghe et al. 2011).	In PCEHR context this translate to successful cost effective and efficient healthcare delivery, such as for GPs treating patient by having a capability of sharing health information with other service providers and for patients who are on long term medications having capability to print prescription from home issued by their doctor through the PCEHR portal. It is important to understand that to maintain the integrity of the network at all times is very important because if wrong information is passed through the network, the errors would be devastating and can propagate quickly and will multiply.
Punctualisation: The concept of ANT is pretty much based of punctualisation. Within the domain of ANT every actor in the web of relations is connected to others and as a whole it will be considered as a single object or concept same as the concept of abstraction is treated in Object Oriented Programming. These sub-actors are sometime hidden from the normal view and only can be viewed in case of the network break-down; this concept is often referred as a depunctulisation. Because ANT require all actors or sections of network to perform required tasks and therefor maintain the web of relations. In case of any actor cease to operate or maintain link the entire Actor-Network would break down resulting in ending the punctualisation. Punctualisation is a process and cannot be achieved indefinitely rather is a relational effect and is recursive that can reproduce itself (Law 1997).	For example, a computer on which one is working would be treated as a single block or unit. Only when it breaks down and one needs help with spare parts can reveal the hidden chain of network consist of different actors made up of (People, Computer parts and organisations). Similarly in a PCEHR context, uploading the health record of a patient is in reality a consequence of the interaction and coordination of many sub-tasks. This will only reveal itself if some kind of breakdown at this point occurs and depunctualisation of the network happens and all sub-tasks then would need to be carefully examined.
Obligatory Passage Point (OOP): Broadly refers to a situation that has to occur in order for all the actors to satisfy the interests that have been attributed to them by the focal actor. The focal actor defines the OPP through which the other actors must pass through and by which the focal actor becomes indispensable (Callon, 1986).	In PCEHR context, we can illustrate this by taking the example of access rights. The interface of the system is developed in a way that no service can access any record without using their IHI, which in this case constitute an obligatory passage point through which they have to pass for their everyday activities.
Irreversibility: Callon (1986, p. 159) states that the degree of irreversibility depends on (i) the extent to which it is subsequently impossible to go back to a point where that translation was only one amongst others and (ii) the extent to which it shapes and determines subsequent translations.	In the context of a very complex nature of healthcare operations irreversibility is very less likely to occur and would be more dependent on social networks and the nature of the interaction between human and non-human actors in the network. Here it is important to remember though the chain of events needs to be monitored carefully so the future events can be addressed in the best possible manners.

Figure 3.

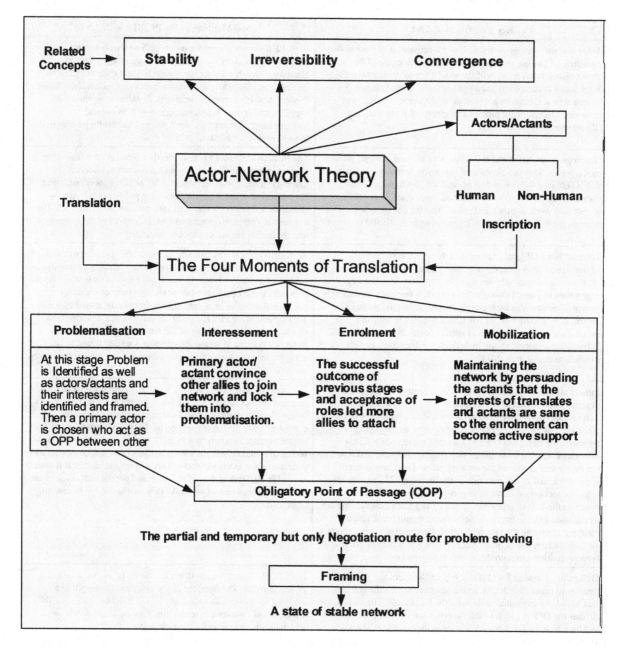

and communications in different mediums including but not limited to journal articles, conference papers and presentations, grants proposals and patents. The idea of Inscription also relates to the notion of durability; for instance a general discussion would be less durable as compared to a recorded meeting. Therefore, the idea behind Inscription is to enhance the durability of the network by associating them with durable material.

Actors can use Inscription as a path to gain credibility in enrolment and the co-optation process during translation.

Stage 2: Translation

Translation is a very important and vital concept in ANT. This term is used to explain the process of creation of Actor-Networks and the formation of ordering effects (Callon 1986; Law 1992). This stage can help researchers in providing the insight into how the software system (PCEHR) can be integrated into the very complex environment of healthcare. The process of Translation can also be called the process of negotiation because after the creation of the network in the presence of many actors a strong or primary actor would translate interests of other actors into his/her own by negotiating with them. At this stage all actors decide to be part of network if it is worth while to build it (Wickramasinghe & Bali 2009).

The process of Translation of Actors/ Actants is achieved through a series of four moments of translations (Callon, 1986). Figure 3 depicts the key ANT concepts along with these four moments of Translation.

Stage 3: Framing

Framing is an operation that can help to define actors and distinguish different actors and goods from each other (Callon, 1986). This last and final stage in the ANT process can help network to stabilise. At this stage key issues occurred throughout the PECHR should already have been negotiated within the network and technologies can become more stable over time (Wickramasinghe & Bali 2009).

THE ACTOR-NETWORK THEORY APPROACH TO EVALUATE IT IN HEALTHCARE

ANT is considered an appropriate choice to analyse the PCEHR evaluation study because it can identify and acknowledge any impact of human and non-human social or policy issues within the healthcare setting (Latour et al. 1996). Moreover, it is robust enough to accurately capture all the complexities, nuances and richness of healthcare operations. In so doing, it can also help to investigate and theorise the question of why and how networks come into existence, what sort of associations and impact they can have on each other, how they move and change their position in a network, how they enrol and leave the network and most importantly how these networks can achieve stability (Doolin & Lowe 2002; Callon 1986; McLean & Hassard 2004). ANT's assumption is that if any new actor is enrolled in the network or an old actor leaves the network it would affect whole network (Cresswell et al. 2010; Doolin & Lowe 2002). These considerations are naturally relevant in the context of the PCEHR.

In addition, ANT also can help to understand the active role of objects in shaping social realities by challenging assumptions of the separation between non-human and human worlds (Walsham 1997; Greenhalgh & Stones 2010; Tobler 2008; Law & Hassard 1999; Rydin 2010). This helps researchers to study the complexities of the relationships between human and non-human actors, the sustainability of power relationships between human actors and what kind of influence artefacts can have on human actors relationships in transforming healthcare (Cresswell et al. 2010).

The rationale to choose ANT to evaluate the PCEHR system is thus the strength of ANT to identify and explore the real and perceived complexities involved in Australian healthcare service delivery. Although ANT has been applied in the implementation and adoption of different healthcare innovation studies (Berg, 2001; Cresswell et al. 2010; Cresswell et al. 2011; Bossen, 2007; Hall, 2005); It is important to note however, that ANT has also been criticised for its limitations (Williams, 2007; Walsham, 1997; Cresswell et al. 2010; Cresswell et al. 2011; Greenhalgh and Stones, 2010).

Some of the key limitations identified in the literature include: ANT's lack of ability to pay proper attention to broader social structures (Walsham, 1997), its lack of ability to pay attention to macro-environmental factors (McLean and Hassard, 2004), ANT's inability to explain the relationship formations between actors and changes of events in network (Greenhalgh and Stones, 2010; Cresswell et al. 2011; Kaghan and Bowker, 2001). To overcome the ANT's limitations, many researchers have suggested that ANT can be combined with other theoretical lenses such as Structuration theory (ST), Strong Structuration Theory (SST), Theory of Practice (ToP) and Social Shaping of Technology (SSoT) (Trudel, 2010; Cresswell et al. 2011; Greenhalgh and Stones, 2010; Walsham, 1997).

CONCLUSION

The need for IT based interventions in the healthcare services delivery to improve information and communication flow is well recognised all around the globe. Different e-health solutions are being implemented with mixed success to address this challenge(Protti & Smit 2006; Basch 2005; DesRoches et al.

2008; Greenhalgh and Stones, 2010). It is, therefore important to evaluate these technologies with theoretically informed approaches in an attempt to enjoy more successful outcomes.

We believe it is important to develop a deeper understanding constraint to the development, adoption, implementation and diffusion of these various ehealth solutions. Specifically, we suggest that a socio-technical ANT based approach can inform and facilitate such evaluations and we illustrate this by presenting an initial analysis and conceptual model for the PCEHR development in Australia. We are confident that this approach can be beneficial to both practitioners and researchers.

This paper has served to outline the key concepts of ANT that are relevant in the context of the PCEHR adoption and implementation and discussed the appropriateness of an ANT based theoretical lens for the evaluation of PCEHR in very complex environment of healthcare service delivery. We have also noted that ANT has been criticised by several scholars on the basis of its appropriateness as an ontology and/or epistemology (Latour, 2005). Therefore we recommend that to reduce the negative impact of these limitations the use of Structuration Theory be incorporated. To further strengthen our study, we plan to combine ANT with structuration theory to analyse the PCEHR in Australia. We close by calling for more confirmatory follow up research in this vital area.

REFERENCES

Aaronson, J. W., Murphy-Cullen, C. L., Chop, W. M., & Frey, R. D. (2001). Electronic medical records: the family practice resident perspective. *Family Medicine, 33*(2), 128–132. PMID:11271741

Aarts, J., Doorewaard, H., & Berg, M. (2004). Understanding Implementation: The Case of a Computerized Physician Order Entry System in a Large Dutch University Medical Center. *Journal of the American Medical Informatics Association, 11*, 207–216. doi:10.1197/jamia. M1372 PMID:14764612

Aarts, J., & Koppel, R. (2009). Implementation of computerized physician order entry in seven countries. *Health Affairs (Project Hope), 28*(2), 404–414. doi:10.1377/hlthaff.28.2.404 PMID:19275996

Abbott, T. (2005). *The adoption of eHealth in Australia*. Paper presented at the Health Informatics Conference Melbourne 2005. Melbourne, Australia.

Agarwal, R., & Prasad, J. (1997). The Role of Innovation Characteristics and Perceived Voluntariness in the Acceptance of Information Technologies. *Decision Sciences, 28*(3), 557–582. doi:10.1111/j.1540-5915.1997.tb01322.x

Al-Azmi, S., Al-Enezi, N., & Chowdhury, R. (2009). Users' attitudes to an electronic medical record system and its correlates: A multivariate analysis. *The HIM Journal, 38*(2), 33–40. PMID:19546486

Ammenwerth, E., Iller, C., & Mahler, C. (2006). IT-adoption and the interaction of task, technology and individuals: A fit framework and a case study. *BMC Medical Informatics and Decision Making, 6*. PMID:16451720

André, B., Ringdal, G. I., Loge, J. H., Rannestad, T., Laerum, H., & Kaasa, S. (2008). Experiences with the implementation of computerized tools in health care units: A review article. *International Journal of Human-Computer Interaction, 24*(8), 753–775. doi:10.1080/10447310802205768

Armstrong, B. K. et al. (2007). Challenges in Health and Health Care For Australia. *The Medical Journal of Australia, 187*(9), 485–489. PMID:17979607

Ashish, K. (2009). Use of electronic health records in U.S. hospitals. *Medical Benefits*. doi: 101056NEJMsa0900592, 25

Avgerou, C. (2008). Information systems in developing countries: A critical research review. *Journal of Information Technology, 23*, 133–146. doi:10.1057/palgrave.jit.2000136

Bahensky, J. A., Jaana, M., & Ward, M. M. (2008). Health care information technology in rural America: Electronic medical record adoption status in meeting the national agenda. *The Journal of Rural Health, 24*(2), 101–105. doi:10.1111/j.1748-0361.2008.00145.x PMID:18397442

Basch, P. (2005). Electronic Health Records and the National Health Information Network: Affordable, Adoptable, and Ready for Prime Time? *Annals of Internal Medicine, 143*(3), 227–228. doi:10.7326/0003-4819-143-3-200508020-00009 PMID:16061921

Bates. (2005). Physicians And Ambulatory Electronic Health Records. *Health Aff, 24*, 1180 - 1189.

Bath, P. (2008). Health informatics: Current issues and challenges. *Journal of Information Science, 34*(4), 501. doi:10.1177/0165551508092267

Berg, M. (2001). Implementing information systems in health care organizations: Myths and challenges. *International Journal of Medical Informatics, 64*(2-3), 143–156. doi:10.1016/S1386-5056(01)00200-3 PMID:11734382

Bernstein, M. L., McCreless, T., & Côté, M. J. (2007). Five constants of information technology adoption in healthcare. *Hospital Topics*, *85*(1), 17–25. doi:10.3200/HTPS.85.1.17-26 PMID:17405421

Boonstra, A., & Broekhuis, M. (2010). Barriers to the acceptance of electronic medical records by physicians from systematic review to taxonomy and interventions. *BMC Health Services Research*, *10*(1), 231. doi:10.1186/1472-6963-10-231 PMID:20691097

Bossen, C. (2007). Test the artefact – Develop the organization: The implementation of an electronic medication plan. *International Journal of Medical Informatics*, *76*(1), 13–21. doi:10.1016/j.ijmedinf.2006.01.001 PMID:16455299

Callon, M. (1986). *Some elements of a sociology of translation: Domestication of the scallops and the fishermen of St. Bricue Bay.* Academic Press.

Car, J., Anandan, C., Black, A., Cresswell, K., Pagliari, C., & McKinstry, B. et al. (2008). *The Impact of eHealth on the Quality & Safety of Healthcare: A Systematic Overview and Synthesis of the Literature.* NHS Connecting for Health Evaluation Programme.

Cash, J. (2008). Technology can make or break the hospital-physician relationship. *Healthcare Financial Management*, *62*(12), 104–109. PMID:19069330

Catwell, L., & Sheikh, A. (2009). Evaluating eHealth Interventions: The Need for Continuous Systemic Evaluation. *PLoS Medicine*, *6*(8), e1000126. doi:10.1371/journal.pmed.1000126 PMID:19688038

Cohn, K., Berman, J., Chaiken, B., Green, D., Green, M., Morrison, D., & Scherger, J. (2009). Engaging physicians to adopt healthcare information technology. *Journal of Healthcare Management*, *54*(5), 291. PMID:19831114

Coiera, E. (2004). Four rules for the reinvention of health care. *BMJ (Clinical Research Ed.)*, *328*(7449), 1197–1199. doi:10.1136/bmj.328.7449.1197 PMID:15142933

Cresswell, K., Worth, A., & Sheikh, A. (2011). Implementing and adopting electronic health record systems: How actor-network theory can support evaluation. *Clinical Governance: An International Journal*, *16*(4), 320–336. doi:10.1108/14777271111175369

Cresswell, K. M., Worth, A., & Sheikh, A. (2010). Actor-Network Theory and its role in understanding the implementation of information technology developments in healthcare. *BMC Medical Informatics and Decision Making*, *10*, 67. doi:10.1186/1472-6947-10-67 PMID:21040575

Culnan, M., & Armstrong, P. K. (1997). *Information Privacy Concerns, Procedural Fairness and Impersonal Trust: An Empirical Investigation.* Retrieved from http://citeseer.ist.psu.edu/viewdoc/summary?doi=10.1.1.40.5243

Davidson, E., & Heslinga, D. (2006). Bridging the IT - Adoption Gap for Small Physician Practices: An Action Research Study on Electronic Health Records. *Information Systems Management*, *24*(1), 15. doi:10.1080/10580530601036786

DePhillips, H. A. (2007). Initiatives and Barriers to Adopting Health Information Technology: A US Perspective. *Disease Management & Health Outcomes, 15*(1). doi:10.2165/00115677-200715010-00001

DesRoches, C. M., Campbell, E. G., Rao, S. R., Donelan, K., Ferris, T. G., & Jha, A. et al. (2008). Electronic health records in ambulatory care–A national survey of physicians. *The New England Journal of Medicine, 359*(1), 50–60. doi:10.1056/NEJMsa0802005 PMID:18565855

Detmer, D., Bloomrosen, M., Raymond, B., & Tang, P. (2008). Integrated Personal Health Records: Transformative Tools for Consumer-Centric Care. *BMC Medical Informatics and Decision Making, 8*(1), 45. doi:10.1186/1472-6947-8-45 PMID:18837999

DoHA. (2010). *Building a 21st Century - Primary Health Care System - Australia's First National Primary Health Care Strategy*. Australian Government Department of Health and Ageing. Retrieved from http://www.yourhealth.gov.au/internet/yourhealth/publishing.nsf/Content/Building-a-21st-Century-Primary-Health-Care-System-TOC

Doolin, B., & Lowe, A. (2002). To reveal is to critique: Actor-network theory and critical information systems research. *Journal of Information Technology*, 69–78. doi:10.1080/02683960210145986

Elrod, J., & Androwich, I. M. (2009). Applying human factors analysis to the design of the electronic health record. *Studies in Health Technology and Informatics, 146*, 132–136. PMID:19592822

Fisher, J. S. (1999). *Electronic Records in Clinical Practice*. Cinical Diabetes.

Frame, J., Watson, J., & Thomson, K. (2008). Deploying a culture change programme management approach in support of information and communication technology developments in Greater Glasgow NHS Board. *Health Informatics Journal, 14*(2), 125–139. doi:10.1177/1081180X08089320 PMID:18477599

Greenhalgh, T., & Stones, R. (2010). Theorising big IT programmes in healthcare: Strong structuration theory meets actor-network theory. *Social Science & Medicine, 70*(9), 1285–1294. doi:10.1016/j.socscimed.2009.12.034 PMID:20185218

Halamka, J. D., Mandl, K. D., & Tang, P. C. (2007). Early experiences with personal health records. *Journal of the American Medical Informatics Association, 15*(1), 1–7. doi:10.1197/jamia.M2562 PMID:17947615

Hall, E. (2005). The 'geneticisation' of heart disease: A network analysis of the production of new genetic knowledge. *Social Science & Medicine, 60*(12), 2673–2683. doi:10.1016/j.socscimed.2004.11.024 PMID:15820579

Hanseth, O. (2007). Integration-complexity-risk: The making of information systems out-of-control. In *Risk, Complexity and ICT*. Oslo: Edward Elgar. doi:10.4337/9781847207005.00005

Hartman, M. et al. (2009). National health spending in 2007: Slower drug spending contributes to lowest rate of overall growth since 1998. *Health Affairs, 28*(1), 246. doi:10.1377/hlthaff.28.1.246 PMID:19124877

Häyrinen, K., Saranto, K., & Nykänen, P. (2008). Definition, structure, content, use and impacts of electronic health records: A review of the research literature. *International Journal of Medical Informatics*, 77(5), 291–304. doi:10.1016/j.ijmedinf.2007.09.001 PMID:17951106

Hennington, A. H., & Janz, B. (2007). *Information systems and healthcare XVI: Physician adoption of electronic medical records: Applying the UTAUT model in a healthcare context.* Academic Press.

Heslop, L. (2010). *Patient and health care delivery systems in the US, Canada and Australia: A critical ethnographic analysis.* LAP LAMBERT Academic Publishing.

HFMA. (2006). *Overcoming Barriers to Electronic health Record adoption: Results of Survey and Roundtable Discussions Conducted by the Healthcare Financial Management Association.* Retrieved from http://mhcc.maryland.gov/electronichealth/mhitr/EHR%20Links/overcoming_barriers_to_ehr_adoption.pdf

Hillestad, R. et al. (2005). Can electronic medical record systems transform health care? Potential health benefits, savings, and costs. *Health Affairs*, 24(5), 1103. doi:10.1377/hlthaff.24.5.1103 PMID:16162551

Hoffinan, L. (2009). Implementing electronic medical records. *Communications of the ACM.* doi: 10114515927611592770

Hoffinan, S., & Podguski, A. (2008). *Finding a cure: The case for regulation and oversight of electronic health record systems.* Academic Press.

Hordern, A., Georgiou, A., Whetton, S., & Prgomet, M. (2011). Consumer e-health: An overview of research evidence and implications for future policy. *The HIM Journal*, 40(2), 6–14. PMID:21712556

Huang, Z., & Palvia, P. (2001). ERP implementation issues in advanced and developing countries. *Business Process Management Journal*, 7(3), 276–284. doi:10.1108/14637150110392773

Jones, D. A., Shipman, J. P., Plaut, D. A., & Selden, C. R. (2010). Characteristics of personal health records: Findings of the Medical Library Association/National Library of Medicine Joint Electronic Personal Health Record Task Force. *Journal of the Medical Library Association*, 98(3), 243–249. doi:10.3163/1536-5050.98.3.013 PMID:20648259

Kaghan, W., & Bowker, G. (2001). Out of machine age? Complexity, sociotechnical systems and actor network theory. *Journal of Engineering and Technology Management*, 18, 253–269. doi:10.1016/S0923-4748(01)00037-6

Kaplan, B., & Harris-Salamone, K. D. (2009). Health IT success and failure: Recommendations from literature and an AMIA workshop. *Journal of the American Medical Informatics Association*, 16(3), 291–299. doi:10.1197/jamia.M2997 PMID:19261935

Karsh, B.-T., Weinger, M. B., Abbott, P. A., & Wears, R. L. (2010). Health information technology: Fallacies and sober realities. *Journal of the American Medical Informatics Association*, 17(6), 617–623. doi:10.1136/jamia.2010.005637 PMID:20962121

Kennedy, B. L. (2011). *Exploring the Sustainment of Health Information Technology: Successful Practices for Addressing Human Factors.* Northcentral University.

Kimaro, H., & Nhampossa, J. (2007). *The challenges of sustainability of health information systems in developing countries: Comparative case studies of Mozambique and Tanzania.* Academic Press.

Kralewski, J., Dowd, B. E., Zink, T., & Gans, D. N. (2010). Preparing your practice for the adoption and implementation of electronic health records. *Physician Executive, 36*(2), 30–33. PMID:20411843

Krohn, R. (2007). The consumer-centric personal health record—It's time. *Journal of Healthcare Information Management, 21*(1), 21.

Latour, B. (2005). *Reassembling the social: An introduction to Actor-Network Theory.* Oxford University Press.

Latour, B., Harbers, H., & Koenis, S. (1996). *On actor-network theory: A few clarifications.* Academic Press.

Latour, B., & Law, J. (1986). *The Powers of Association: Power, Action and Belief.* Academic Press.

Latour, B., Mauguin, P., & Teil, G. (1992). A note on socio-technical graphs. *Social Studies of Science, 22*(1), 33. doi:10.1177/030631279202200102

Law, J. (1991). *A Sociology of monsters: Essays on power, technology, and domination.* Routledge.

Law, J. (1997). *Heterogeneities, published by the Centre for Science Studies.* Lancaster University. Retrieved from http://www.biomedcentral.com/1472-6947/10/67/

Law, J., & Hassard, J. (1999). *Actor network theory and after.* Wiley-Blackwell.

Leila. (2009). The Bug's Blog: Bruno Latour, Action-Network-Theory, and Dismantling Heirarchies: Actor-Network-Theory: Terms and Concepts. *The Bug's Blog.* Retrieved from http://latourbugblog.blogspot.com/2009/01/actor-network-theory-terms-and-concepts.html

Leslie, H. (2011). *Australia's PCEHR Challenge | Archetypical.* Retrieved from http://omowizard.wordpress.com/2011/08/30/australias-pcehr-challenge/

Liu, L. S., Shih, P. C., & Hayes, G. R. (2011). Barriers to the adoption and use of personal health record systems. In *Proceedings of the 2011 iConference* (pp. 363–370). Academic Press.

Lorenzi, N. et al. (2009). How to successfully select and implement electronic health records (EHR) in small ambulatory practice settings. *BMC Medical Informatics and Decision Making, 9,* 15. doi:10.1186/1472-6947-9-15 PMID:19236705

McLean, C., & Hassard, J. (2004). Symmetrical Absence/Symmetrical Absurdity: Critical Notes on the Production of Actor-Network Accounts. *Journal of Management Studies, 41*(3), 493–519. doi:10.1111/j.1467-6486.2004.00442.x

McReavy, D., Toth, L., Tremonti, C., & Yoder, C. (2009). *The CFO's role in implementing EHR systems: the experiences of a Florida health system point to 10 actions that can help CFOs manage the revenue risks and opportunities of implementing an electronic health record.* Retrieved December 16, 2011, from http://findarticles.com/p/articles/mi_m3257/is_6_63/ai_n35529545/

Mitchell, V., & Nault, B. (2008). *The emergence of functional knowledge in sociotechnical systems.* Academic Press.

Mohd, H., & Mohamad, S. M. S. (2005). Acceptance model of electronic medical record. *Management, 2*(1), 75.

Mort, M., Finch, T., & May, C. (2007). Making and unmaking telepatients: identity and governance in new health technologies. *Science, Technology & Human Values, 34*(1), 9–33. doi:10.1177/0162243907311274

Nagle, L. M. (2007). EHR versus EMR versus EPR. *Nursing Leadership, 20*(1), 30–32. doi:10.12927/cjnl.2007.18782 PMID:17472137

NEHTA & DoHA. (2011). *Concept of Operations: Relating to the introduction of a Personally Controlled Electronic Health Record System*. Retrieved from http://www.yourhealth.gov.au/internet/yourhealth/publishing.nsf/Content/CA2578620005CE1DCA2578F800194110/$File/PCEHR%20Concept%20of%20Operations.pdf

Neves, J. et al. (2008). Electronic Health Records and Decision Support Local and Global Perspectives. *WSEAS Transactions on Biology and Biomedicine, 5*, 8.

NHHRC. (2009). *A Healthier Future for All Australians: National Health and Hospitals Reform Commission - Final Report June 2009*. Retrieved from http://www.health.gov.au/internet/main/publishing.nsf/Content/nhhrc-report

Protti, D., & Smit, C. (2006). The Netherlands: Another European Country Where GPs Have Been Using EMRs for Over Twenty Years. - Google Scholar. *Healthcare Information Management & Communications, 30*(3).

Rhodes, J. (2009). Using Actor-Network Theory to Trace an ICT (Telecenter) Implementation Trajectory in an African Women's Micro-Enterprise Development Organization. *Information Technologies & International Development, 5*(3), 1–20.

Rogers, E. (2003). *Diffusion of innovation*. New York: Free Press.

Rosebaugh, D. (2004). Getting ready for the software in your future. *Home Health Care Management Practice*. doi: 101177108482230325987 7

Rosenbloom, S. T. et al. (2006). Implementing Pediatric Growth Charts into an Electronic Health Record System. *Journal of the American Medical Informatics Association, 13*, 302–308. doi:10.1197/jamia.M1944 PMID:16501182

Ross, S. E., Schilling, L. M., Fernald, D. H., Davidson, A. J., & West, D. R. (2010). Health information exchange in small-to-medium sized family medicine practices: Motivators, barriers, and potential facilitators of adoption. *International Journal of Medical Informatics, 79*(2), 123–129. doi:10.1016/j.ijmedinf.2009.12.001 PMID:20061182

Rydin, Y. (2010). Actor-network theory and planning theory: A response to Boelens. *Planning Theory, 9*, 265–268.

Showell, C.M. (2011). Citizens, patients and policy: A challenge for Australia's national electronic health record. *Health Information Management Journal OnLine, 40*(2).

Tang, P. C., Ash, J. S., Bates, D. W., Overhage, J. M., & Sands, D. Z. (2006). Personal health records: Definitions, benefits, and strategies for overcoming barriers to adoption. *Journal of the American Medical Informatics Association, 13*(2), 121–126. doi:10.1197/jamia.M2025 PMID:16357345

Thompson, T. G., & Brailer, D. J. (2004). *The decade of health information technology: Delivering consumer-centric and information-rich health care. Framework for Strategic Action, Office for the National Coordinator for Health Information Technology (ONCHIT)*. Department of Health and Human Services, and the United States Federal Government.

Thweatt, E. G., & Kleiner, B. H. (2007). New Developments in Health Care Organisational Management. *Journal of Health Management*, 9(3), 433–441. doi:10.1177/097206340700900308

Timpka, T. et al. (2007). Information infrastructure for inter-organizational mental health services: An actor network theory analysis of psychiatric rehabilitation. *Journal of Biomedical Informatics*, 40(4), 429–437. doi:10.1016/j.jbi.2006.11.001 PMID:17182285

Tobler, N. (2008). *Technology, organizational change, and the nonhuman agent: Exploratory analysis of electronic health record implementation in a small practice ambulatory care.* The University of Utah.

Torda, P., Han, E. S., & Scholle, S. H. (2010). Easing the adoption and use of electronic health records in small practices. *Health Affairs (Project Hope)*, 29(4), 668–675. doi:10.1377/hlthaff.2010.0188 PMID:20368597

Trudel, M. C. (2010). *Challenges to personal information sharing in interorganizational settings: Learning from the Quebec Health Smart Card project.* The University of Western Ontario.

Vitacca, M., Mazzù, M., & Scalvini, S. (2009). Socio-technical and organizational challenges to wider e-Health implementation. *Chronic Respiratory Disease*, 6(2), 91–97. doi:10.1177/1479972309102805 PMID:19411570

Walsham, G. (1997). Actor-network theory and IS research: Current status and future prospects. In *Information Systems and Qualitative Research: Proceedings of the IFIP TC8 WG 8.2 International Conference on Information Systems and Qualitative Research.* Academic Press.

Weimar, C. (2009). Electronic health care advances, physician frustration grows. *Physician Executive*, 35(2), 8–15. PMID:19452840

Westbrook, J. I., & Braithwaite, J. (2010). Will Information and Communication Technology Disrupt the Health System and Deliver on Its Promise? *The Medical Journal of Australia*, 193(7), 399–400. PMID:20919970

Wickramasinghe, N., & Bali, R. K. (2009). The S'ANT Imperative for Realizing the Vision of Healthcare Network Centric Operations. *International Journal of Actor-Network Theory and Technological Innovation*, 1(1), 45–58. doi:10.4018/jantti.2009010103

Wickramasinghe, N., Bali, R. K., & Lehaney, B. (2009). *Knowledge Management Primer.* Taylor & Francis.

Wickramasinghe, N., & Schaffer, J. (2010). *Realizing Value Driven e-Health Solutions.* IBM centre for the Business of government.

Wickramasinghe, N., Tatnall, A., & Goldberg, S. (2011). The Advantages Of Mobile Solutions For Chronic Disease Management. *PACIS 2011 Proceedings*. Retrieved from http://aisel.aisnet.org/pacis2011/214.

Williams, R. (2007). Managing complex adaptive networks. In *Proceedings of the 4th International Conference on Intellectual Capital, Knowledge Management & Organizational Learning* (pp. 441–452). Academic Press.

Yusof, M., Stergioulas, L., & Zugic, J. (2007). *Health information systems adoption: Findings from a systematic review.* Academic Press.

KEY TERMS AND DEFINITIONS

EMR (Electronic Medical Record): A medical record that is written, stored and maintained in a digital format usually by a clinician.

EPR (Electronic Patient Record): An electronic record of patient related facts maintained and managed by healthcare organisations at large.

HER (Electronic Health Record): A comprehensive health record that consists of information collected from different service providers and point-of-service systems such as pharmacies and diagnostic centres for all healthcare encounters an individual may have with them.

Individual Health Identifiers: Unique numbers given to each user of the PCEHR so that their record remains unique at all times.

PCEHR (Personally Controlled Electronic Health Record): The Australian national e-health solution that has an unique shared governance model.

Chapter 3
Climate Information Use:
An Actor–Network Theory Perspective

Maryam Sharifzadeh
Yasouj University, Iran

Ezatollah Karami
Shiraz University, Iran

Gholam Hossein Zamani
Shiraz University, Iran

Davar Khalili
Shiraz University, Iran

Arthur Tatnall
Victoria University, Australia

ABSTRACT

This chapter employed an interdisciplinary attempt to investigate agricultural climate information use, linking sociology of translation (actor-network theory) and actor analysis premises in a qualitative research design. The research method used case study approaches and purposively selected a sample consisting of wheat growers of the Fars province of Iran, who are known as contact farmers. Concepts from Actor-Network Theory (ANT) have been found to provide a useful perspective on the description and analysis of the cases. The data were analyzed using a combination of an Actor-Network Theory (ANT) framework and the Dynamic Actor-Network Analysis (DANA) model. The findings revealed socio political (farmers' awareness, motivation, and trust) and information processing factors (accuracy of information, access to information, and correspondence of information to farmers' condition) as the key elements in facilitating climate information use in farming practices.

INTRODUCTION

Climate information has become recognized as a basic production factor affecting agricultural systems (Harrison and Williams, 2007). This is while, despite significant improvements in the climatic information production in the last decade (Subbiah*et al.*, 2004; Ziervogel*et al.*, 2005; Hu *et al.*, 2006; Artikov*et al.*, 2006),

farmers as focal decision makers of farm systems and main users of uncertain Agricultural Climate Information (ACI), have not altered management decisions to take advantage of this type of information (Artico*et al.*, 2006; Hu *et al.*, 2006; Nazemos'sadat*et al.*, 2006).

Decision making in complex dynamic environments, like the field of agricultural process management, takes place in a network of dif-

DOI: 10.4018/978-1-4666-6126-4.ch003

ferent actors with their own interests, concerns, and control of a part of the resources needed for successful farming practices (Hermans, 2005). The issue of not considering climate information in farming decisions was scholarly investigated from different perspectives.

Existing literature on economic and management research related to climate science shows increasing attention to necessity of stakeholder interactions to both contribution to and learning from climate applications and indicates that the methodologies commonly used in modeling decision-making are based on the assumption that users have an idealized response to the information (Sherrick *et al.*, 2000). Each methodology is appropriate to (Hansen and Sivakumar, 2006): 1) understand how uncertain information can be incorporated into decisions, 2) serve as a bridge between physical and social sciences and decision-makers, and 3) reveal the conditions under which climate information can have value. However, as human actors are not always optimizing, rational and idealized decision-makers, smallholder farmers will not all make the same decisions in the same way and in isolation (Ziervogel, 2004).

In contrast to an earlier wave of optimism in economic and management science regarding the use of climate information, psychological theories aim to explain the relation between intention and action in particular contexts and in relation to specific practices. The scholarly literature reveals that these theories have also been criticized by some psychologists (Richetin *et al.*, 2008) for relying on analyses of correlation, rather than causes, and for assuming too much about the instrumental relation between attitude and intention (May and Finch, 2009). Therefore, in the 1980s, a growing interest in the social foundations of behaviour within the Information Systems (IS) field led to a shift towards a broader perspective with the continuum of theories, all of which examine the relationships of individuals and technology, the varying influences that affect these relationships and how together they influence technology adoption and use. Giddens' structuration theory, Bijker's social construction of technology, Orlikowski and Gash's technology frames, and Latour and Callon's sociology of translation (Actor-Network Theory – ANT) are amongst this stream of attempts (Mac Leod, 2001), which are now used much more extensively in investigating information systems and IS adoption. The ANT approach by posing opportunities to function as a go-between for the two extremes of actor and system perspectives, has inspired a number of IS researchers to conceptualise the interrelation of actors in the socio-technical contexts of systems development during the past decade (Walsham, 1997; Tatnall, 2000; Dunning-Lewis and Townson, 2004; Bakshaie, 2008; Everitt-Deering, 2008).

Policy science, as another body of scholarship, focuses on different approaches available to study the characteristics of actors and networks. Stakeholder analysis (Cameron, 2005; Rubas *et al.*, 2006; Lybbert *et al.*, 2007), social network analysis (Ziervogel, 2004; Ziervogel and Downing, 2004; Ziervogel *et al.*, 2005), and actor perception analysis (Bots *et al.*, 2000; Hermans, 2008; Hermans and Thissen, 2009) are probably the most widely used approaches by environmental management in policy science literature.

A comparison of the aforementioned theoretical approaches suggests that perceptions, values and interests, and resources are the three basic dimensions in multi-actor decision making processes (Hermans, 2005; Hermans and Thissen, 2009). Perceptions (causal beliefs, cognitions or frames of reference) are the image that actors have of the world around them, both of the other actors and networks, and of the substantive characteristics of a problem (Bots *et al.*, 2000; Enserink *et al.*,

2010). Values (specific 'objectives', 'goals', 'interest' and 'targets') provide the directions in which actors would like to move. Resources are the practical means over which actors have control and in which they have some interest'. Such resources enable actors to influence other actors, relations and rules in a network (Hermans and Thissen, 2009; Enserink *et al.*, 2010). In other words, farmers' perceptions can be a result of constrained or bounded decisions due to limited knowledge, interests, cognitive capacity, resources and the structure of the environment in which decisions occur. Besides, climate information use as an important interest in the farming context is not issued in a 'neutral environment'. Many non-weather factors may preclude the uptake of climate information (Klopper *et al.*, 2006). Furthermore, some users are in a better position than others to capitalize on the information provided through climate information. These factors together with actor associations and networks are seen as filters in determining what the responses to new information will be.

Preferably, implications of actor-network theory with actor-analysis models could be integrated in our study for their relevance and fitness in explaining the reasons underlying farmers' decision processes regarding climate information in the network of dynamically complex and uncertain agricultural environments.

Accordingly, this article uses a causal mapping study to present farmers' viewpoints regarding agricultural climate information use. Using the terminology of actor-network theory, along with details of the translation mechanism in the process of agricultural climate information use, provides a useful approach. Furthermore, this paper contributes to enhance understanding of the linking between actor-network theory and actor analysis models, and its influence in shaping processes underlying climate information use

in agricultural practices. It has been built using dynamic actor-network analysis method, specifically the DANA software program (Pieter W.G. Bots, Delft University of Technology, Netherlands). DANA combines elements from stakeholder analysis and strategic decision analysis (Bots *et al.*, 1999; Bots *et al.*, 2000). It expands upon both in that it requires the analyst to make explicit how each of the actors views this decision issue in terms of what is desirable, what actions are possible, what exogenous influences are to be expected, and how these aspects are causally related (Bots, 2008).

The following section of the paper outlines the background and concepts of actor-network theory. The subsequent sections provide a description of the methodology. The paper ends with an analysis of the farmers' perceptions and beliefs, and their role in processes of agricultural climate information use.

BACKGROUND THEORY

Historically, the diffusion of innovation frameworks (Aikat, 2001), which focused on 'early adopters' (Rogers, 1995), 'lead users' (Von Hippel, 1988), and 'cyclical technological progress' (Toffler, 1970), has relied on technological determinism and innate properties of technologies as important components of diffusion process. Actor- network theory has been proposed as an alternative view, as it suggests that many complex social factors are involved in the interaction of society and technology (Latour, 1991) and that any process of technological adoption must inevitably involve multiple, parallel and divergent activities with a set of complex negotiations between all those involved (Tatnall, 2000; Everitt-Deering, 2008).

The theoretical framework of this paper is based on actor-network theory as a useful approach to modelling innovation and understanding the process of introduction of a certain technology (involving translation, dissemination and evolution of technological innovations) in society (Hanseth and Monteiro, 1997; Knights *et al.*, 1997; Aanestad and Hanseth, 2000; McGrath, 2001; Alcouffe *et al.*, 2008). The suggested framework is attuned to emphasize the symmetry between actors and their technologies, and to focus attention on the translations that occur as knowledge and resources are exchanged in networked interactions (Lamb, 2006). The essential tenet of ANT concentrates on issues of network formation and investigates the human and non-human alliances (Feldman *et al.*, 2006; Elbanna, 2009) by looking at the decisions that were made, who made them and on what basis they were made. This implies that any changes might be described as learning new ideas, as changes in behaviour or as transformations that emerge through networks of actors (Fenwick, 2006).

ANT theorists have developed a particular vocabulary with which to identify the participants in a network and to conceptualise the means by which these participants are manoeuvred, coordinated, aligned and rendered stable. As the name suggests, the concept of actor/ actant (non- human actors) is key to ANT. Typically actors are the participants in the network which includes both the human and non-human objects and/or subjects. A network is an interactive assembly of actors, groups, or 'strings of actions' involving a number of potential mediators (Dolwick, 2009). In actor-network theory the extent of a network is determined by actors that are able to make their presence individually felt by other actors (Wickramasinghe *et al.*, 2007).

It is necessary to distinguish between stakeholders (actors, those who have an interest affected when certain changes occur in the process regardless of whether these changes occur due to either one's actions, the actions of other actors, or external developments) and agents (actors with the capability to cause changes in the process but who have no stake in the issue). Actors not only have an interest in the process, they also have to have the ability to influence the process "to bend space around themselves" (Goodwin *et al.*, 2005). However, both actors and networks are interchangeable: an actor may be viewed as a network, and a network may be viewed as an actor (Dolwick, 2009).

Central to ANT is the concept of translation (Latour, 1996) as the *woven fabric* of tight or loose relations and forces connecting the *actors* in a stable or unstable network of actions. ANT proposes the concept of "translation" to replace "diffusion" and sees innovation spreading as a result of how actors "translate" (or adapt) the interest of others so that they become aligned with their own interests. Successful translation of the interests of both human and non-human entities to create effective alliances contributes to effective "transfer of technology" (Everitt-Deering, 2008).

According to Callon as stated by Shoib *et al.* (2006) and Sophonthummapharn (2008), translation has four stages: *problematization*– the network is established through a consensus on the problems to be fixed by the relevant actors, *intressement*– the effort to get the actors interested in the network and to determine the level of their involvement, *enrollment*– role acceptance by the actors within the network, and *mobilization of allies* –coordination of active actors who build a strong network. When these translations of one's interests are put into material form they result in inscriptions (Gao, 2005).

In other words, translation leads to the alignment of the actors' interests and consequently the formation of a stabilised actor-network (Gustad, 2006). According to Callon as stated by Li and Zhang (2009), "the identity of actors, the possibility of interaction and the margins of manoeuvre are negotiated and delimited" in the translation process.

Guided by actor-network theory, this study explores the role of agricultural climate information use as a heterogeneous network of human and non-human actors in "summing up" farmers' capacities (Latour, 1999). The answer to whether the agricultural climate information adoption initiative may succeed or fail, according to ANT, lies in: following the actors, "reading" the networks, describing the "translations" and estimating the strength in the inscriptions. Further, this means that the success of an Agricultural Climate Information System (ACIS) lies within the result of a stabilised network (Gustad, 2006). Based on the developed framework, the main focus of the present research is to understand: a) how can the translation process define the dynamics by which farmers recruit others into agricultural climate information system?, b) what aspects are considered relevant to the climate information use in farming practices?, c) how may changes in some factors cause

changes in other factors?, d) which changes are expected to occur due to external developments?, and e) how do farmers value the changes that may occur?

METHODOLOGY

The case study approach was selected because of its capacity to capture knowledge gained through vicarious experiences and its suitability to study people engaged in real –life activities, in context. This approach enabled the examination of the complexities of climate information adoption and its interactions among farmers, including their insights into barriers of climate information use in farm practices.

Case Study Site: Fars Province, Iran

Fars province with an area of about 133 299 km^2, is located in the southwest of Iran (Figure 1). This province has 8.6 million hectares of

Figure 1. Iran map and Study Area

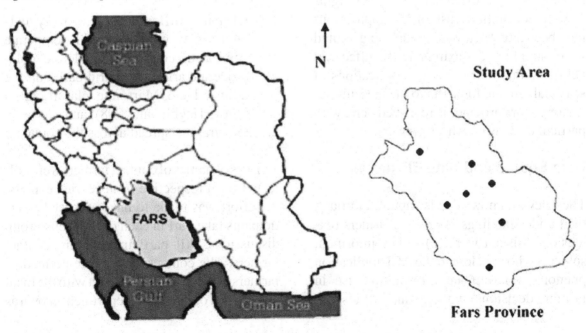

rangeland, 1.212 million hectares of forest and 1.6 million hectares of cropland. The annual rainfall is 150-200mm. The mean annual temperature is 17°C, with hot summers and cold winters (Moameni, 1999). Agriculture is of great importance in Fars province. The main agricultural crops include cereals (wheat and barley), citrus fruits, dates, sugar beets and cotton. The province is known for its high production rate in wheat crop compared with other parts of the country, with a total annual wheat production of 13.44 million tons (Karimi *et al.*, 2009).

Farming has been faced with water stresses due to climatic variability and drought challenges in recent years. These are the main sources of agricultural risks in the study region. The dwindling supply of irrigation water in different regions across the province has resulted in regional declines in the grain area and much reduced crop yield potential in the affected areas. Most farming regions land use properties have changed due to recent water stresses and many of the previously advantaged farmers in the area have begun decreasing the size of their production. Given the devastating impacts of the recent droughts in the region, initial thrusts of ACI applications have been directed towards addressing limited water supply and maximising the effectiveness of rainfall. The case of applications of seasonal climate forecasts to the agricultural sector in Fars province is reviewed here, with particular focus on wheat growers.

Data Source and Data Elicitation

The relevant knowledge is captured through tacit understandings of contact farmers (extension-assistant-workers) as key informant subjects who are close to the ACI application phenomenon (climate information use in farming decisions) in the manner of discovery. In this way, a more narrative approach is developed to identify concepts that emerge from the interview transcripts. The mapping of farmers' cognition is based on team mental models in reference to mental representation of knowledge, assumptions, interests, values and beliefs, to make sense of farmers and to communicate information about their farm practices in terms of their own frame regarding the meaning of the situation (Mohammed *et al.*, 2000; Kolkman *et al.*, 2005). As focus group interviews allow for a better understanding of the domain of inquiry through the use of appropriate probes (Patton, 2002), discovery interviews were held, aimed at capturing the knowledge of agricultural climate information use and the reasoning behind the decisions to use agricultural climate information and other related purposeful farming actions.

Focus group discussions were organised in seven villages in different wards in the study area. The regions were selected following primary studies and expert discussions with the related organisations in order to encompass a typical view of the farming zones experiencing recent droughts on the basis of:

1. Climatic differences – broadly categorised to cold semi-arid, temperate semi-arid, cold arid, moderate cold, moderate arid, warm arid based on the adjusted De Martonne classification index (Haghighat, 2008); and
2. Geographic spread across the county.

Lists of names of contact farmers were collected from respective rural service centers. An effort was made to have at least 15 participants take part in each of the focus group discussions. All participants were contact farmers. The contact farmers are defined as farmers who have direct contact with the local agricultural organisation department known as rural service centers which are the main extension organisations in each *Dehestan* (district or a collection of villages). The various rural service centers target contact farmers for dissemination of agricultural information in the villages, and then rely on the contact farmers to disseminate the information (technologies) to other, non-contact farmers. Relative to other types of farmers, contact farmers are of particular interest because, in addition to their improved confidence in applying new information and improved skills in farm management; they enhanced recognition as farmers. Working with contact farmers was considered desirable because it overcame some of the problems of getting farmers together in an agreed location, at an agreed time. Attendance rates were very high after the purpose was explained to invitees (i.e., greater than 70% of those contacted, average group size of ten farmer participants). This suggested the topic was important to wheat growers.

The participants were predominantly male with two-thirds over 40 years of age. Most were wheat growers, who experienced recent drought in the region. Their farms averaged about 5-10 hectares in size. In each case a large part of their farms were irrigated to ensure their production would meet market quality demands. The participants' average level of education was 10 years. The sessions were held between September 2009 and February 2010.

Elicitation of the relevant cognitions of farmers and casting of their cognitions into appropriate structural representations was provided by means of an Interactively Elicited Causal Mapping (IECM) technique. IECM implies reliance on the terms, language, and concepts used by the people being studied (Mohammed et al., 2000; Narayanan and Armstrong, 2005) and probes to collect in-depth qualitative information about groups' experiences on the use of climate information. In an open exchange of ideas, participants were asked of their successes and failures in relation to this type of information, their constraints and ideas on how the information could be improved and how it should be implemented in future.

A focus group protocol was used to guide the process. The interview process consisted of semi-structured interviews. An interview guide was developed by the researcher to facilitate the interview process. Each session lasted about two hours, as suggested by Patton (2002). All questions were open-ended. These interviews were then transcribed into a document format.

In each case, the participants were instructed to consider the extent to which they believed the concepts mentioned during IECM were causally related to one another using the following 7-point scale: -3 -2 -1 0 $+1$ $+2$ $+3$, where -3 denoted a very strong perceived negative causal relationship (i.e., an increase in the causal variable directly decreases the effect variable) and $+3$ denoted a very strong perceived positive causal relationship (i.e., an increase in the causal variable directly increases the effect variable). Zero denoted the fact that a neutral relationship was perceived to exist between the variables in question.

The point of redundancy or convergence among the subjects represents the point at which further interviews or data collection does not lead to additional concepts or coding categories (Nelson and Nelson, 2000; Narayanan and Armstrong, 2005). This point also serves as a way of establishing the adequacy of the sample which was operationalized by gradually cumulating the group causal mappings 1, 2, 3 … up to 6.

Non-Human Actors

One of the underlying principles of actor-network theory is to give due recognition to the contributions of all actors, regardless of whether they are human or non-human, and to treat these contributions equally. In addition to the important human actors, this study involved collecting data and studying the influences of a large number of non-human actors including: Government policy, the weather and climate information, and farming information (crop type, planting information, seed variety, planting area, and irrigation). A researcher obviously cannot collect data directly from non-human actors in the same way as for human actors by using interviews and surveys, but they can collect this data nevertheless. Techniques for data collection from non-human actors include questioning the agricultural climate information system's human actors (related climate producers, extensionists, and farmers), study of documents (and archives of related organisations – Agri-Jihad Orgnisation, Meteorological Organisation, etc.) and observation.

Coding

An exploratory approach was used to identify conceptual categories drawn from the texts or narratives. The process typically involves participants in brainstorming a large set of statements relevant to the topic of interest. A list of concepts is drawn from the narratives to be examined dynamically. Going through the texts, new concepts were continually identified. Code-based analysis, or thematic coding method, was used to reduce text data into manageable summary categories. Some concepts were the exact words used by the respondents (farmers' fatalism) while some others resulted from coding of these words into generalized concepts (correspondence of forecasts information to farmers' condition). Filtering or aggregation, which is the process of determining which part of the text to code (Jackson and Trochim, 2002) and

what words to use in the coding, was used to increase the level of comparability across the resultant maps and aggregate phrases in the raw causal maps to avoid misclassification of concepts due to peculiar wording on the part of individuals.

The above multiple coding process, by creating its own categories, helps ensuring that the categories are exhaustive and reliable. Construct semantic validity (Jackson and Trochim, 2002) with an internal component in terms of how the units of analysis are broken down as well as an external component in terms of how well coders understand the symbolic meaning of respondents' language confirms this study. In addition, the validity of the concepts and constructs was checked through dialogues with the original respondents at both stages of coding and verifying the final maps.

Data Analysis

A major development for qualitative research methods is efficient and flexible qualitative database software (Abernethy *et al.*, 2005) to serve two purposes: first, to create a database through systematic coding of qualitative data, and second, to analyze relations in the database of interview transcripts. Therefore, this study used Dynamic Actor-Network Analysis (DANA) software program (Pieter W.G. Bots, Delft University of Technology, Netherlands) to create and test relational maps that aid in developing theoretical models and assess the relative strength of the relations among the variables of the models (Bots *et al.*, 1999).

The DANA workbench is based on the assumption that the situations by which actors are influenced and to which they adapt themselves do not stem from the 'objective' world of the analyst, but from their own subjectively perceived world (Bots, 2007). This method represents the views of the actor in cognitive maps, which are conceptual models that show the perceptions of actors as a combination of objectives, factors and instruments, which are linked by arrows that depict the assumed influences among the elements (Hermans and Muluk, 2008). These views are important to understand the dynamics of actor-networks (Hermans and Muluk, 2002).

When working with DANA, the perceptions of the actors in terms of relevant factors and actor specific instruments and goals, are made explicit in a qualitative, conceptual language in terms of factual, causal and teleological assumptions. Factual assumptions make a statement about the state of a particular factor, e.g. equipment is inadequate. Causal assumptions represent the logic that an actor attributes to chains of events, e.g., "*if* governmental policy programs support farmers through moderating fluctuation of prices *then* planting patterns will be changed according to the experts' advice", "*if* planting patterns develop *then* climate information use will probably refine", and so on. Formally, a link has a cause part and an effect part, where cause and effect are considered as two different operational steps in DANA. Links may be uncertain, which is expressed by a hedge like 'possibly', 'probably' or 'definitely', typically associated with some value between 0 and 1. A link should be read as "*if* cause *then* certainty effect". Teleological assumptions represent an actor's true (or stated) objectives, e.g., "governmental support is inadequate". The actor's *insistence* on a goal expresses how strongly he desires a factor to change, typically on some ordinal scale, e.g., from 'preferably' to 'definitely'.

Borrowing from dynamic actor-network analysis (Bots *et al.*, 2000), DANA software focuses on problem situations as modelled by actors called 'arenas'. Within an arena, actors have their own perception of this situation. The perceptions of actors are modelled in terms of factors (variables of different types) and the causal relations between these factors. Factors that are particular to one actor are called 'attributes' of that actor (Table 1).

In a model, a standard factor coloring convention was used to represent the objectives and constraints: blue indicates a desired decrease, and orange a desired increase. The color intensity reflects the actor's insistence on a change. For constraints, the fill style of the oval is vertical bars, rather than solid fill. Factual assumptions are highlighted by showing the factor name in bold face. The vector determines the direction of causal relations with a solid arrow head indicating an increase, and an open arrow head a decrease. Furthermore, the width of an arrow reflects the ratio of effect over cause – the modifier in – and as uncertainty increases, a link takes on a lighter shade of purple.

The unit of analysis, or the level of aggregation of causal maps employed was 'farmers groups', and the level of aggregation was also the 'group'. The results were modelled into diagrams, which could then later be combined in an overview diagram that show the "group perspective" as a holistic perspective. DANA software provides the opportunity to merge actor perceptions from different case files into a single case to model the issue in such a way that it best reflects the combined views of all stakeholders.

RESULTS AND DISCUSSION

According to the results, focus groups suggested that farming decisions could be categorized into four groups corresponding to different stages of crop production: 1) agronomic decisions through the planting season (including choice of crop type, seed variety, planting area, and planting date), 2)

Table 1. DANA legend

DANA Element	Symbols As	Definition	Sample
Factor	-	A factor is a problem or a cause of a problem	Planting pattern
Factor / Actor	🕴	Factor that is connected to an actor	Trust / Farmers
Action / Actor	🕴	Solution given by an actor	Data collection / Farmers

growing-season decisions (including applications of pesticides, herbicides, fertilizers, and irrigation), 3) harvest and post-harvest decisions (including harvest date, autumn irrigation and tillage for the next crop), and 4) economic decisions (i.e., crop insurance, marketing decisions and cost-benefit optimization decisions). However, farmers believed that climate information was only applicable to a part of the growing season. The types of decisions that farmers mentioned they might undertake in response to climate information were mostly short-term tactical decisions such as sowing less wheat, planting earlier or irrigation tactics, rather than long-term strategic decisions. These decisions are a form of agricultural risk management (Carberry *et al.*, 2000). The focus on short-term decisions is expected when using new information. While these decisions seem to be an appropriate way of responding to climate information, the question is: why are more long-term decisions neglected by the farmers? Which factors affect the preferences of decision makers regarding use of climate information?

As stated before, in ANT translation involves all the strategies through which an actor identifies other actors (human and non-human) and arranges them in relation to each other (Tatnall, 2000). This research speaks of how chains of translation can transform (or translate) the agricultural climate information use problem. By using ANT as a cognitive lens, the focal actor (farmers) would view the entire scenario as an actor-network. In this perspective, climate information has to be integrated into farmers' practices. Such information is similar to the quasi objects exchanged between the actors in an actor-network. Information application in a real context (farm situation) is successfully integrated into farmers' decisions if it passes the expected quasi objects. Any passage of an unexpected object will result in the breakdown of the application and hence will result in the breakdown of the network. Results of another study (Sharifzadeh et al.

2012) pointed to the inefficient use of climate information in farming decisions. As making decisions in a real situation takes place in a condition in which the most demanding and the troubling task is to define the nature of the problem, rather than its solutions (Giordano *et al.*, 2007), the first step, therefore, is to focus on *problematization* to understand how the network is established through a consensus on what problems need to be fixed and who the relevant actors are. This process was facilitated by DANA to provide a useful lens to understand the process of translation in ACIS.

Problematization: Farmers' Perception on Agricultural Climate Information Use

As stated before, the results of an actor analysis are lists of relevant multi-actors or rather actor groups, and factors affecting climate information use from the farmers' perceptions. Results of the research show that utilization of farmers' perceptions in terms of interests, aims and problem definitions can greatly facilitate the problematization of the ACI utilization and possible related conflicts and alliances. Figure 2 shows the perceptions of respondents regarding factors affecting climate information use in farming decisions. Ovals represent factors affecting climate information use in farming practices. By default, a factor is a *system attribute*, i.e., it is a property of the system as a whole, rather than a property of a specific actor. As indicated in Figure 2, "access to information" is clearly relevant for the case, but it is not an actor itself. Therefore, "access to information" is a system attribute. The "equipment", on the other hand, is modelled as an *actor attribute* because there are several actors whose equipment is relevant to the case, and clearly the equipment for the Meteorological Organisation is not the same factor as the equipment for farming practices. The same

argument also holds for *actions*: the action "identifying communication channels", that can be taken by the Meteorological Organisation is not the same factor as the action "identifying communication channels", that can be taken by the Agri-Jihad Organisation. Note that, rectangles are used for representation of actions, including a reference to the actor who controls them.

A goal represents a desired change in a factor. The actor represented in Figure 2, for example, desires a strong decrease in the factor "insufficient government support". Causal links from one factor to another are defined by positive and negative correlations and can be described by seven different "change multipliers" (from a strong decrease – a big minus – through "no change" to a strong increase – a big plus). Figure 2 reflects farmers' belief that the use of climate information should improve farming decisions, that farmers' trust in climate forecasts and information cause this application to ameliorate (arrow labeled '✚'), that providing an appropriate market for products, which is under indirect control of governmental efforts, facilitates accretion of climate information use in farming practices ('+'), and that farmers' awareness, which is a result of access to information, will promote information use in farming decisions, although its influence is limited (small '+'). These results reveal that a complex combination of human and non-human actors control the utilization of ACI in farming decision making through their direct and indirect effects.

Interessement: Relevant Actors and Agricultural Climate Information Use

As ANT proposes, the key to information system use is the creation of powerful enough consortiums of actors (McMaster *et al.*, 1997) to carry the network through, and in the case of information failure, it could be a reflection

Figure 2. DANA diagram of farmers' perception of factors' affecting climate information use

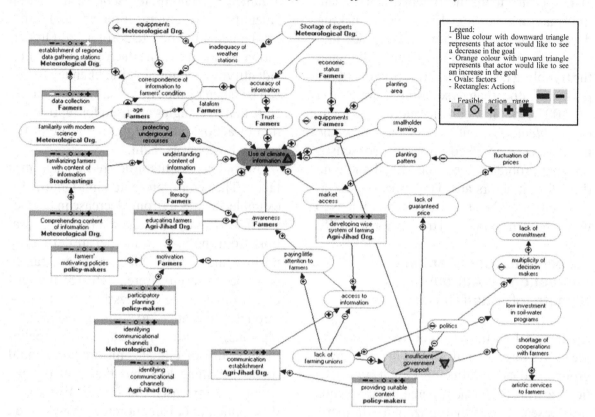

of the inability of those involved to construct the network of alliances. Therefore, a better understanding of the problem at hand lies in finding out which actors (acting units) are involved. Such acting units can be an individual person or a collective, like an organisation or an institute and also public or private, or semi-public (Bots *et al.*, 2000).

In seeking to identify whose objectives and interests are at stake (*stakeholders*) in the ACIS, Table 2 lists the actors indicated by farmers. This Table summarizes the results from DANA to identify the relevant actors (actors affected by the problem situation or by solutions considered; actors having an influence on the problem situation and its devel-

opment; actors needed for implementation of solutions; and other actors formally involved in decision making in the field).

As suggested by farmers, the Agri-Jihad Organisation (*Sazeman-e Jihad-e-Keshavar-zi*) which is known as the executive hand of the Minister of Jihade-e Agriculture in Iran, is committed to provide a safe environment for farming activities and is responsible for providing the latest news on climatic events for farmers. Media, especially public broadcasts, are the other communication channels that play a vital role in information transfer to farmers through TV and radio programs. Farmers believe that the Meteorological Organisation is another axis of climate information production

Table 2. Main stakeholders involved in climate information management cycle from farmers' viewpoint

Stakeholders	Level of Involvement in Climate Information Management Cycle
Farmers	Depend on amount and source of irrigation water and/or irrigation system; type of products and other related technical issues, increasing competition for products and their livelihood survival.
Meteorological Organisation	National-level government institution responsible for weather information processing, involved mainly through county agencies.
Media: • TV news, • Radio news, • News paper.	National and County level institution responsible for providing informing programs and related warning massages.
Minister of Jihade-e Agriculture	National-level agency responsible for agricultural development, involved mainly through county organisations.
Policy makers	National-level actors (planners) formally involved in decision making in the field, e.g. public authorities.

and transfer. Farmers believe that the main climate information stakeholders are farmers themselves (from small holding farming to plantation systems). Finally, farmers view national-level politicians (policy makers) as *agents* of the climate information management cycle responsible for agricultural development enactment.

When adopting climate information as an innovative technology, various actors including humans and technology working together to form a dynamic but robust network. Each actor plays a unique role in this network. Farmers argue that the Meteorological Organisation plays an important role in the initiating, organising, and leading of the climate information network. They believe that this organisation together with other extensionists (Agri-Jihad organisation and media) should use various ways such as persuasion and incentives to enroll new actors to join the network.

Suggested by the interessement notion, the focal actor attempts to assign other actors' specific roles in the problem solving process, impose and stabilize the other actors' identities, and is useful for seeing how groups of actors are brought together around a particular goal (Li and Zhang, 2009). Investigating this

process through focal actors' views revealed that respondents were agreed on an actors' identity in the network. In other words, the role of other actors as seen by farmers is truly defined. This issue proved that aligning the goals of information producers and extensionists are embedded in the network, which could be seen by end users (farmers). Figure 2 reveals how significant the actors' influence is from the perspective of achieving the farmers' objectives (the arrow functions as a change multiplier).

Non-Human Actants and Agricultural Climate Information Use

Actor-network theory suggests that non-human actors are also essential components of any human activity system (Pouloudi and Whitley, 2000). ACI application behaviour cannot only be explained as a result of human interaction or exclusive human communications, but non-human elements are central to explain the way in which an action is performed. Human actors in their ACI application behaviour interact not only with other human beings but also with other material objects. In the complex agricultural environment, the planting information, weather and its related components, and governmental policies are all elements that interact with the human actors of the network and have an active role in its development

Investigating planting information in Fars province revealed that crop insurance, crop type (planting pattern), seed variety, planting area, and irrigation issues are affected by ACI while at the same time influencing climate information use in the farmers' perceptions. However, results of an unpublished survey study revealed that farmers resist changing their farming strategies (changing planting pattern, seed varieties, irrigation techniques, planting area, etc.) with regard to in-advance climate information (Sharifzadeh et al. 2012). Salinity and drought are among the most important environmental stresses that limit crop production in farming systems of Fars province, Iran. These challenges make the policy makers interested in investigating government policies regarding crop production.

A review of government strategies to reduce the impact of environmental stresses on crop production shows that government policy has been directed toward controlling economic activities and markets in order to support either producers or consumers through input subsidies, credit programs, guaranteed price, and distribution of coupon, supplying higher yielding seeds, improving machinery services, augmenting fertilizer usage, enhancing water systems and pest management practicesresulting in wheat self-sufficiency (Najafi and Bakhshoodeh, 2002; RahimiBadr and Moghaddasi, 2009). Another supportive strategy was an overarching institutional and public policy framework of the development of the agrometeorological office in the Meteorological Organisation in 2007, and three agrometeorology research centers in three different sites of Fars province. These activi-

ties are a positive development to combat the impacts of recent drought. They are already having a major impact on producing up-to-date information on climatic events.

Although, analysis of farmers' beliefs (Figure 2), reveals that farmers weren't satisfied with these supports, the government strategy to develop and implement aforementioned programmes identify the most urgent adaptation activities to reduce and control the challenges. However, a major effort is required to scale up the institutional, financial, and technical capacity to address the climate challenges which, embody positive supportive behaviour of farmers regarding performance of rational use of climatic information.

Enrolment: Factors Affecting Agricultural Climate Information Use

The relevance of factors was also assessed in investigating farmers' viewpoints regarding climate information use. In the context of actor perceptions, 'relevance' denotes whether an actor does or does not perceive a factor as being relevant to the problem (Bots *et al.*, 2000). Factor relevance is measured as occurrence (the number of actor perceptions in which a factor appears) and importance, where the number of actor goals stated for a factor and the interest these goals represent are relevance indicators. Results showed that, actors put a strong emphasis on socio-political (14 occurrences), economic (8 occurrences), and information processing factors (6 occurrences), respectively, tofacilitate the use of climate information use in farming practices. Utilization of the above factors from farmers' perspectives suggests that by no means would it be possible to arrive at one unique perception. This implies that climate information use in farming practices could not refer to its official standard to convince all actors that they should apply this information, as these actors would encounter problems that were equally valid.

Results revealed that farmers' trust influences the use of climate information due to the negative effects of factors controlling the accuracy of climate information, which would affect the farmers trust directly. Consequently, most farmers found it far easier to rely on their traditional farm practices irrespective of climate information and based on their experience rather than what they saw as 'modern risky information'. Correspondence of forecast information to farmers' condition did not make it any easier to convince sceptical farmers to use this risky information. Most actors did not expect that this would change in the near future, unless significant changes occurred in resolving impediments of information use, which could be done through equipping the Meteorological Organisation (*Sazeman-e Havashenasi*) and its staff with the modern devices and scientific knowledge to promote quality of forecasts to adapt to each climatic region.

To further investigate enrolment, dynamic actor-network analysis suggests that, the extent to which a factor may be controlled depends on the existence of both actions and prospects (external influences) that can influence the factor. As suggested by farmers, factors affecting use of climate information in farming practices are under control of some external factors. Prospects (e.g. "multiplicity of decision makers" in Figure 2) represent independent changes in a factor (i.e. not caused by some action in the perception graph) the actor expects to occur (Bots, 2008). Policy issues, multiplicity of decision makers, and lack of suitable equipment facilitating farm practices are perceived as prospects that generate "noise" which reduces controllability over use of climate information.

Furthermore, actions can be taken to cause a factor to change, or to encounter external influences (the rectangles in Figure 2). The minuses and pluses in the headers of boxes refer to seven possible "change levels" which range from a strong decrease of the action (a big minus) through "no change" to a strong increase (a big plus), as compared to the current course of action. While minuses and/or pluses shown in black colour represent change levels that the actor considers to be possible to occur, change levels displayed in white colour are considered to be impossible. For example, the actor 'farmer' regards a constant degree of participatory planning by the policy makers as possible (Figure 2).

The results (Table 3) indicate that access to information is widely considered to be a very effective measure in this regard, and one which is under control of other factors. Familiarity with appropriate communication channels which convey information to farmers is one of the important factors subject to the Meteorological Organisation and the Agri-Jihad Organisation extension programs, whose shortage provides the users with an excuse for access and finally using climate information. Table 3 summarizes the main factor groups that are affected directly and indirectly by external factors controlling climate information use from a farmers' points of view. These results are confirmed by Carmen Lemos *et al.* (2002) who argue that socio-economic, political, and cultural conditions can compromise the use of seasonal forecasts by both farmers and policymakers.

Inscriptions describe both a process and a product. As a process it may be seen as information format, and seen as a product it may be studied in practice, in situations where it heavily influences the behaviour of the network, or where it is rejected or used in farming practices. In this perspective, farm-

Table 3. Number of prospects and actions influencing climate information use

Factors	# Prospect	# Action
Economic factors:		
• Economic status.	-	
• Equipment shortage.	-	-
• Fluctuation of prices.	-	-
• Lack of guaranteed prices.	1	-
• Low investment in soil-water issues.	1	-
• Market access.	-	-
• Smallholder farming.	-	-
Information processing factors:		
• Access to information.	2	4
• Accuracy of information.	1	1
• Correspondence of information to farmers' condition.	1	1
• Familiarity with information content.	-	1
• Familiarity with modern science.	-	-
• Timeliness informing.	2	-
Organisational factors:		
• Accountability.	2	-
• Bureaucratic formalities.	-	-
• Shortage of experts.	-	-
Socio- political factors:		
• Age.	-	-
• Artistic services to farmers.	1	-
• Awareness.	2	5
• Fatalism.	-	-
• Insufficient government support.	1	-
• Lack of commitment.	2	-
• Lack of farming unions.	1	-
• Literacy.	-	-
• Motivation.	1	2
• Multiplicity of decision makers.	1	-
• Paying little attention to farmers.	1	-
• Politics.	0	-
• Shortage of cooperation with farmers.	1	-
• Trust-building among farmers.	1	1
Technical factors:		
• Inadequacy of weather stations.	1	-
• Planting area.	-	-
• Planting pattern.	-	-
• Protecting underground resources.	4	9
• Use of climate information.	4	9

ers' perceptions on barriers to use of climate information were mentioned earlier. In this way, explicit programmes of actions for producers and extensionists and role definitions for the system are proposed. It is obvious that by inscribing a particular pattern (provision of information and extension through appropriate channels) onto climate information, the information then becomes an actor and imposes its inscribed actions onto its users.

Suggested by farmers, extension activities are widely recognized as useful instruments, which should be provided by the regional Agri- Jihad organisations and mass media programmes. Knowledge production - a timely and accurate knowledge of weather conditions - and familiarizing extension agents with the content of the information facilitates the application of climate information from the farmers' viewpoint. In other words, understanding climate information (wordings and content of information, uncertainties underlying the probabilistic seasonal climate forecasts, and know-how to decide based on probabilistic climate information) is perceived as one of the factors constraining climate information use. Perhaps understanding relevant content by extension agents would facilitate users' comprehension of the information.

Findings of this study reveal that, the agricultural climate information system is a conceptual space maintained by actor-networking, "under" which actor-worlds such as farmers, the Meteorological Organisation, the Agri-Jihad Organisations, policy-makers, broadcasts and media based organisations are engaged in a self-reinforcing process of translation. According to ANT's enrollment and mobilization concepts, farmers specifically have involved Meteorological Organisation, Agri-Jihad Organisation and mass media (broadcasting programmes) into the climate information network. Respondents pointed to the producer-extensionists coalition with the intention to strengthen user relations and act as an instrument for further contact. Although interest in the solution proposed is quite high, neither "influencing power" nor strong public appearance of this coalition seems to contribute to the extension of the network. These ideas are used to analyze how climate information issues are conceptualized as networks of interconnected heterogeneous human and nonhuman actors.

Results also reveal that, ANT is useful for accounting for the introduction of climate information to the farming society. It can help map the agricultural climate information network and consider certain aspects of how power flows (actor dependencies) within it. The emphasis on the dynamic and relational aspects of the problem (agricultural climate information use in farming decisions) is a useful lens for studying nonlinear change

and the unintended outcomes of ACIS which is provided by ANT. Besides, ANT usefully reveals the highly complex setting of ACI and work practices, the highly emergent and evolutionary use of ACI, and the complexity of the network of human and non-human alliances that needed to be recruited to agricultural climate information use.

For researchers who seek to answer questions related to the constraints of agricultural climate information use from an actors' perspective, ANT may provide conceptual tools and inspiration, but not a sophisticated theory of either human actors or the generative causality of related structures. This study showed that, a combination of actor analysis models with selected features of ANT could explain adoption of climate information by farmers in complex environments, like farming systems.

CONCLUSION

Advances in the ability to predict climate information months in advance suggest the opportunity to improve the climatic risk management in agriculture, but only if particular conditions take place. This paper outlined factors affecting patterns of climate information use from the farmers' viewpoint. The study used ANT to explain how such patterns of use accorded with the aligned interests of farmers in farming processes. The approach described here rests on DANA as a potentially promising way to investigate the multi-actor decision making setting in which farmers find themselves.

The results summarized by DANA contain a sorted list of five major categories of economic, information processing, organisational, socio-political, and technical factors to utilize ACI in farming decision making. An impression of the relevant factors affecting human and non-human actors was obtained by reviewing the frequency with which a factor was mentioned by respondents. Results shed light on the importance of socio political factors (farmers' awareness, motivation, and trust of information), and information processing factors (accuracy of information, timeliness access to information, correspondence of information to farmers' conditions, and familiarity with content of information) to facilitate information use enrolment and mobilization in farm practices. Despite the importance of the impact of uncertainty in ACI use, the methodology used in this paper does not provide much insight to this issue. Further research is needed to better understand the impact of uncertainty and provide suggestions to overcome resistance to information that is perceived to be uncertain.

Economic factors and organisational issues were also often mentioned, referring to problems such as the limited instruments and staff and the influence government support has on the development and implementation of climate information related policies and regulations. Findings contend that a process of learning must ensue in order to determine appropriate ACIS. At the same time, the presentation of the climate information and its mode of communication to farmers are critical for successful application and mobilization.

Results revealed that ANT provides a comprehensive vocabulary for exploring the problematization in ACIS. Findings showed that drawing on ANT can serve to deepen and sharpen a detailed analysis of why farmers use or misuse climate information in the farming context and how ACIS facilities and elements integrate and evolve, how work routines and collaborative relationships should adjust, and how positions and skills of "actors" should undergo change over a period of time. ANT's ability to incorporate ACI through the inscription of human actors' interests provides an

understanding of the complex social interactions involved in ACI use.

By performing DANA analysis of the perceptions, values, resources and tactics of farmers in the negotiating arena a better understanding of the effects of these in the processes of interessement and enrolment was provided. The actions suggested by farmers were focused on facilitating enrolment. The most "popular" solutions that were mentioned by farmers to enrol and mobilize ACI in farm practices were financial support for providing related instruments necessary for up taking climate information, co-ordination between organisations, improved policies of motivating and supporting the agricultural sector and farmers through developing farmer unions, and addressing the policies by means of participatory planning. Providing education and increasing farmers' and extensionists' awareness to get familiar with information content were also mentioned often as part of a solution, but it should be added that these activities cover a wide range of subjects and a wide range of actors' support. Results showed that there was no one actor responsible for this – it was the accumulation of influences by different actors (the Meteorological Organisation, the Agri-Jihad Organisation and mass media).

Although, analysis of all actors with a special stake in the climate information network (climate information producers, and policy makers, and extensionists) is not reported in this paper, the insights produced by farmers as an important client of ACIS translated into practical recommendations, suggesting a sequence of activities that the decision makers should consider in the development of a new policy regarding climate information. This indicates that the information produced by DANA could be useful, although the final test for this aspect is still to come, provided by its actual use in the climate information management cycle.

However, this study is subjected to some limitations due to its research methodology. Despite the convenience and time saving that focus groups provide to the research, focus group discussions do not provide a statistically representative sample of respondents. However, the participants in the discussions were carefully selected to represent all the climatic variation groups in the province. The other constraint with this technique may be lack of participants' confidentiality and shyness due to avoiding answering some questions directly (for fear of divulging embarrassing information on the community or even people around) or being too shy to speak on some issues.

REFERENCES

Aanestad, M., & Hanseth, O. (2000). *Implementing open network technologies in complex work practices: A case from telemedicine.* Paper presented at the IFIP 8.2. Aalborg, Denmark.

Abernethy, M. A., Horne, M., Lillis, A. M., Malina, M. A., & Selto, F. H. (2005). A multi-method approach to building causal performance maps from expert knowledge. *Management Accounting Research, 16,* 135–155. doi:10.1016/j.mar.2005.03.003

Aikat, D. D. (2001). Pioneers of the early digital era: Innovative ideas that shaped computing in 1833-1945. *International Journal of Research into New Media Technologies, 7*(4), 52–81. doi:10.1177/135485650100700404

Alcouffe, S., Berland, N., & Levant, Y. (2008). Actor-networks and the diffusion of management accounting innovations: A comparative study. *Management Accounting Research, 19,* 1–17. doi:10.1016/j.mar.2007.04.001

Artikov, I., Hoffman, S. J., & Lynne, G. D., PytlikZillg, L. M., Hu, Q., Tomkins, A. J., ... Waltman, W. J. (2006). Understanding the influence of climate forecasts on farmer decisions as pllanedbehavior. *Journal of Applied Meteorology and Climatology, 45,* 1202–1214. doi:10.1175/JAM2415.1

Bakhshaie, A. (2008). *Testing an actor-network theory model of innovation adoption with econometric methods.* (Thesis). Virginia Polytechnic Institute and State University, Blacksburg, VA.

Bots, P. W. G. (2007). Analysis of multi-actor policy contexts using perception graphs. In *Proceedings of IEEE/WIC/ACM International Conference on Intelligent Agent Technology* (pp. 160-167). IEEE.

Bots, P. W. G. (2008). *Analyzing actor networks while assuming frame rationality.* Paper presented at the conference on Networks in Political Science (NIPS). Cambridge, MA.

Bots, P. W. G., Van Twist, M. J. W., & Van Duin, J. H. R. (2000). Automatic pattern detection in stakeholder networks. In J.F. Nunamaker & R.H. Sprague (Eds.), *Proceedings 33rd Hawaii International Conference on System Sciences.* Los Alamitos, CA: IEEE Press.

Bots, P. W. G., van Twist, M. J. W., & van Duin, R. (1999). Designing a power tool for policy analysts: Dynamic actor network analysis. In *Proceedings of Thirty–Second Annual Hawaii International Conference on System Sciences.* IEEE.

Cameron, T. A. (2005). Updating subjective risks in the presence of conflicting information: An application to climate change. *Journal of Risk and Uncertainty, 30*(1), 63–97. doi:10.1007/s11166-005-5833-8

Carberry, P., Hammer, G. L., Meinke, H., & Bange, M. (2000). The potential value of seasonal climate forecasting in managing cropping systems. In G. L. Hammer, N. Nicholls, & C. Mitchell (Eds.), *Applications of seasonal climate forecasting in agricultural and natural ecosystems: The Australian experience*. Dordrecht, The Netherlands: Kluwer Academic Publishers. doi:10.1007/978-94-015-9351-9_12

Carmen Lemos, M., Finan, T. J., Fox, R. W., Nelson, D. R., & Tucker, J. (2002). The use of seasonal climate forecasting in policy making: Lessons from North East Brazil. *Climatic Change*, *55*, 479–507. doi:10.1023/A:1020785826029

Dolwick, J. S. (2009). The social and beyond: Introducing actor-network theory. *Journal of Maritime Archaeology*, *4*(1), 21–49. doi:10.1007/s11457-009-9044-3

Dunning-Lewis, P., & Townson, C. (2004). *Using actor network theory ideas in information systems research: A case study of action research*. Lancaster University.

Elbanna, A. (2009). Actor network theory and IS research. In Y. K. Dwivedi, B. Lal, M. D. Williams, S. L. Schneberger, & M. Wade (Eds.), *Handbook of research on contemporary theoretical models in information systems*. Hershey, PA: IGI Global. doi:10.4018/978-1-60566-659-4.ch023

Enserink, B., Hermans, L., Kwakkel, J., Thissen, W., Koppenjan, J., & Bots, P. (2010). *Policy Analysis of Multi-Actor Systems*. The Hague, The Netherlands: Lemma.

Everitt-Deering, P. (2008). *The adoption of information and communication technologies by rural general practitioners: A socio-technical analysis*. (Ph.D dissertation). Faculty of Business and Law, Victoria University, Melbourne, Australia.

Feldman, M. S., Khademian, A. M., Ingram, H., & Schneider, A. S. (2006). Ways of knowing and inclusive management practices. *Public Administration Review*, *66*(1), 88–89.

Fenwick, T. (2006). Toward enriched conceptions of work learning: Participation, expansion, and translation among individuals with/in activity. *Human Resource Development Review*, *5*(3), 285–302. doi:10.1177/1534484306290105

Gao, P. (2005). Using actor-network theory to analyse strategy formulation. *Information Systems Journal*, *15*, 255–275. doi:10.1111/j.1365-2575.2005.00197.x

Giordano, R., Mysiak, J., Raziyeh, F., & Vurro, M. (2007). An integration between cognitive map and casual loop diagram for knowledge structuring in river basin management. In *Proceedings of International Conference on Adaptive & Integrated Water Management (CAIWA), Coping with Complexity and Uncertainty*. Basel, Switzerland: CAIWA.

Goodwin, P., Fildes, R., Lee, W. Y., Nikolopoulos, K., & Lawrence, M. (2005). *Unearthing some causes of BPR failure: An actor-network theory perspective*. Bath University. Retrieved from http://www.bath.ac.uk/management/research/pdf/2007-14.pdf

Gustad, H. (2006). *Implications on system integration and standardisation within complex and heterogeneous organisational domains: Difficulties and critical success factors in open industry standards development.* (Master of Science thesis). Norwegian University of Science and Technology, Department of Computer and Information Science.

Haghighat, M. (2008). *Determining the best date of cultivating rain fed wheat considering climatic parameters and soil characteristics in appropriate places for rain-fed agriculture in Fars Province.* (M.Sc. Thesis). Islamic Azad University, Science and Research branch.

Hansen, J. W., & Sivakumar, M. V. K. (2006). Advances in applying climate prediction to agriculture. *Climate Research*, *33*, 1–2. doi:10.3354/cr033001

Hanseth, O., & Monteiro, E. (1997). Inscribing behaviour in information infrastructure standards. *Accounting. Management and Information Technology*, *7*(4), 183–211. doi:10.1016/S0959-8022(97)00008-8

Harrison, M., & Williams, J. B. (2007). Communicating seasonal forecasts. In A. Troccoli et al. (Eds.), *Seasonal Climate: Forecasting and Managing Risk*. Springer.

Hermans, L. M. (2005). *Actor analysis for water resources management: Putting the promise into practice.* Delft, The Netherlands: Eburon Publishers.

Hermans, L. M. (2008). Exploring the promise of actor analysis for environmental policy analysis: Lessons from four cases in water resources management. *Ecology and Society*, *13*(1), 21.

Hermans, L. M., & Muluk, Ç. B. (2002). Actor analysis for the implementation of the European Water Framework Directive in Turkey. In R.S.E.W. Leuven, A.G. van Os, & P. H. Nienhuis (Eds.), *Proceedings NCR-Days 2002 Current Themes in Dutch River Research*, (pp. 74-75). Academic Press.

Hermans, L. M., & Muluk, Ç. B. (2008). *Facilitating policy learning about the multi-actor dimension of water governance.* Paper for the XIII World Water Congress. Montpellier, France.

Hermans, L. M., & Thissen, W. A. H. (2009). Actor analysis methods and their use for public policy analysts. *European Journal of Operational Research*, *196*(2), 808–818. doi:10.1016/j.ejor.2008.03.040

Hu, Q., PytlikZillg, L. M., Lynne, G. D., Tomkins, A. J., Waltman, W. J., Hayes, M. J., ... Wilhite, D. A. (2006). Understanding farmers' forecast use from their beliefs, values, social norms, and perceived obstacles. *Journal of Applied Meteorology and Climatology*, *45*, 1190–1201. doi:10.1175/JAM2414.1

Jackson, K. M., & Trochim, W. M. K. (2002). Concept mapping as an alternative approach for the analysis of open-ended survey responses. *Organizational Research Methods*, *5*, 307–336. doi:10.1177/109442802237114

Karimi, M., Kheiralipour, K., Tabatabaeefar, A., Khoubakht, G. M., Naderi, M., & Heidarbeigi, K. (2009). The Effect of Moisture Content on Physical Properties of Wheat. *Pakistan Journal of Nutrition*, *8*, 90–95. doi:10.3923/pjn.2009.90.95

Klopper, E., Vogel, C. H., & Landman, W. A. (2006). Seasonal climate forecasts- potential agricultural- risk management tool? *Climatic Change, 76,* 73–90. doi:10.1007/s10584-005-9019-9

Knights, D., Murray, F., & Willmott, H. (1997). Networking as knowledge work: A study of strategic interorganizational development in the financial service industry. In B. P. Bloomfield, R. Coombs, D. Knights, & D. Littler (Eds.), *Information technology and organizations: Strategies, networks, and integration.* Oxford, UK: Oxford University Press. doi:10.1093/acprof:oso/9780198289395.003.0007

Kolkman, M. J., Kok, M., & Van der Veen, A. (2005). Mental model mapping as a new tool to analyse the use of information in decision-making in integrated water management. *Physics and Chemistry of the Earth, 30,* 317–332. doi:10.1016/j.pce.2005.01.002

Lamb, R. (2006). Alternative paths toward a social actor concept. In *Proceedings of the Twelfth Americas Conference on Information Systems.* Acapulco, Mexico: Academic Press.

Latour, B. (1991). Technology is society made durable. In J. Law (Ed.), *A sociology of monsters: Essays on power, technology and domination.* Routledge.

Latour, B. (1996). *Aramis, or, The Love of Technology* (C. Porter, Trans.). Cambridge, MA: Harvard University Press.

Latour, B. (1999). On recalling ANT. In J. Law, & J. Hassard (Eds.), *Actor Network Theory and After.* Oxford, UK: Blackwell.

Li, X., & Zhang, Y. (2009). Analysis of the 3G diffusion and adoption in China. *Journal of Academy of Business and Economics, 9*(3), 175–187.

Lybbert, T. J., Barrett, C., McPeak, J. G., & Luseno, W. K. (2007). Bayesian herders: Updating of rainfall beliefs in response to external forecasts. *World Development, 35*(3), 480–497. doi:10.1016/j.worlddev.2006.04.004

Mac Leod, M. A. (2001). *Actor network theory: Examining the role of influence on technology adoption.* University of Hawai'i at Manoa.

May, C., & Finch, T. (2009). Implementing, embedding, and integrating practices: An outline of normalization process theory. *Sociology, 43*(3), 535–554. doi:10.1177/0038038509103208

McGrath, K. (2001). The golden circle: A case study of organizational change at the London Ambulance Service. *ECIS 2001 Proceedings.* Retrieved from http://aisel.aisnet.org/ecis2001/52

McMaster, T., Vidgen, R. T., & Wastell, D. G. (1997). Towards an understanding of technology in transition: Two conflicting theories. In *Proceedings of Information Systems Research in Scandinavia, IRIS20 Conference.* Hanko, Norway: University of Oslo.

Moameni, A. (1999). *Soil quality changes under longterm wheat cultivation in the Marvdasht plain, south-central Iran.* (Ph.D. Thesis). Ghent University, Belgium.

Mohammed, S., Klimoski, R., & Rentsch, J. R. (2000). The measurement of team mental models: We have no shared schema. *Organizational Research Methods, 3*(2), 123–165. doi:10.1177/109442810032001

Najafi, B., & Bakhshoodeh, M. (2002). Effectiveness of government protective policies on rice production in Iran. In *Proceedings of the Xth EAAE Congress on 'Exploring Diversity in the European Agri-Food System.* Zaragoza, Spain: EAAE.

Narayanan, V. K., & Armstrong, D. J. (2005). *Casual mapping for research in information technology*. Hershey, PA: Idea Group Publishing Inc.

Nazemos'sadat, S. M. J., Kamgar-Haghighi, A. A., Sharifzadeh, M., & Ahmadvand, M. (2006). Adoption of Long – Term Rainfall Forecasting: A Case of Fars Province Wheat Farmers. *Iranian Agricultural Extension and Education Journal, 2*(2), 1–16.

Nelson, K. M., & Nelson, H. J. (2000). Revealed causal mapping as an evocative method for information systems research. In *Proceedings of the 33rd Hawaii International Conference on System Sciences*. IEEE.

Patton, M. Q. (2002). *Qualitative research and evaluation methods* (3rd ed.). Sage Publications, Inc.

Pouloudi, A., & Whitley, E. A. (2000). Representing Human and Nonhuman Stakeholders: On Speaking with Authority. In R. Baskerville, J. Stage, & J. I. DeGross (Eds.), *Organizational and Social Perspectives on Information Technology*. Boston: Kluwer Academic Publishers. doi:10.1007/978-0-387-35505-4_20

RahimiBadr. B., & Moghaddasi, R. (2009). Iran wheat price forecasting performance evaluation of economic. In *Proceedings of 4th International congress on Aspects and Visions of Applied Economics and Informatics (AVA)*. Debrecen, Hungary: AVA.

Richetin, J., Perugini, M., Adjali, I., & Hurling, R. (2008). Comparing leading theoretical models of behavioral predictions and post-behavior evaluations. *Psychology and Marketing, 25*(12), 1131–1150. doi:10.1002/mar.20257

Rogers, E. M. (1995). *Diffusion of innovations* (4th edn.). New York: Free Press.

Rubas, D. J., Hill, H. S. J., & Mjelde, J. W. (2006). Economics and climate applications: Exploring the frontier. *Climate Research, 33*, 43–54. doi:10.3354/cr033043

Sharifzadeh, M., Zamani, G. H., Khalili, D., & Karami, E. (2012). Agricultural climate information use: An application of the planned behaviour theory. *Journal of Agricultural Science and Technology, 14*, 479–492.

Sherrick, B. J., Sonka, S. T., Lamb, P. J., & Mazzocco, M. A. (2000). Decision-maker expectations and the value of climate prediction information: Conceptual considerations and preliminary evidence. *Meteorological Applications, 7*(4), 377–386. doi:10.1017/S1350482700001584

Shoib, G., Nandhakumar, J., & Jones, M. (2006). Using social theory in information systems research: A reflexive account. In A. Ruth (Ed.), *Quality and Impact of Qualitative Research, Proceedings of the 3rd International Conference on Qualitative Research in IT & IT in Qualitative Research*. Brisbane, Australia: Institute for Integrated and Intelligent Systems, Griffith University.

Sophonthummapharn, K. (2008). *A comprehensive framework for the adoption of techno-relationship innovations: Empirical evidence from eCRM in manufacturing SMEs*. (University Dissertation). Handelshögskolan vid Umeåuniversitet.

Subbiah, A. R., Kalsi, S. R., & Yap, K. (2004). *Climate information application forehancing: Resilience to climate risks*. Paper presented at the the International Committee of the Third International Workshop on Monsoons (IWM-III). Hangzhou, China.

Tatnall, A. (2000). *Innovation and change in the information systems curriculum of an Australian University: A socio-technical perspective.* (Dissertation). Central Queensland University.

Toffler, A. (1970). *Future shock.* New York: Bantam Books.

Von Hippel, E. (1988). *The sources of innovation.* New York: Oxford University Press.

Walsham, G. (1997). Actor-network theory and IS research: Current status and future prospects. In A. S. Lee, J. Liebenau, & J. I. DeGross (Eds.), *Information Systems and Qualitative Research.* Chapman and Hall. doi:10.1007/978-0-387-35309-8_23

Wickramasinghe, N., Tumu, S., Bali, R. K., & Tatnall, A. (2007). Using actor-network theory (ANT) as an analytic tool in order to effect superior PACS implementation. *International Journal of Networking and Virtual Organisations*, *4*(3), 257–279. doi:10.1504/IJNVO.2007.015164

Ziervogel, G. (2004). Targeting seasonal climate forecasts for integration into household level decisions: The case of smallholder farmers in Lesotho. *The Geographical Journal*, *170*(1), 6–21. doi:10.1111/j.0016-7398.2004.05002.x

Ziervogel, G., Bithell, M., Washington, R., & Downing, T. (2005). Agent-based social simulation: A method for assessing the impact of seasonal climate forecast applications among smallholder farmers. *Agricultural Systems*, *83*(1), 1–26. doi:10.1016/j.agsy.2004.02.009

Ziervogel, G., & Downing, T. E. (2004). Stakeholder networks: Improving seasonal climate forecasts. *Climatic Change*, *65*, 73–101. doi:10.1023/B:CLIM.0000037492.18679.9e

KEY TERMS AND DEFINITIONS

Actant: Non-human actor.

Actor-Network Theory: ANT is a sociological theory that emerged as a useful vehicle to study the intertwining of the social and technical agency to constitute a performing network.

Actor: An human and non-human entity that is made to act affect by the support of other actors.

Agricultural Climate Information: Agricultural climate information is one among many sources of inputs that can be used by decision makers to reduce risks and to optimize gains in agricultural systems. Part of climatic information is forecast of climate fluctuations with a seasonal (i.e. several months) lead-time, which appears to offer significant potential to contribute to the efficiency of agricultural management. Accessibility, interpretability, credibility, and relevance (lead time and forecast period, covering different climate variables, finer spatial resolutions, and inclusion of historical analogs) are among factors facilitating climate information use.

Agricultural Information System: Agricultural information system links people and institutions. It integrates farmers, agricultural educators, researchers and extensionists to harness knowledge and information from various sources to promote mutual learning and generate, share and utilize agriculture-related technology, knowledge and information for better farming and improved livelihoods.

Case Study: A case study methodology attempts to answer the why and how questions about a contemporary phenomenon that is complex.

Causal Map: A causal map is a diagnostic tool that illustrates the structure of individuals' idiosyncratic belief systems in a particular domain and cause and effect relationships among these concepts.

Interactively Elicited Cognitive Mapping: Also known as IECM, is an idiographic

approach of measuring mental models, in which participants themselves provide the content of the knowledge to be mapped.

Network: Network is a conceptual tool which describes and makes sense of the simultaneously social and technical character of any social arrangements. It is concerned with mapping how actors define and distribute social, political, technical, or bureaucratic roles, and mobilise or invent others to play these roles.

Theoretical Saturation: The point at which analysis of additional data provides no new insights into the substantive theory of action generated from the data.

APPENDIX 1: ABBREVIATIONS

- **ACI:** Agricultural climate information.
- **ACIS:** Agricultural climate information system.
- **ANT:** Actor-network theory.
- **DANA:** Dynamic actor-network analysis.
- **IECM:** Interactively elicited causal mapping.
- **IS:** Information system.

Chapter 4
A Fluid Metaphor to Theorize IT Artifacts:
A Post–ANT Analysis

Rennie Naidoo
University of Pretoria, South Africa

Awie Leonard
University of Pretoria, South Africa

ABSTRACT

This chapter extends existing metaphors used to conceptualise the unique features of contemporary IT artifacts. Some of these artifacts are innately complex, and current conceptualisations dominated by a "black box" metaphor seem to be too limited to further advance theory and offer practical design prescriptions. Using empirical material drawn from a longitudinal case study of an Internet-based self-service technology implementation, this chapter analyses various aspects of an artifact's fluidity. Post-actor network theory concepts are used to analyse the artifact's varying identities, its vague boundaries, its unexpected usage patterns, and its resourceful designers. The successes and failures of the artifact, its complex and elusive relations, and the unintended ways user practices emerged, are also analysed. This chapter contributes by extending orthodox metaphors that overemphasise a stable and enduring IT artifact—metaphors that conceal the increasingly unpredictable and transitory nature of IT artifacts—with the distinctive characteristics of fluidity. Several prescriptions for the design and management of fluid IT artifacts are offered.

INTRODUCTION

Despite the increasing breadth and scale of IT artifacts and the pervasive role they play in our day to day lives, IT research continues to be dominated by studies that focus on the psychological and socio-psychological context of IT acceptance and use, the social and managerial contexts of IT, and systems analysis and design techniques that form an essential part of the construction of IT. The conceptual significance of the IT artifact itself has often been overlooked, either by treating it as a simple measurable variable, by decou-

DOI: 10.4018/978-1-4666-6126-4.ch004

pling it from the social context, and by even excluding it in IT studies (Akhlaghpour, Wu, Lapointe & Pinsonneault, 2009). The shift in many firms from adopting custom built applications to adopting packaged applications may partly explain why the IT artifact remains an under researched area. Many tertiary institutions have likewise shifted their curriculum to focus on managerial aspects such as enterprise architecture, IT governance, project management, service and operations management, systems analysis and design, and the teaching of packaged applications with lesser attention being paid to arguably the core subject matter, the IT artifact (Topi, Valacich, Wright, Kaiser, Nunamaker, Sipior & Vreede, 2012).

Orlikowski and Iacono (2001) analysed 188 articles published in the *Information Systems Research* (ISR) journal in the decade beginning in 1990 and ending in 1999 to understand the prevalent conceptualizations of the IT artifact. According to these authors, 13 different views of IT artifact could be distinguished but the black box metaphor remains the favoured and dominant conceptualisation as a foundation for inquiry. Although they identified 13 clusters of IT in these articles, they classified these into 5 general clusters. The clusters suggest that researchers focused on: "the building of IT artifacts with particular capabilities (the computational view of technology), their intended uses (the tool view of technology), technology as a variable (the proxy view of technology), and the interaction between people and technology (the ensemble view of technology) (Gregor & Juhani, 2007:5)". Despite a recent similar analysis by Akhlaghpour et al. (2009) that reviewed published research over the years 2006-2008, and carried out on two more journals including the *Management Information Systems Quarterly* (MISQ), and the *Journal of the Association of Information*

Systems (JAIS), little evidence was found to suggest that researchers have been deeply engaging with the conceptualization of the IT artifact. One of the potential problems with the current orthodox conceptualization of the IT artifact is that it may be misleading and also impeding the theoretical development of the IT discipline and consequently a body of knowledge useful to IT practitioners.

Metaphors can be described as useful "symbolic fictions" to describe and construct knowledge about the world in which we live (Morgan, 1980). Morgan (1980) has suggested an important link between metaphors and the way scientific theory is constructed and argues that they can exert an important influence on a discipline's body of knowledge as well as practices. Contemporary IT artifacts are innately complex and the current metaphorical orthodoxy seems to be too limited to productively theorise and research them. Our principal aim therefore in attempting to understand the IT artifact is to depart from a predictable 'black box' conceptualization of the technological artifact to a more complex and 'fluid' conceptualization. Our major contention is that the contemporary IT artifact cannot be viewed only as predictable and seamless. We therefore glean from an alternative theoretical framework that is concerned with fluid nature of the IT artifact. As such we turn to post-ANT, and the more recent feminist work to highlight the fluid nature of an IT artifact (Moser & Law, 2006; de Laet & Mol, 2000).

In this chapter, we analyse the 'fluid' nature of an IT artifact observed in a case study of a self-service technology (SST) implementation at a large multinational private healthcare insurance firm, based in South Africa. The argument of this chapter is presented as follows. In the next section, the literature on the technological artifact is critically reviewed.

Next, we provide an overview of the case study research context and a justification for the research methods used. The subsequent and most substantive section reports on the fluid nature of the SST under investigation. In the final two sections, we offer an interpretation and discuss the implications of our findings in the context of the IT artifact literature.

LITERATURE REVIEW

Conventional Perspectives of the IT Artifact

The dominant research paradigm in IT research assumes that the IT artifact is a man made object that is capable of having a predetermined impact on social systems and therefore technology is typically assigned as the independent variable (Simon, 1996). A number of the models used to describe IT artifacts based on rational decision theory models of user choice and attribution theory (Pandya & Dholakia, 2005; Venkatesh, Morris, Davis, & Davis, 2003; Goodhue & Thompson, 1995). Ciborra and Lanzara (1994) have expressed concerns regarding the notion of the IT artifact has purely closed and instrumental in character. Nevertheless mainstream research assumes that an IT artefact has a static goal and continues to focus on determining stable user characteristics when using a stable technology medium (Venkatesh, Thong & Xu, 2012). Current understandings of IT artifacts using natural science approaches are too narrow to provide an understanding that can be used to develop a more holistic view of the technology artifact (Gregor & Livari, 2007). From a design science perspective, researchers interested in designing IT artifacts to solve organizational and social problems have also adopted a somewhat narrow perspective of the IT design artifact viewing it in abstract terms

ranging from stable constructs, models, and methods, to stable material manifestations such as their actual instantiations (Hevner, March, Park & Ram, 2004; Jones, 2007). To view the IT artifact as a definitive end product is also an oversimplification (Gregor & Livari, 2007). By studying how an IT artifact comes into being and evolves over time, we share Orlikowski and Iacono's (2001:121) critique of mainstream research as 'conceptualizations of IT artifacts as relatively stable, discrete, independent, and fixed' and respond to their call to theorise how IT artifacts are designed, constructed, and used by people and shaped by the interests, values, and assumptions of a wide variety of communities.

Early ANT Perspectives of the IT Artifact

ANT's view of society as a socio-technical web has enjoyed broad appeal in the IS community's efforts to understand the IT artifact (Cho, Mathiassen & Nilsson, 2008; Heeks & Stanforth, 2007; Hanseth & Aanestad, 2004; Faraj, Kwon & Watts, 2004; Tatnall & Lepa, 2003; Monteiro, 2000). From a traditional ANT perspective, stability and social order are continually renegotiated as a process of aligning the interests of symmetrical human and non-human actors (Callon, 1986; Law, 1992). According to ANT, translation is required to maintain stability in socio-technical systems (Callon, 1982; Latour, 2005). In this process one actor, typically a focal actor represents the interests of another actor, to the focal actor's own. Translation also relies on some kind of medium for it to take place – a material embodiment, such as text or a machine, into which it can be inscribed. The unchanging nature of immutable mobiles assumed by these perspectives is viewed here as too restrictive for contemporary IT artifacts. More specifically, conventional actor-network approaches have been criticized for being too rigid (Law & Singleton, 2005; de Laet & Mol, 2000). Consequently, there is a considerable scope for researchers to explore the range of relations that make up an IT artifact. In the next sections, we review alternative conceptual tools that Post-ANT perspectives offer to understand complex a specific type of mutable technology artifact.

Post-ANT perspectives of Mutable Artifacts

In contrast to conventional and early ANT studies of the IT artifact, Post-ANT studies suggest that we need to depart from a predictable view of contemporary IT and offers a framework that is sensitive to the more complex and 'fluid' character of IT (Moser & Law, 2006; de Laet & Mol, 2000). For Post-ANT work, contemporary technological artifacts are characterised by mutability. Unlike immutable mobiles that hold themselves rigid, mutable mobiles appear to change gently and fluidly. In this perspective, the network analogy is 'loosened' up, the role of control is downplayed, and as opposed to being rigid, network relations are seen to be fleet-ing (Moser & Law, 2006; Law & Singleton, 2005; Whitley & Darking, 2006). Similar to immutable mobiles, there is still a core of stability characterising a mutable object (Naidoo & Leonard, 2012; Nielsen & Hanseth, 2007). In other words, even though the artifact changes, it is still part of what subsisted previously. The notion of a mutable IT artifact however is not new. Prior work such as those done by Orlikowski and Gash (1994) have already recognised and described the mutable behaviour of the IT artifact. However these works have not described and analysed these mutable behaviours using the more recent fluid concepts. What is unique in this study is the general fluidity of relations that is assumed to make up a mutable technological artifact. According to de Laet & Mol (2000), artifacts do not exist as a single spatial type. They view an object from a topological perspective - that is by the space that it occupies as opposed to a region or a network. Accordingly fluid technological artifacts can be depicted by six properties: boundaryless, multiple identities, mixtures, robustness, continuity, and dissolving ownership (See Table 1).

Table 1. The six characteristics of 'Fluid' IT Artifacts

Characteristic	Definition
Boundaryless	Boundaries are unclear. There is no line separating where one variant of the technology stops and the other begins. Attempts to fix a boundary tend to falter.
Multiple Identities	Identities come in a diverse array. Identities are variant and cannot be predetermined.
Mixtures	Composed of a multitude of mixtures with variable degrees of viscosity. They are not solid or stable. There is no central element that binds these mixtures.
Robustness	Since they are not clingy to anything in particular, the bonds within fluid spaces are always unstable. However they cannot collapse easily, as they are quite adaptable in altering their form.
Continuity	Exhibits mutability within an unpredictable world they subsist in. Transformations that occur while they flow occur without discontinuity.
Dissolving ownership	Over time different actors can take the ownership role, make a contribution, and then release their ownership

Adapted from de Laet & Mol (2000) and Nielsen & Hanseth (2007)

METHODOLOGY

Research Approach

In this research we were mainly concerned with social, technical and human relations, more specifically practices which carry and produce these complex relations that can only be understood over time. Unlike most interpretive researchers who target human actors as their focal point (Orlikowski & Iacono, 2001), we assume that there is no clear dividing line between the social and technical, and accordingly adopted a mild social constructionist approach. We are therefore generally willing to attribute the properties and effects of technology in social shaping (Brey, 1997). Thus, our overarching philosophical perspective for this empirical study was informed by a practical orientation, led largely by an interpretive approach. As a way of improving the quality of research we adopted Klein and Myers (1999) proposed set of principles based on the hermeneutic orientation. In the remainder of this section we provide a review of the research context, followed by a discussion of how the fieldwork was staged, including sampling and data analysis procedures.

Research Context

This study is attuned to a holistic case study approach since it observes a global programme or initiative (Yin, 1999a; 1999b). The unit of analysis consisted of the HIC1 organisation, a pioneer in consumer-driven healthcare, and its SST initiative (eHIC). The case study focuses on understanding the implementation of HIC's SST, an initiative that was aimed at facilitating the online electronic interactions between its clients and the organisation. Although HIC had service portals dedicated to providers, employers and brokers, clients in this case study refer to HIC's largest and top online priority at the time, the member-related portion of the self-service website (See Figure 1).

There were several reasons for why this particular private healthcare insurance sector represents an ideal context to study the social, technical, and human. The first is that healthcare costs globally are continually rising. As a result private insurance firms are seeking novel ways to control the cost of financing healthcare. Some healthcare insurance firms are now modelling their business on the so-called "consumer-driven healthcare" concept. They frame this emerging concept

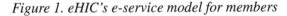

Figure 1. eHIC's e-service model for members

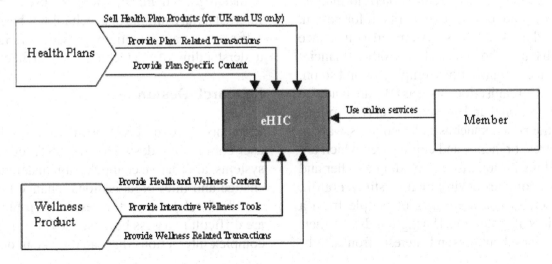

as one were consumers play a greater role in making decisions on their healthcare, have better access to information to make informed decisions, and share more in the costs (Cannon and Tanner, 2005). Although the South African (RSA) economy is perceived to be an emerging market, the RSA private healthcare insurance industry is widely recognised as being highly developed (Benatar and Fleischer, 2003). HIC is also viewed globally as one of the pioneers in consumer-driven healthcare (CDH), and was in a fairly advanced phase of implementing its online SST to complement this strategy. At the turn of 2000, HIC attempted to capitalise on joint ventures (JVs) and trends towards consumer-driven healthcare in the international market, particularly in the UK. With an ageing population, and one in four UK citizens being obese, a tremendous burden was being placed on the National Health Service (NHS). The UK product range was being built on HIC's 'consumer-driven healthcare' experience, even though it was not a full replacement scheme, as in RSA. Nevertheless, this case has global relevance.

A second appeal of this context, is that private healthcare insurance organisations such as HIC perform an information-intensive activity, including processing a member's personal details, claim details, service provider details, procedures and health conditions information, tariffs and coverage benefits information, and the like. HIC's systems also interface with banks, hospitals, clinics, other financial services organisations, employers and so on. Third, organisations such as HIC are made up of traditional and alternative service options for their clients such as intermediaries, walk-in centres, branches and call centres which are well institutionalised, providing a richer and more complex environment of study. Fourth, the organisation consists of people from a variety of professional backgrounds with their peculiar subcultures and interests, from health-

care practitioners, actuaries, call centre staff, information technologists, and accountants to broker consultants and underwriters, making for a dynamic organisational context. Fifth, the healthcare insurance industry touches on the daily lives of a broad section of citizens and institutions. The on-going inequity between those with access to private medical care and those dependent on the public sector remains one of the biggest challenges for the RSA health system. In the UK approximately 6.5 million UK citizens use some form of private healthcare insurance to supplement the NHS funding.

Sixth, the RSA healthcare insurance market for historical reasons is arguably more mature than some of its counterparts in first world countries, who have only recently turned their attention to more innovative private healthcare funding mechanisms as a means of managing healthcare (Costello and Tuchen, 1998). This novel approach has been taken by consumer-driven healthcare insurance firms on the assumption that they can curb costs by focusing on wellness rather than on illness, and the resultant design of their products on the basis that 'prevention is better than cure'. Seventh, an Internet-based self-service technology context was selected because an integrative theory in this area is lacking, despite its phenomenal growth. Furthermore, the use of the Internet as an example of a technology-based self-service can potentially provide a richer understanding of the technological artifact.

Research Design

Elements of a complex IS artifact such as the user community, designers, the SST, legacy systems, alternative channels, organisational change and inter-organisational change that are the focal point of the research problem are difficult to assess because they represent complex interactions that can only be understood over time. The case study strategy was therefore chosen because of its advantages in creating novel and profound insights and its focus on examining the rich social, cultural, political an technical influences on the implementation of SST initiatives in the context of a health insurance services organisation. Yin (1999b) argues that a case study is an ideal form of empirical enquiry to investigate a contemporary issue or event within its real-life context, especially where the boundary between such issues or events and its context is not clearly defined. This is equally relevant in novel areas where few theories have been applied (Cornford & Smithson, 1996). Yin (1999b) also makes a particular case for the use of case study methods in health services research and asserts that other empirical methods are at a distinct disadvantage in developing our understanding of contemporary developments, specifically in healthcare systems that are linking multiple components in new ways and producing 'mega-systems' of great complexity.

Yin (1999b) also adds that the system rules in healthcare are in a high state of flux, continually and rapidly changing and endorses the study of single facilities such as health centres, hospitals, and community mental health centres, explaining that a single case can often produce a more penetrating study. A case study offers a means of investigating complex social aspects in which multiple variables are intertwined. In fact, a number of researchers have demonstrated the effectiveness of single case studies in bringing about a broader understanding of the IT artifact, a particular focus of this study, and the potential to perhaps improve practice (Whitley & Darking; Schultze & Orlikowski, 2004). Meanwhile a number of researchers have now suggested a set of clearly defined methodological guidelines for interpretive case study research. As a way of improving the quality of research

conducted from the interpretive perspective, Atkins and Sampson (2002) comprehensive guideline for the conduct of a single case study will be used. Their guidelines emerged through a synthesis of leading research work of case studies, particularly in the IS field (Klein & Myers, 1999; Walsham, 1995). The guidelines are organised in a framework which suggest five focus areas: way of thinking; way of working; way of controlling; way of supporting; and way of communicating.

Data Collection

Semi-structured interviews and secondary source analysis were the main data collection mechanisms. Individual interviews carried out on site were the primary technique used to elicit information from the HIC's informant. The duration of these interviews varied from 1 to 2 hours. The fieldwork for the case study took take place at various intervals during the period from February 2004 to November 2008. From June 2004 to July 2005, our main focus was directed at creating a historical reconstruction of the SST implementation from 2000 to 2004 (see Table 3). During this period one of the researchers was immersed in a large number of public and confidential reports that had been provided by several of the managers and members of the implementation team. The reports included management reports, weekly operation reports, call-centre and online user feedback, strategic plans, news articles, forms, fliers, prior research reports, presentations and so on (See Table 2).

The use of the website feedback from the inception of the technology was shown to be a very innovative and useful approach to collecting and analysing information to understand user perception and evaluation of Internet-based SST. The SST feedback data retained information about the user and the date of the feedback.

Table 2. Sources of empirical evidence

Data Sources
250 hours of participant observation including call centre queries and weekly meetings
Over 100 pages of field notes
55 formal interviews (planned)
9 informal interviews (planned and opportunistic)
Record of online user feedback log (over 5 000 responses since inception of SST)
Approximately 80 documents
Listening to member calls at main client call centre
Management reports, annual reports, emails, strategic plans, magazines, news articles, forms, fliers, surveys, previous research
1 research diary

This then gave us a rich historical perspective of user experiences. Users were expressing themselves in a 'real context', and not an artificial context created by researchers.

The field research for the case study was carried out in four main periods, consisting of three months in early 2004, two months in early 2006, two months in late 2007, and another month in late 2008. One of the researchers conducted a total of 55 formal interviews during this period (see Table 3). All 55 interviews were tape-recorded, and extensive research notes were taken. This practice ensured that everything said was preserved for analysis. A number of documents were prepared for the research by participants after the study had begun. The specific purpose for generating these documents was to learn more about the events being investigated. Data collection and analysis by and large consisted of an iterative process, and this approach assisted with subsequent collection of data ensuring richer and deeper interpretation. The interviewees were chosen for their relevance to the conceptual

Table 3. Summary of interviews conducted with design team

Nature of Group	Number of Interviews				Totals	
	Field trip 1	Field trip 2	Field trip 3	Field trip 4	Number of interviews	Number of respondents
Management team	1	6	5	2	15	11
Business/systems analysts	11		3	1	15	13
Usability analysts	1				1	1
Java developer	4	2			7	6
System architects	5		1	1	7	7
Graphic designers		6			6	6
Subject matter experts*		1	1	1	2	2
Marketing	1	1			2	2
Other		1			1	1
Total	23	17	10	5	55	49

questions rather than their representativeness. Initial participants (at the first group interview) were asked to suggest names of other actors involved in the topic of the case study, and general networking through personal contacts expanded the sample. The total number of respondents to interview was reached heuristically, that is, the decision to stop adding respondents was taken when nothing new was being learned from the interviews and a state of theoretical saturation was achieved (Eisenhardt, 1989).

Although there were no set boundaries for selecting the interviewees, we favoured pursuing respondents who had a longer history with the SST implementation initiative. A few 'newcomers' were invited to obtain a more balanced perspective. All interviews were conducted in English and transcribed in 'Word' format. Interview transcripts and written notes were analysed systematically through iterative and repeated re-reading. This made it possible to gain an increasingly profound understanding of each interviewee's viewpoint and perspective, of links and contradictions within and across interviews, of complex contextual factors emerging from these interviews and of the many relationships between the relevant concepts. During the transcription of the interviews, key themes from fluid ANT were identified and new perspectives and questions generated. These themes subsequently acted as inputs to discussions with interviewees and guided further analysis and interpretation of the transcripts. These interviews, together with the large collection of rich, thick qualitative information from a number of sources, played an important role in revealing the complexity of organisational processes and of the context studied. This triangulation of data was important in counteracting any biases in the collection and analysis of data (Darke, Shanks, & Broadbent, 1998). Key participants were given a chance to check the results of the analysis by reviewing transcripts of their interviews (Nandhakumar & Jones, 1997). Discussions were also held with key informants in order to give them a chance to reflect on the output of the case.

Data Analysis

We used the relevant ANT theoretical concepts and themes from Post ANT to guide the analysis process. Version 5 of ATLAS.ti was used to code and store these themes and categories at the textual and conceptual level. The textual level included activities such as coding text and writing memos. The conceptual level focused on model-building activities. ATLAS.ti was also used for the overall management of the research project and its associated data. This archive consisted of the case study field notes, case study documents, quantitative data and other electronic files generated during the case study. These 'files' were catalogued, indexed chronologically as the research process unfolded, and filed for easy access and retrieval (Bassey, 1999). Using ATLAS.ti for easy cross-referencing assisted in maintaining a chain of evidence to support the case study conclusions (Muhr and Friese, 2004; Darke, Shanks & Broadbent, 1998). In the next section, themes from ANT, Post-ANT, and feminist theories on objects will be used to express the findings of the study.

RESEARCH FINDINGS

The following selection of episodes will be drawn upon to represent the story of the SST implementation: from information portal to channel of choice; from channel of choice to a wellness tool; and from wellness tool to a complementary channel. While the order of the flow of these episodes will maintain their sequence during this analysis, the choice of certain incidents that will be featured, and the length of discussion accorded to these, are chosen to illuminate and make explicit the key fluid concepts discussed in previous sections. As depicted in Figure 2, the SST is about a mutable rather than immutable mobile. Thus, the SST needed to change its shapes as it reformed itself from ideals of a dominant channel to eventually emerging as a 'complementary channel'. These taken together will describe the particular flow of the SST object.

FLUID TRANSITION 1: FROM INFORMATION PORTAL TO CHANNEL OF CHOICE

Until 1999, disseminating "static content" dominated HIC's website implementation. As the CIO recalls there was no clear point of departure: 'Initially it was just to be part of the space. And no one really could draw a more rational reason than that. You have got to be part of this play. The whole world was going to go online.' Nevertheless, a new identity for the SST to become the "Channel of Choice" was emerging as the ability to transact

online became more plausible and as capital expenditure on call centre technology continually increased. This new identity was also strengthened by the dotcom hype created by various steering mechanisms. An organising vision emerging from a heterogeneous collective consisting of the academic world, media, consultants, software vendors, and dotcom start-ups who bestowed a lot of appeal upon the substitution claim and other "efficiency" inscriptions (Hannemyr, 2003). Swanson and Ramiller (2004) assert that this "bandwagon" phenomenon is especially prevalent where an innovation achieves a high public profile, as with the Internet and e-commerce. Planned action was dismissed by the urgency to join the 'stampeding herd', despite the high costs and apparent risk.

The primary intent of HIC's e-business designers was to substitute the call centre consultant with the Web front-end. This major shift in identity from information purveyor to take on the role of the call centre consultant was based on the assumption that the SST would be able to reduce calls by having answers to key call reasons programmed and made available on the website. To show support for the new channel, the leadership

Figure 2. The fluid transitions of an IT artifact over time[2]

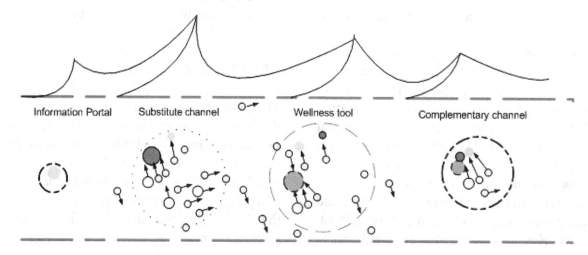

of the organisation provided autonomy (It was being granted a separate strategic business unit status) to the SST, to transform the way customers interacted with the firm and the way they were to be serviced. One of the senior managers described the growing optimism at the time: 'HIC itself was flying having redefined health insurance in South Africa so all in all there was an abundance of goodwill and positive energy. Pretty much anything Tom (referring to the Head of eHIC) and I wanted to do with eHIC we could and we had the support. Our only limiting factor was capacity and time'. Consequently, the firm invested significantly in hardware and software. However, the predominantly batch mode of processing between internal systems, the use of a non-standard database platform, poor data quality and integration issues made for weak ties. The internal system, a robust client-server application used by the firm's operational areas for high volume data capture was at odds with the Java-based application for developing Web applications. Furthermore a number of the functionally-driven applications were built as one monolithic piece of software code and therefore could not be easily adopted for a component-based Web environment. One of the senior managers recalled: 'HIC was and always has been a mess of technology types, especially in the development environment. We tried to follow the industry standards and use what at that stage would have been best of breed technology offerings, which we did, but not without huge and consistent resistance from other key players'. Meanwhile there were fears in several of the functional areas that the role of the SST was to 'squash' legacy practices embedded in traditional channels.

For users, interessement was positioned in advertising and promotions using email, call centre promotions and the customer magazine, by positioning 'convenience, secure and real-time information benefits', in various media.

However the poor ICT infrastructure supplied by the South Africa's monopoly telecommunications firm, the speed and cost of bandwidth, the corporate standard to verify user name and password through a call-centre validated process became major barriers to using the new system. This following complaint is indicative of how users were impacted: "Even with a dedicated 64K ISDN line, I have not managed to get to any page I tried to access. Please ensure that there is either more bandwidth for access or the system does not bump you off in too short a time. Note: Most people access with a normal Telkom (referring to the monopoly firm) phone line at 14,400 K and must find it impossible to access the site'. During the same period a minority of users reported contradictory experiences as the SST worked its ways through barriers in the national ICT infrastructure. One user submitted the following compliment: 'Thanks for an excellent site! What a pleasure to browse the site. The content, the speed, the layout, etc. are well planned and executed'. As a result of the failure to cope with these mixture of technologies and the unsteady support from both system and business departments, the implementation of the project was completed much later than envisaged. However, despite the weaknesses in the RSA ICT infrastructure at the time and the absence of converting a majority of users away from traditional channels, the SST did not simply collapse.

Conventional ANT suggests that successful translations depend on how faithful key actors are towards their alliances. From this discussion, it might be concluded that the translation of the actors failed to achieve the desired outcome as the preferred channel. While the initial network and its loosely formulated goal – "to become the channel of choice" – was readily accepted internally by a few of the key senior executives, it remained too weak to mobilise a sufficiently strong enough network to become the dominant channel. One of the reasons for this unintended outcome was that SST's goals were not sufficiently convincing for those managers and staff that represented the traditional channels the SST was attempting to substitute. Achieving synergy between departments proved to be particularly problematic largely due to the various actors having to facilitate multiple and often conflicting agendas. As a result negotiations were often beset by 'clashes' of interest and these conflicts often became irresolvable. More specifically, the interactions with both human and non-human actors supporting the traditional channels were during this part of the implementation journey more contentious than collaborative. Certainly, the local ICT infrastructure was not supportive of a self-servicing environment for a majority of the users. Furthermore standards and security were impeding the Web channel when compared to alternative channels. For example one customer voiced the following dissatisfaction when key functionality was removed from the channel as a result of fraud and issues related to securing online transactions: 'I am extremely disappointed that you have discontinued your online claims submission. Come on, guys, surely there are better solutions than stopping a useful service!'. In addition, the poor interoperability with systems designed to support internal processes and the lack of technical skills of the newly appointed development e-business team also translated into unsuccessful and unstable translations between actors. As a result, attempts to mobilise, expand and stabilise the majority of potential users turned out to be complex and frustrating. For consumers, the telephone clearly had a better inscription than the web as suggested by this online feedback: 'dealing with your call centre guys is just great ... once you give them the membership number they have the information on their fingertips.

Also the folks just seem to be enjoying what they're doing ..a pleasure to deal with'. Allied to the telephone was the customer's membership card, inscribed with a membership number and telephone number, as opposed to a user name and password which remained a cognitive challenge for the typical user. Importantly the evolving consumer-driven healthcare product was laden with jargon. Apart from understanding the Health Savings Account (HSA) mechanism and how it works, customers needed to understand a plethora of unfamiliar concepts related to the mechanics of the HSA. On the other hand, in servicing a customer, a call centre consultant was able to retranslate the perplexing consumer-driven healthcare terminology to facilitate client understanding. Whereas banking transactions are frequent and the 'language' of banking is standard across banking institutions, the complexity of the health insurance product and its associated jargon were serving to constrain the use of the online channel. Embedded in the products are jargon terms like medical savings account, self-payment gaps, above-threshold balance, cryptic and abbreviated claims reason codes, all of which are beyond the understanding of the average user. A remark by a user demonstrates the frustration that is felt because of the product complexity and its associated language: 'could you please dilute your language ... not every member is highly educated to can understand your oxford English. the purpose here is to transmit info and not to impress via language protocol or style' However, banking terms such as withdrawals, transfers, borrowing rates, lending rate, deposits and overdraft are well accepted and are part of the individual's practical consciousness. On the other hand, consumer-driven health insurance concepts are unique terms and for most members are not institutionalised in their social practice. In reviewing the medical savings account concept, while it sounds as if it should work like a bank account, it is subject to limits such as the tariff rate that the healthcare provider charges, the condition or procedure one is being treated for, or the health plan one has chosen. In addition, these products contain a host of other actuarial, clinical and commercially loaded language. Furthermore, while bank products remain fairly stable, health plans are subject to significant changes in terms of benefits structures and so on. These remarks by a senior manager allude to some of the differences between online banking and e-health insurance services: 'But I also think the nature of the product. Because the cheque account doesn't really change. You know what I mean? You understand how it works. So, where we introduce different things along the way, the product rules changes, limits changes, legislation changes. So there is always, you know, your family changes, you have different claim requirements, your needs. The way the doctors behave differently, differ over time. So, the doctor might be inconsistent in the way he handles the claims, and sometimes the treatment.. There's many unknowns. Many changes.' Consequently, weak inscriptions demonstrated by the allies of the SST and their inability to act in ways that maintained the network, led to the majority of customers persisting with the use of the traditional call-centre channel. Given the properties of irreversibility demonstrated by the traditional channel, there was this realisation among the senior management team that the "substitution claims" may have been farfetched. The CIO responsible for this initiative stated: When the dot bomb started happening, you know, the realisation of what happened is that, maybe it really wasn't this tsunami'. Similarly, another Senior Manager stated: 'From a landscape point of view we may have been a bit blind to the fact that connectivity levels and even on dial-up at that stage were abysmal'. It would

soon emerge that the e-business was more suited as a channel for the HIC business, rather than a separate business unit. Shortly after the launch of eHIC, the world was hit by the dotcom crash. Silicon Valley companies came crashing down around the globe. HIC senior management expediently absorbed eHIC back into the business as one of the systems department. No more 'chill room', no more fringe benefits. The staff of eHIC were now working for a financial services company and were integrated into the IT department. However, this dissolution of ownership did not alter the continuity of the SST. Instead signs of the SST's robustness were emerging as new boundaries were being defined to make the technology work. Ironically, the once solidified structure of an information portal was giving way to new forms not unlike flow patterns reminiscent of a tidal wave.

FLUID TRANSITION 2: FROM CHANNEL OF CHOICE TO WELLNESS TOOL

The wellness episode describes the second major shift in the continuity of the online SST. The role of the SST was fundamentally driven by the firm's new focus on its wellness offering. Within this identity the emphasis was on hedonic aspects as opposed to merely health plan transactional features during the "preferred channel" episode. Customers were now being incentivised to stay healthy and the wellness team was interested in establishing whether the Web could be an appropriate mechanism to promote a healthy lifestyle. Loyalty points and other forms of rewards were offered to customers for following a healthy lifestyle with the aid of online tools. At the same time, the wellness business unit was powerfully positioned and was becoming a major part of the firms' actor-network.

Consequently wellness was being viewed by business proponents of the SST as an important ally to their success. As one of the firms wellness experts stated: 'The problem with human beings is you need to give them some reward for their effort. And the whole concept of Wellness is that you get a reward or an incentive, but by then you would feel so good that, that it is sort of, you know, an added bonus, 'the reason why I am doing it is because I am actually feeling good by exercising or if I meditate twenty minutes every day'. So you catch people, either with, you know, scaring them, showing them the huge risk, or you catch them with the incentives, which are just so attractive, that in the end, you have them stick with it long enough to feel it in their bodies and mind that they are actually benefiting'. At the time, both the wellness business unit as well as the e-business department shared similar views on the "cost savings" argument as the rhetorical device to enrol users to self-service using online wellness applications as opposed to using more costly wellness practitioners.

Translating the wellness innovation and HIC's engaging style to the online world appealed to only a minority of users. It appeared

that after the novelty effect of active participation in the Wellness program subsided, so did the user's participation online. Over time, as much as 60% of the users who registered never returned to the website (see Figure 2). This is in stark contrast to the health member base churn or lapse rate of 3–4%. In fact, only 25% of the registered users continued to remain loyal and habitual users of the online channel. Another segment of the user base showed sporadic use. On the other hand, the use of the call centre remained relatively high, with repeat calls accounting for a significant component of the call volume. Internal statistics showed that 40% of members were phoning more than once a month.

The wellness programme itself while proving to be an effective product differentiator for the health insurer and an attractive selling point for brokers, was not effective at enrolling a majority of its customer base online and effecting actual behavioural change. Instead, many users used the incentive points for online use in an unanticipated way. One of the users provided the following feedback: 'I have been trying to complete the four exams for the nutrition section of the web site to get the 500 Wellness points. I passed the last three

Figure 3. Leaky bucket syndrome – losing users over time.[3] Source: Internal Report: Statistical analysis of retention (2008).

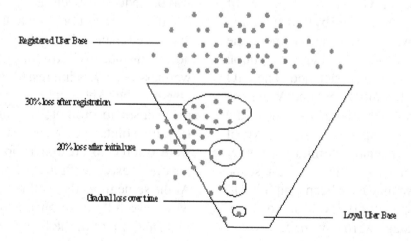

but can't get the 1st (basic) exam to display. My girlfriend logs on and sees (and completed) all four so it can't be my PC it must be something to do with what happens when I log on to the web site. Please advise how I can get the 500 points'. Rather than follow the assigned way of using the online channel as an obligatory approach to "improve their health", the anti-program of "points chasers" emerged as a result of the online incentives. Many users appeared to be more interested in earning points through the Web with minimal behavioural changes to their lifestyles. The designers had been betrayed by many of the users they thought they were representing. Meanwhile with the appointment of a new head for the wellness programme, there was a "push" to drive customers towards the firm's wellness partners. As the majority of customers showed a preference for using the call centre to resolve administrative related queries, similarly of the two networks, face-to-face consultation would prevail over the use of virtual wellness diagnosis and consultations. One of the firm's dieticians who assisted in the design of the wellness tool made the following remark: 'trying to figure out everything a dietician would want to know from a person and trying to put that in some kind of tool was actually a mistake because we really didn't intend to become or replace the services of a registered dietician. There is absolutely no way that we could possibly do that and yet we were trying so hard to get to that point of being an online dietician.'

Meanwhile, internally senior management used the SST as an ally to convince joint venture (JV) partners of their 'innovativeness'. Following this identity, the attention of the development team shifted to the international operations and their online requirements. The initial assumption was that the team could just 'plug and play' local services into the UK context. However, in many areas, the local inscriptions did not apply to the international initiatives. The excerpt by a business analyst below points to a local contradiction that influenced the shaping of the SST in the UK: 'You see, whereas, in our UK market we needed to get customers, in South Africa, we needed to maintain customers. So they were two fundamentally different things, which I used to always get frustrated with, because I would say, why we are basing a new Website, for a new company on an existing company's Website that services existing members'. Some of the key members of the UK SST development team expressed concerns that the UK initiative was seen by management as a project where the team could simply leverage off the RSA SST. Consequently, the UK site inherited a lot of the logic of the RSA website. However, the UK market had to perform functions such as 'quote submissions' and 'health plan comparisons' to enable direct-to-consumer sales, which were not available in the RSA SST. Whereas the assumption was that the team could reuse the components developed for the RSA SST, a number of new applications and features had to be built from scratch for the UK site. In this case, the Internet shrinks the world, because JV partners located overseas can access skilled developers based in South Africa. The different local requirements were constraining the RSA development team and as a result, relationships between the local and UK design teams were often strained. 'Well, in the beginning, there was a very much of an 'us' and 'them. so, we are the RSA developers and now we are doing the UK Website a favour by helping them, developing stuff for them. We were not a priority. And, yes, I used to get a, a definite feeling that we were the enemy, coming in to use up their time, you know'.

Some interesting challenges were associated with creating a 'UK-friendly' online nutrition centre. For instance, it soon emerged that the dietetic products available in the UK market far exceeded what was available in South Africa – for example wheat-free and soy-based products. Upon reflecting on these major differences, a senior member of the nutrition team viewed this as a mistake: 'A major nutritional issue in the UK is intolerance or allergy to wheat and there are many more vegans than in South Africa. We had not catered for wheat-free and vegan meal plans on the SA Nutrition Centre and the UK office requested that we design such options to suit their market. This involved the UK dietician supplying us with the names of products available in the UK that could be used as substitutes for wheat and animal protein foods'. The other issue that came to light was that the content and educational articles needed to be checked and modified for the UK market, owing to certain differences in how the UK national dietary guidelines and health systems operate. Furthermore, while the RSA dietary guidelines were designed specifically for that nation and its nutritional issues, the UK had its own set of dietary guidelines. The team had to change the content to reflect these cultural practices instead of the RSA recommendations. In addition, any reference to contacting a registered dietician in South Africa was updated to reflect the process in the UK, which is completely different, because of the way the National Health Service (NHS) works in the UK. The design of the UK nutrition site initially had the RSA nutrition rules embedded in it. In a sense, the UK system of nutrition was initially ignored by designers. Despite their absence, various UK actors came together to make their presence felt, such as UK nutrition standards and guidelines, UK nomenclature, a qualified UK nutritionist, and the UK language and metric system. This narrative demonstrates the tension in developing SST systems between wishing to standardise applications for efficiency and imposing the same applications on different local contexts. In a sense then, the SST could not gently flow but had to make certain 'jumps and discontinuities' to cater for the different sets of relations and context of practices of UK users.

In the end, despite moderate use by a minority of end users both in the RSA and UK, its was acknowledged by the wellness team that the SST was an oversimplification of what wellness practitioners do in the "real world" and that only wellness practitioners can deal with the full complexity of the wellness practice. The lead technical designer made the following remark: 'And Wellness is kind of moving away from just the Web, you know. I think we have been fairly Web centric. Now they have said ... no, for nutrition, you have got to go and see a nutritionist, which I agree with. There are certain things that we are not very good at, for the Web. You are not going to go to the gym on the Web. You are going to go to the gym near you, physically, you know, physical'. While the SST arguably did not succeed in its identity either as a substitute for the call centre agent, or as a comprehensive wellness tool for replacing a wellness practitioner, the way it fluidly dissolved into a complementary channel emerged to rescue the technology that had been thrown into crisis yet again.

FLUID TRANSITION 3: FROM WELLNESS TOOL TO COMPLEMENTARY CHANNEL

This final episode in our story emerged as current social, technical and political contexts made the inscriptions for other channels stronger. During this phase there was a notable shift in alliances with the wellness team away from the SST to networks of healthcare professionals. By 2006, despite attempts by the self-service's team management to persuade other key actors that wellness online "worked", the third major identity that emerged was one in which the self-service channel was regarded as a complementary channel. There was a shift from 'click and point' towards more incentives for using a network of practitioners. At the same time, a new discourse about the role of the SST began to emerge. The SST was once again legitimised by the swift modifying of its purpose to ward off the challenge by dissenting groups within the organisation, as described below: 'I think, if we look today at where we have come, our initial objective was to convert a channel (call centre) into another channel [online self-service]. And lessons are learnt, you know, this is a social environment, okay. There is no dominant channel. It's apparent to me that the channels are interlinked, merged, and one will use whatever is closest in proximity'. The SST management had conceded that it was not within their scope to alter the behaviour of their members to use the most 'cost-effective channel' in their engagement with the health insurer.

There was also a prevailing view that the majority of users in the UK had problematised the SST as research tool. Although the majority of UK users preferred purchasing health insurance directly from brokers in the UK, the role of the Web in the sales process was not totally dismissed. Instead it underwent another 'smart' retranslation as a 'sales research tool'. The CIO mentioned: 'we have seen strong evidence to support the notion that the online channel used in conjunction with the interactive channels probably results in the highest levels of success. So, even if you are buying very high-end products, and very complex products, you will use the online channel to do research and understand the products, and do the evaluations even though you might not execute in that channel'. In other words, the

SST was rearticulated as playing an effective role in supporting the sales process. The newly appointed head of the SST also elaborated on the technologies adaptability for research as another new justification for its existence: 'We are starting to see some feedback now in terms of the differences between the conversion rates in the UK environment, to the people that visit more than one channel and they've got higher conversion rates. The ability to be able to research, doesn't necessarily lead to online sales, but it certainly supports the sales process. And that also, I think, can definitely translate into servicing as well. But it has almost become something that people need to have now ... It is not optional in the long run'. As observed, the path of the translation of technology is seldom predictable. The initial organisational metaphor guiding the identity of the self-service channel implementation was framed as the "channel of choice". Initially it was conceived that the SST would replace the call-centre consultant for a broad population of the firm's customers. When contradictory facts emerged this identity was retranslated to a "Wellness Tool" were the technology was used for more hedonistic purposes. By early 2008, the SST was being framed as a complementary channel.

In this most recent mutation, the designers became a stock of usability experts that could deliver front-end development expertise for the rest of the firm. Consequently their identities, in this case, were threatened rather than reinforced. As the CIO remarked: 'One of the things that H-World does particularly well, which maybe more an art than a science is the softer elements of the channel. How to position, how to message, how to design the user interfaces. Now all of those are steeped in something that is not well understood by the traditional environment'. Eventually the online channel itself was fully integrated in the firm's business functions. Not surprisingly the use of this technological frame evoked a less grand conclusion about the SST's capabilities and led the SST team to become anxious about their roles in the organisation. The CIO in hindsight now had this to say: 'Our members can continue to use the call centre and rely on our regular postal mailings, but if they find the Web more convenient, we wanted them to have that option as well. We also recognized that many members would want to communicate with us across multiple touch points, including Web self-service, phone, e-mail, or even mobile devices. So making sure that the experience was seamless and consistent across all those service points was critical. We want to be able to recognise our members as individuals on contact and ensure that we are using everything we know about them to deliver the most fulfilling service experience possible'. By mid-2008, the SST department was disbanded. Indeed, apart from maintenance and product changes, the SST was being managed on an ad hoc basis by various system departments in the IT department. The SST technology itself had become completely immersed in the fabric of the organisation, and simply became one of many channels that the business units would use to engage or interact with stakeholders. The transition of eHIC from what was envisaged at its birth to what transpired had clearly moved in another unintended direction. Its latest mutation compared with how it was conceived in 1999 departed in yet another interesting way.

Nevertheless, this particular configuration as a complementary channel has maintained the SST for now. With this redefinition it was acknowledged that the SST would continue to have many identities. Furthermore, by drawing upon the inclusion of previous identities in this broader 'complementary redefinition', the SST was able to resist many of the commonly

held beliefs that the SST had not delivered on previous identities. For instance, it was now acceptable that a user could use the phone, engage directly with a wellness practitioner or broker, as well as the self-service website. The new identity provided the SST with more flexibility and in this way became immersed in the firm's total service offering. In fact, in its latest conception, the SST was becoming a proxy that tangibly represented the firm's 'innovativeness' and as such, would assist in enrolling JV partners willing to form an alliance with a 'progressive health insurer', but this is part of another episode.

DISCUSSION

The study contributes to previous research by confirming the analytical generalisability of the fluidity concept to a specific IT artifact case study (Nielsen & Hanseth, 2007; Ruddin, 2006). The results corroborate the propositions that major fluid characteristics such as boundaryless, multiple identities, mixtures, robustness, continuity and dissolving ownership matched the IT artifact we studied (Nielsen & Hanseth, 2007; de Laet & Mol, 2000). By departing from the orthodox black box metaphor of the IT artifact to a more complex and fluid metaphor, this study has provided unique perspectives about contemporary IT artifacts. One of the most defining characteristic of the IT artifact in the case was its shifting identities and unclear boundaries that were needed for the technology to work. Any attempts to fix its identity would have most likely faltered. When exploring the changing nature of the IT artifact, the first dominant identity that designers attempted to lock into the artifact was the notion of substituting the call centre consultant. While the use of the SST as a replacement channel was in some respect a rational decision based on internal efficiency goals, it neglected the socially-technical rich context of the external user and their already intersecting routines with the existing and more traditional channels. Unlike internal users who are subjects of the governing structures of the organisation, external users appear to possess substantial discretion in their use of self-service channels. This dynamic context made the process of attracting, converting and retaining users a major challenge for the firm. This fluid object in particular existed in a world of mixtures of both self-service and traditional services. Failing to see the intricacies of interacting with traditional channels resulted in a misconception that the SST could have somehow taken on the role of traditional channels.

Similarly, during the next wave of change we observed how the use of the channel as a wellness practitioner dismissed the interpersonal roles that are essential in a wellness practitioner-patient interaction. Thus, SSTs are subject to their social-technical contexts for their continuous adaptations. Designers of SSTs shape SSTs but cannot control them in a deterministic way and SSTs that do not match the demands of their social contexts are unlikely to evolve in ways inscribed by designers. Furthermore SST designers have to face external users as well as internal business and traditional systems staff interests and alignments that are often contingent and unstable. A fluid perspective reveals the unpredictability that emerges when one mixes the grand vision of channel domination by designers and their practices on the one hand and the rather complex and situated realities of channel practices of users on the other. Nevertheless, despite failing to enrol a majority of users away from the call-centre channel or to online and remote wellness, the SST was not compelled to maintain its stability and linkages in these previous identities. Since it did not cling to any particular identity, the IT artifact was able to change adaptably. Nevertheless, it had to let go notions of channel domination and settle into a complementary role.

Despite the fact that the identity of the IT artifact was unstable and changed over time its boundaries remained unclear as it absorbed these multiple identities, it still remained robust. As pointed out by Monteiro (2000), this seemingly self-perpetuating process whereby a solution spreads, gathers momentum or picks up speed is in an on-going effort to survive. In other words the IT artifact is never complete or final. Nevertheless we can consider it to be a success even if only some of its identities are working or even partially working. In the case, the plasticity of the SST enabled it to be consistently reinvented by both designers and users. At every implementation shift, there were distinctive and fluid transformations of the technology, the organisation, and the environment. And each set of changes provided a new context for later set of identities.

The case reveals that the IT context is not some clearly bounded unity. The practices of the users and designers are vital in reproducing the SST, but traditional service channels, the media, software vendors, regulators, and other steering mechanisms are equally implicated in its reproduction as a service channel. Furthermore, traditional channels are well institutionalised in day-to-day practices and show strong symptoms of irreversibility. While SSTs are applicable and appropriate in certain contexts, this study reveals that the subtle roles of service consultants and wellness practitioners such as dieticians and stress counsellors that are acted out in face-to-face or interpersonal encounters in this healthcare insurance context cannot be completely delegated to technology. Furthermore, the study revealed that many users still adopt institutionalised practices such as the call centre, and prefer interpersonal engagements for complex or sensitive matters. Granted, there are some users who prefer to use only the online channel, but in this case they did not form a critical mass that is sufficient to justify the spend on the technology. Even then, some service encounters are simply more effectively serviced by traditional channels. As such, the SST did not radically alter the way services were provided in the case study. At best, the SST was more effective at supporting other channels than replacing them.

The main practical contribution of this study is the finding that in this particular healthcare insurance services context, SSTs are most appropriately positioned as a complementary service channel. The study showed that existing embedded relationships between stakeholders such as the health insured member and the call-centre agent were in the main irreversible or resistant to change. Many health insured members prefer to use the human interface, as opposed to the Web interface, to manage complex or sensitive queries. Users modified the way in which they appropriated the SST, and the modified use often did not resemble the intentions of the designers. Based on their servicing need, users very often tailored the use of the SST channel in the context of other channels (Germonprez et al. 2007).

This case also reveals that a seemingly endless improvisation with its role characterises this IT artifact. It appears that in this particular case no particular role was permanent specifically in an environment of alternative service channels. As such ongoing negotiations characterised the identity of this SST better than the 'black box' metaphor. However the most recent translation as a complementary channel is likely to remain a special kind of fluid, an alloy. After all, within this broader conception of a complementary channel, the interests of other actors could fall in with the

SST's schemes without too much controversy. For practitioners, this means that they should not try to set fixed boundaries to contemporary complex IT. Moreover, they should acknowledge that they cannot anticipate the success of every service that can be produced online. Instead of being both ambitious and rigid, practitioners themselves should adopt a flexible strategy for an SST implementation. Instead of mastering the fluid features that characterise contemporary IT, perhaps one should be prepared for tinkering, experimenting, tailoring, and serendipitous outcomes (Ciborra, 2002; Hovorka & Germonprez, 2009). Practitioners need to recognise that processes of implementation of contemporary IT cannot be managed by following a recipe book as very often the identity of such an IT artifact is still in a "discovering" stage. While practitioners should nurture and "cultivate" the identities of a fluid object, they should also be aware that these objects are often "out of control", and design and implementation tactics are consequently much more 'subtle and limited' (Ciborra & Hanseth, 1998).

Furthermore, despite their beliefs and related inscriptions, designers in the case study were shown to be flexible and contingent in the way in which they followed their own interests. We witnessed that management – be it by acquiescence or coercion – were not 'deterministically' or dogmatically bound to their planned identities and related inscriptions of developing the 'channel of choice' or the 'wellness tool'. Another important point is that post ANT concepts of spatiality paid careful attention to the localities of the particular SST contexts, as opposed to offering generalised predictors of change. As a result, the theory demonstrated that universal solutions for fluid IT artifacts are unlikely to be immediately successful in multiple user social groups in multiple locations spanning different social, political, institutional, strategic and complex servicing contexts. These analytical devices confirm that the implementations fluid IT artifacts are indeed context dependent. Ultimately, we are not dismissing the use of Internet-based self-service technology or even implying that traditional service channels are better than electronic channels. We are merely suggesting that the metaphor on thinking about some IT artifacts and their implementation needs to change to emphasise fluidity and not just the black box.

CONCLUDING REMARKS

A major issue in this study was to enhance our understanding of the complexities of contemporary IT artifacts by devoting more attention to one of the central concepts in IS – the IT artifact (Orlikowski & Iacono, 2001). However, by using conventional and early ANT concepts alone typical of this kind of analysis, one runs the risk of concealing the fact that IT artifacts can enact additional and more dynamic spatial forms. We argued that de Laet & Mol's (2000) fluid metaphor is a particularly useful for describing and analysing IT artifacts. While some simplification was necessary in this study, by using post-ANT concepts we were able to reveal the multiple nature of the IT artifact. Consideration of fluidity revealed how fragile the IT innovation was during certain stages. The three major waves in the SST's configuration also provided a useful approach to trace the key movements during the SST implementation and the catalytic effect of global shifts in e-commerce and consumer driven healthcare as well as concomitant organisational priorities. Research of this kind is scant in the literature and for SSTs in particular is almost non-existent, making this study significant.

The use of the fluid concept as an analytical device also lends itself to understanding the dynamic nature of innovative IS implementations. By using concepts such as boundaryless, multiple identities, mixtures, robustness, continuity, and dissolving ownership, we were able to tease out important features about this contemporary IT artifact (Nielsen & Hanseth, 2007; de Laet & Mol, 2000). Importantly, the analysis above has demonstrated that there was no clear distinction between central and peripheral actors who feature prominently in a conventional ANT analysis. Instead, the SST was composed of a mixture of different identities that had to continually compose and recompose themselves, as these gained different significance to different owners. During movements from one wavelike episode to another, the SST for the most part was robust enough to maintain what is was part of before and because of its innovative nature, the SST could include many participants, in its various configurations. However this does not imply that all IT artifacts are fluid. It appears that SSTs depending on their context can either be mutable or immutable – For instance it can be argued that SSTs that support online banking are more immutable than healthcare insurance.

Fluid IT artifacts in general are becoming important objects in our lives. However the lack of understanding of these kinds of technological artifacts by both practitioners and academics may be able to partly account for the extravagant and unrealistic claims we make about them, our failure to make better predictions about their adoption and use, inflexibility in our implementation approaches, and the protracted uptake of these technologies in some markets. It is very unlikely that we can and whether we should even attempt to curtail the fluid trajectory of certain IT artifacts. Even more disconcerting for those consultants and managers that offer prescriptions to manage IS implementation are the implications of the fluid nature of IT artifacts for IS management techniques. How is one to be normative about aspects such as formal system development methods and investment appraisal techniques, when no other fluid artifact can possibly compare to another in its implementation? One thing is for certain, that a rigid view of the particular SST in the case and being faithful to unyielding design and implementation prescriptions, would have led to its demise. It appears that despite their intrinsic risk, we may need fluid IT artifacts since their novel nature could potentially inspire new business models, and new users and stakeholders. In fact, in turbulent and unpredictable environments, due to new business ideologies and organisational relationships, IT artifacts should be flexible. Perhaps we should get used to the feeling of anxiety associated with the implication that that no single identity to a fluid IT artifact will necessarily be solid and lasting, and embrace trial and error as a risky yet exciting feature of their design, implementation and use. Certainly early implementers of fluid IT artifacts should perhaps have the corresponding risk appetite and entrepreneurial flair to let these artifacts 'loose'.

Using a fluid metaphor, our complex technological artifact still remains somewhat of a mystery. Certainly the conceptual tools for the fluid metaphor demonstrates that technological artifacts are sometimes not coherent objects, and clearly the more traditional managerial lens and early ANT concepts are likely to sharply conceal its unfinished status. What new perspectives can be opened about the movement of fluid IT artifacts? While a conventional fluid analysis can perhaps account for the gentle ebb and flows of artifacts, it reveals little about surges in their movement activity. Movements of fluids also appear to occur in waves. In addition,

it reveals very little about their viscosity as well – their immensity (thin or thick fluid) or even their stickiness. Furthermore, while existing 'fluid' concepts are particularly useful for describing what is bound up in the body of the IT artifact, it does relatively little for describing the dynamic behaviours that are especially crucial for examining the fluidity of IT artifacts. We therefore need to promote a new agenda for theorising fluid IT artifacts, an agenda that explores their spatiality, an agenda of mobility. We also need to deepen our implementation, design and use theories to account for fluid artifacts; artifacts that are more sharply amenable to designer and user experimentation, and serendipity; artifacts that adapt and change to survive.

REFERENCES

Akhlaghpour, S., Wu, J., Lapointe, L., & Pinsonneault, A. (2009). Re-examining the Status of IT in IT Research-An Update on Orlikowski and Iacono (2001). In *Proceedings of AMCIS 2009*. AMCIS.

Atkins, C., & Sampson, J. (2002). Critical appraisal guidelines for single case study research. In *Proceedings of the Xth European Conference on Information Systems (ECIS)*. Gdansk, Poland: ECIS.

Bassey, M. (1999). *Case Study Research in Educational Settings*. Open University Press.

Benatar, S. R. (1993). Reforming the South African health care system. *Physician Executive, 19*(5), 51–57. PMID:10130951

Brey, P. (1997). Philosophy of technology meets social constructivism. *Society of Philosophy & Technology, 2*, 3–4.

Callon, M. (1986). Some elements of the sociology of translation: Domestication of the scallops and the fisherman of St Brieuc Bay. In Power, Action and Belief: A New Sociology of Knowledge (pp. 196-223). Routledge.

Callon, M., & Law, J. (1982). Of interests and their transformation: Enrolment and counter-enrolment. *Social Studies of Science, 12*(4), 615–625. doi:10.1177/030631282012004006

Cannon, M. F., & Tanner, M. D. (2005). *Healthy Competition: What's Holding Back Health Care and How to Free IT*. Cato Institute.

Cho, S., Mathiassen, L., & Nilsson, A. (2008). Contextual dynamics during health information systems implementation: An event-based actor-network approach. *European Journal of Information Systems, 17*, 614–630. doi:10.1057/ejis.2008.49

Ciborra, C. U. (2002). *The Labyrinths of Information: Challenging the Wisdom of Systems*. Oxford University Press.

Ciborra, C. U., & Hanseth, O. (1998). From Tool to Gestell Agendas for Managing the Information Infrastructure. *Information Technology & People, 11*(4), 305–327. doi:10.1108/09593849810246129

Ciborra, C. U., & Lanzara, G. F. (1994). Formative Contexts and Information Technology: Understanding the dynamics of innovation in organisations. *Accting., Mgmt., &. Info. Tech., 4*, 61–86.

Cornford, T., & Smithson, S. (1996). *Project Research in Information Systems: A Student's Guide*. Macmillan.

Costello, G. I., & Tuchen, J. H. (1998). A comparative study of business to consumer electronic healthcare within the Australian insurance sector. *Journal of Information Technology, 13*, 153–167. doi:10.1080/026839698344800

Cresswell, K.M., Worth, A., & Sheikh, A. (2010). Actor-Network Theory and its role in understanding the implementation of information technology developments in healthcare. *BMC Medical Informatics and Decision Making, 10*, 1–12. doi:10.1186/1472-6947-10-67 PMID:20067612

Darke, P., Shanks, G., & Broadbent, M. (1998). Successfully Completing Case Study Research: Combining Rigor, Relevance and Pragmatism. *Information Systems Journal, 8*, 273–289. doi:10.1046/j.1365-2575.1998.00040.x

de Laet, M., & Mol, A. (2000). The Zimbabwe bush pump: Mechanics of a fluid technology. *Social Studies of Science, 30*, 225–263. doi:10.1177/030631200030002002

Eisenhardt, K. M. (1989). Building theories from case study research. *Academy of Management Journal, 14*(4), 532–550.

Faraj, S., Kwon, D., & Watts, S. (2004). Contested artifact: technology sensemaking, actor networks, and the shaping of the Web browser. *Information Technology & People, 17*(2), 186–209. doi:10.1108/09593840410542501

Germonprez, M., Hovorka, D., & Callopy, F. (2007). A Theory of Tailorable Technology Design. *Journal of the Association for Information Systems, 8*(6), 351–367.

Goodhue, D. L., & Thompson, R. L. (1995, June). Task-technology fit and individual performance. *Management Information Systems Quarterly*, 213–227. doi:10.2307/249689

Gregor, S., & Jones, D. (2007). The Anatomy of a Design Theory. *Journal of the Association for Information Systems*, 8(5), 312–335.

Gregor, S., & Juhani, I. (2007). *Designing for mutability in information systems artifacts*. Canberra, Australia: ANU Press.

Hannemeyr, G. (2003). The Internet as hyperbole: A critical examination of adoption rates. *The Information Society*, 19, 111–121. doi:10.1080/01972240309459

Hanseth, O., & Aanestad, M. (2004). Actor-network theory and information systems: What's so special? *Information Technology & People*, 17(2), 116–123. doi:10.1108/09593840410542466

Hanseth, O., & Monteiro, E. (1997). Inscribing Behaviour in Information Infrastructure Standards. *Accounting. Management & Information Technology*, 7(4), 183–211. doi:10.1016/S0959-8022(97)00008-8

Heeks, R., & Stanforth, C. (2007). Understanding e-Government project trajectories from an actor-network perspective. *European Journal of Information Systems*, 16, 165–177. doi:10.1057/palgrave.ejis.3000676

Hevner, A. R., March, S. T., Park, J., & Ram, S. (2004). Design science in information systems research. *Management Information Systems Quarterly*, 28(1), 75–106.

Hovorka, D. S., & Germonprez, M. (2009). Reflecting, Tinkering, and Bricolage: Implications for Theories of Design. In *Proceedings of the Fifteenth Americas Conference on Information Systems*. San Francisco, CA: Academic Press.

Klein, H. K., & Myers, M. D. (1999). A set of principles for conducting and evaluating interpretive field studies in information systems. *Management Information Systems Quarterly*, 23(1), 67–94. doi:10.2307/249410

Latour, B. (2005). *Reassembling the Social: An Introduction to Actor-Network Theory*. Oxford University Press.

Law, J. (1992). Notes on the Theory of Actor-Network: Ordering, Strategy and Heterogeneity. *Systems Practice*, 5(4), 379–393. doi:10.1007/BF01059830

Mahring, M., Holmstrom, J., Keil, M., & Montealegre, R. (2004). Trojan actor-networks and swift translation: Bringing actor-network theory to IT project escalation studies. *Information Technology & People*, 17(2), 210–238. doi:10.1108/09593840410542510

March, S., & Smith, G. F. (1995). Design & natural science research on information technology. *Decision Support Systems*, 15(4), 251–266. doi:10.1016/0167-9236(94)00041-2

Mol, A., & Law, J. (1994). Regions, Networks and Fluids: Anaemia and Social Topology. *Social Studies of Science*, 24, 641–671. doi:10.1177/030631279402400402 PMID:11639423

Monteiro, E. (2000). Actor-Network Theory and Information Infrastructure. In *From Control to drift: the dynamics of corporate information infrastructures* (pp. 71–83). Oxford University Press.

Morgan, G. (1980). Paradigms, metaphors, and puzzle solving in organization theory. *Administrative Science Quarterly*, 25(4), 605–622. doi:10.2307/2392283

Moser, I., & Law, J. (2006). Fluids or flows? Information and qualculation in medical practice. *Information Technology & People*, *19*, 55–73. doi:10.1108/09593840610649961

Muhr, T., & Friese, S. (2004). *User's Manual for ATLAS.ti 5.0* (2nd ed.). Scientific Software Development.

Naidoo, R., & Leonard, A. (2012). Observing the 'Fluid' Continuity of an IT Artefact. *International Journal of Actor-Network Theory and Technological Innovation*, *4*(4), 23–46. doi:10.4018/jantti.2012100102

Nandhakumar, J., & Jones, M. (1997). Too close to comfort? Distance and engagement in interpretive information systems research. *Information Systems Journal*, *7*, 109–131. doi:10.1046/j.1365-2575.1997.00013.x

Nielsen, P., & Hanseth, O. (2007). *Fluid standards: A case study of the Norwegian standard for mobile content services*. Retrieved from http://heim.ifi.uio.no/~oleha/Publications/FluidStandardsNielsenHanseth.pdf

Orlikowski, W. J., & Iacono, C. S. (2001). Research commentary: Desperately seeking IT in IT research - A call to theorizing the IT artifact. *Information Systems Research*, *12*, 121–134. doi:10.1287/isre.12.2.121.9700

Pandya, A., & Dholakia, N. (2005). B2C Failures: Towards an Innovation Theory Framework. *Journal of Electronic Commerce in Organizations*, *3*(2), 68–81. doi:10.4018/jeco.2005040105

Ruddin, L. (2006). You can generalize stupid! Social scientists, Bent Flyvbjerg, and case study methodology. *Qualitative Inquiry*, (4), 797–812. doi:10.1177/1077800406288622

Schultze, U., & Orlikowski, W. (2004). A practice perspective on technology-mediated network relations: The use of Internet-based self-serve technologies. *Information Systems Research*, *15*(1), 87–106. doi:10.1287/isre.1030.0016

Simon, H. A. (1996). *The Sciences of the Artificial* (3rd ed.). MIT Press.

Swanson, E. B., & Ramiller, N. C. (2004). Innovating mindfully with information technology. *Management Information Systems Quarterly*, *28*(4), 553–583.

Tatnall, A., & Lepa, J. (2003). The Internet, e-commerce and older people: an actor-network approach to researching reasons for adoption and use. *Logistics Information Management*, *16*(1), 56–63. doi:10.1108/09576050310453741

Topi, H., Valacich, J. S., Wright, R. T., Kaiser, K. M., Nunamaker, J. F., Sipior, J. C., & Vreede, G. J. (2012). *IS 2010–Curriculum Guidelines for Undergraduate Degree Programs in Information Systems. Association for Information Systems*. AIS.

Venkatesh, V., Morris, M. G., Davis, G. B., & Davis, F. D. (2003). User Acceptance of Information Technology: Toward a Unified View. *Management Information Systems Quarterly*, *27*(3), 425–478.

Venkatesh, V., & Thong, J., & Xu. (2012). Consumer acceptance and use of information technology: Extending the unified theory of acceptance and use of technology. *Management Information Systems Quarterly*, *36*(1), 157–178.

Walsham, G. (1995). Interpretive case studies in IS research. *European Journal of Information Systems*, *4*, 74–81. doi:10.1057/ejis.1995.9

Whitley, E. A., & Darking, M. (2006). Object lessons and invisible technologies. *Journal of Information Technology*, *21*, 176–184. doi:10.1057/palgrave.jit.2000065

Yin, R. K. (1999a). Case study research, design and methods. *Sage (Atlanta, Ga.)*.

Yin, R. K. (1999b). Enhancing the quality of case studies in health services research. *Health Services Research*, *34*(1), 1209–1224. PMID:10591280

KEY TERMS AND DEFINITIONS

Black Box Metaphor: A dominant metaphor that assumes the IT artifact to be closed and instrumental in character. In other words the IT artefact is an immutable mobile that has a static goal. Hence the focus of researchers is on determining stable user characteristics when using a stable technology medium.

Fluid Metaphor: An emerging metaphor that assumes the IT artifact to be characterised by mutability. Unlike immutable mobiles that hold themselves rigid to a goal, IT artifacts appear to change gently and fluidly. Researchers view the IT artifact following these six properties: boundaryless, multiple identities, mixtures, robustness, continuity, and dissolving ownership (Laet & Mol, 2000).

Metaphors: Metaphors can be described as useful "symbolic fictions" to describe and construct knowledge about the world in which we live (Morgan, 1980).

ENDNOTES

1. Health Insurance Company (HIC) and eHIC (e-business arm) refer to anonyms chosen to preserve the confidentiality of the identities of the organisations involved.

2. Figure 2 depicts the SST behaving like waves in a loosely defined fluid rather than as particles in a rigid network. The SST hit three major threshold points at which it radically changed forms. It started in a somewhat stable fashion as an information portal, and then attempted to change sharply to a dominant channel and next to a wellness tool. Despite being permeable and seeping, these waves of change eventually overlapped with a new wave, to form a new, more robust, sticky and alloy-like object - a complementary channel.

3. Figure 3 shows registration to use the online service being followed by discontinued use. This amounted to leakage - an evaporation of the majority of registered users.

Chapter 5
How the Rich Lens of ANT Can Help us to Understand the Advantages of Mobile Solutions

Nilmini Wickramasinghe
Epworth Healthcare, Australia & RMIT University, Australia

Arthur Tatnall
Victoria University, Australia

Steve Goldberg
INET Intl. Inc., Canada

ABSTRACT

The WHO has labelled diabetes the silent epidemic. This is because the instances of diabetes worldwide continue to grow exponentially. In fact, by 2030 it is expected that there will be a 54% global increase. Thus, it behooves all to focus on solutions that can result in superior management of this disease. Hence, this chapter presents findings from a longitudinal exploratory case study that examined the application of a pervasive technology solution, a mobile phone to provide superior diabetes self-care. Notably, the benefits of a pervasive technology solution for supporting superior self-care in the context of chronic disease are made especially apparent when viewed through the rich lens of Actor-Network Theory (ANT), and thus, the chapter underscores the importance of using ANT in such contexts to facilitate a deeper understanding of all potential advantages.

INTRODUCTION

In today's Information Technology Age one area that has yet to embrace the full benefits of ICT (Information Communication Technologies) to facilitate superior operations is healthcare. Yet slowly, this bastion is beginning to be besieged by numerous tools and solutions which at the surface at least appear to offer a panacea to the current challenges of escalating costs and poor quality.

In such a context then, it becomes important to have an appropriate and rich theoretical lens of analysis so that it is possible to judge and

DOI: 10.4018/978-1-4666-6126-4.ch005

evaluate the true advantages of these potential ICT solutions. The following section serves to proffer that ANT (Actor-Network Theory) is indeed such a lens.

ACTOR-NETWORK THEORY (ANT)

Actor-network theory (ANT) provides a rich and dynamic lens of analysis for many socio-technical situations. Essentially, it embraces the idea of an organisational identity and assumes that organisations, much like humans, possess and exhibit specific traits (Brown, 1997). Although labelled a 'theory', ANT is more of a framework based upon the principle of generalised symmetry, which rules that human and non-human objects/subjects are treated with the same vocabulary. Both the human and non-human counterparts are integrated into the same conceptual framework.

ANT was developed by two French social sciences and technology scholars Bruno Latour and Michel Callon and British sociologist John Law (Latour, 1987, 2005; Law and Hassard, 1999; Law, 1992, 1987; Callon, 1986). It is an interdisciplinary approach that tries to facilitate an understanding of the role of technology in specific settings, including how technology might facilitate, mediate or even negatively impact organisational activities and tasks performed. Hence, ANT is a material-semiotic approach for describing the ordering of scientific, technological, social, and organisational processes or events.

Key Concepts of Actor Network Theory

Central to ANT and relevant for this specific context are the six key concepts as follows:

1. **Actor/Actant:** Typically actors are the participants in the network which include both the human and non-human objects and/or subjects. However, in order to avoid the strong bias towards human interpretation of Actor, the neologism Actant is commonly used to refer to both human and non-human actors. Examples include nurses, doctors, thermometers, electronic instruments, technical artifacts and graphical representations.

2. **Heterogeneous Network:** Is a network of aligned interests formed by the actors. This is a network of materially heterogeneous actors that is achieved by a great deal of work that both shapes those various social and non-social elements, and 'disciplines' them so that they work together, instead of 'making off on their own' (Latour, 2005).

3. **Tokens/Quasi Objects:** Are essentially the success outcomes or functioning of the Actors which are passed onto the other actors within the network. As the token is increasingly transmitted or passed through the network, it becomes increasingly punctualised and also increasingly reified. When the token is decreasingly transmitted, or when an actor fails to transmit the token (e.g., the broadband connection breaks), punctualisation and reification are decreased as well.

4. **Punctualisation:** Is similar to the concept of abstraction in Object Oriented Programming. A combination of actors can together be viewed as one single actor. These sub-actors are hidden from the normal view. This concept is referred to as Punctualisation. An incorrect or failure of passage of a token to an actor will result in the breakdown of a network.

When the network breaks down the result is the breakdown of punctualisation and the viewers will then not be able to view the sub-actors making up the actor. This concept can be referred to as depunctualisation.

5. **Obligatory Passage Point (OOP):** Broadly refers to a situation that has to occur in order for all the actors to satisfy the interests that have been attributed to them by the focal actor. The focal actor defines the OPP through which the other actors must pass through and by which the focal actor becomes indispensable (Callon, 1986).

6. **Irreversibility:** Callon (1986) states that the degree of irreversibility depends on
   a. The extent to which it is subsequently impossible to go back to a point where that translation was only one amongst others; and
   b. The extent to which it shapes and determines subsequent translations.

In order to apply these to the context of diabetes self-care it is first necessary to provide a brief background on the current healthcare environment and the application of ICT for supporting diabetes self-care.

THE HEALTHCARE ENVIRONMENT BACKGROUND

Healthcare delivery for all OECD countries is experiencing exponentially increasing costs (OECD 2010a-c). This is of great concern to all and many believe information technology (IT) offers the promise of cost effective healthcare delivery (Geisler and Wickramasinghe, 2009). One area that appears to be particular problematic in this regard is connected with chronic diseases such as diabetes and hypertension, not only because they consume a disproportionate slice of healthcare services but also because there is no foreseeable cure and hence these cost pressures will continue for the life of the patient. Moreover, they are likely to increase, because if the chronic disease is poorly managed secondary complications will inevitably develop (Wickramasinghe and Goldberg, 2003).

Diabetes is one of the five major chronic diseases. It afflicts over twenty million people in the United States and accounts for almost $100 billion in medical costs (Geisler and Wickramasinghe, 2009). Diabetes is also one of the leading chronic diseases affecting Australians and its prevalence continues to rise. The total number of diabetes patients worldwide is estimated to rise from 171 million in 2000 to 366 million in 2030 (Wild et al., 2004). With increasingly growing prevalence this means that an estimated 275 Australians are developing diabetes daily (Diabetes Australia, 2008). Given these alarming trends not only in Australia and the United States but worldwide diabetes has been termed as the silent epidemic by the WHO. Table 1 shows the increases for diabetes in various regions throughout the world.

Given that technology may play a role in contributing to a more efficient and effective delivery of care that may also assist in controlling costs, the following examines the possible potential use of wireless technology in the monitoring of diabetic patients. Specifically, this paper provides insights from a longitudinal exploratory case study that investigated the major research question "how can technology support superior self-care for sufferers of chronic disease"? To understand the role for technology it is first necessary to understand critical issues regarding diabetes self-care.

Table 1. Prevalence of diabetes throughout the world

Region	2010 (Millions)	2030 (Millions)	% Increase
North America	33.9	53.2	42%
South America	18	29.6	65%
Africa	12.1	23.9	98%
Europe	55.2	66.2	20%
Middle East	26	51.5	94%
S.E Asia	58.7	101.0	72%
World	**284.6**	**438.4**	**54%**

Source: IDF, 2009 (http://www.idf.org) accessed January 2011

DIABETES SELF-CARE

Diabetes is a chronic disease that occurs when there is too much glucose in the blood because the body is not producing insulin or not using insulin properly (DA, 2007).

Diabetes management involves a combination of both medical and non-medical approaches with the overall goal for the patient to enjoy a life which is as normal as possible (AIHW, 2008; AIHW, 2007). Critical to this management regimen is the systematic monitoring of blood sugar levels. However, as there is no cure for diabetes, achieving this goal can be challenging because it requires effective lifestyle management and careful and meticulous attention and monitoring by both the patient and health professionals (Britt, 2007). In particular, to be totally successful, this requires patients to be both informed and actively participate in their treatment regimen.

The Role for Technology to Facilitate Superior Chronic Disease Management

Simply stated, the research goal of this project was to design and develop a mobile (wireless) Internet environment to improve patient outcomes with immediate access to patient data and to provide the best available clinical evidence at the point of care. To do this, INET International Inc's research (Goldberg, 2002a-e; Wickramasinghe and Goldberg, 2003, 2004) started with a 30-day e-business acceleration assessment in collaboration with many key actors in hospitals (such as clinicians, medical units, administration, and IT departments). From this it was possible to design a robust and rigorous web-based business model: the INET web-based business model (figure 1). The business model brings together all the vital considerations and in turn provides the necessary components to enable the delivery framework to be positioned in the best possible manner so it can indeed facilitate and support superior chronic disease management. From this business model then the INET mobile solution was developed.

Thus the INET solution represents a pervasive technology enabled solution, which, while not exorbitantly expensive, can facilitate the superior monitoring of diabetes. The proposed solution (Figure 2) enables patient empowerment by way of enhancing self-management. This is a desirable objective because it allows patients to become more like partners with their clinicians in the management of their own healthcare (Radin, 2006; Mirza et al. 2008) by enhancing the traditional clinical-patient

Figure 1. INET-web-based-business model. (Adapted from Wickramasinghe and Goldberg, 2004).

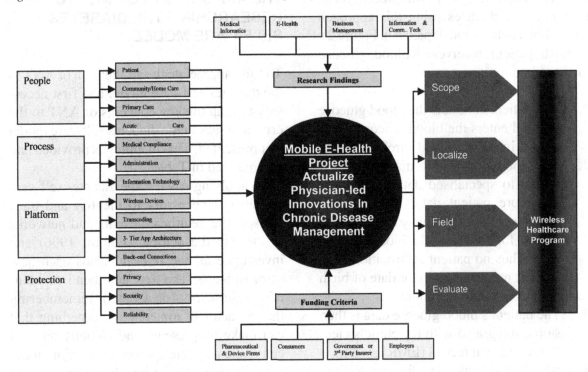

Figure 2. The Wireless Enabled Solution for Diabetes Self-Care. (Reproduced with the permission of INET).

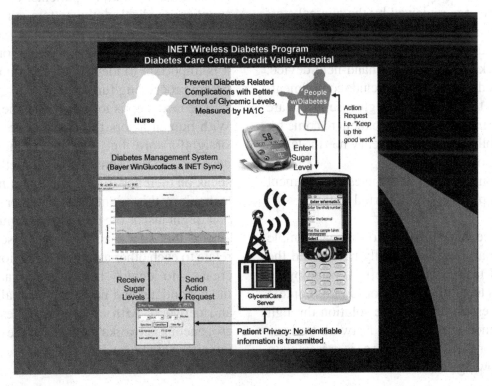

interactions (Opie 1998). The process steps in monitoring diabetes using this approach are outlined below (and depicted in Figure 2).

Each patient receives a blood glucose measurement unit.

1. The patient conducts the blood glucose test and enters the blood glucose information into a hand-held wireless device.
2. The blood glucose information is transmitted to specialised database servers that store patient data. The patient's hand-held device uniquely identifies the patient for recording the blood glucose data. Thus no patient information such as their name, ethnicity or date of birth is transmitted to the clinic.
3. The patient's blood glucose data is then stored/integrated with the clinic's electronic medical record (EMR) system.
4. An alert is generated for the clinical staff with the patient's blood glucose information.
5. The blood glucose information of the patient is reviewed by the clinical staff (physician/nurse).
6. Feedback on glucose levels is transmitted back to the patient's hand-held device. Feedback examples include complimenting the patient when glucose levels are normal or asking the patient to come for a follow-up appointment when the levels are out of norm.
7. Monitor trends in diabetes management for patients over a period of time.

At face value this solution appears simple and trivial; however to fully understand the full and far reaching benefits of employing wireless technology solutions in this context it is necessary to view the solution through a combined lens of Social Network analysis and Actor-Network Theory as is presented in the following sections.

THE APPLICATION ON ANT TO RESEARCHING THE DIABETES SELF-CARE MODEL

To fully appreciate the application of ANT to the diabetes self-care model it is first necessary to map the key concepts of ANT to the critical issues in the diabetes self-care model just presented. This mapping is provided and summarised in Table 2.

In applying ANT to the diabetes self-care model it is necessary to identify and trace the specific healthcare events and networks to "follow the actors" (Latour, 1996) and investigate all the relevant leads each new actor suggests. The first step then is to identify these actors (or actants), remembering that an actor is someone or something that can make its presence individually felt and can make a difference to the situation under investigation. Thus, in healthcare contexts the actors would include: medical practitioners, nurses, medical instruments, healthcare organisations, regulators, patients, equipment suppliers, medical administrators, administrative computer systems, medical researchers, and so on. In a particular operation (or event) it is important to identify all relevant actors before proceeding further.

The next step is to 'interview' the actors. With human actors this is, of course, quite straightforward, but with non-humans it is necessary to find someone (or something) to speak on their behalf. For an item of medical technology this might be its designer or user, or it might just be the instruction manual. The aim of this step is to see how these actors relate to each other and the associations they create – to identify how they interact, how they negotiate, and how they form alliances and networks with each other. These 'heterogeneous networks' consists of the aligned interests held by each of the actors.

Table 2. Key concepts of ANT

Concept	Relevance to Diabetes Self-Care
Actor/Actant: Typically actors are the participants in the network which include both the human and non-human objects and/or subjects. However, in order to avoid the strong bias towards human interpretation of Actor, the neologism 'Actant' is commonly used to refer to both human and non-human actors. Examples include humans, electronic instruments, technical artifacts, or graphical representations.	In the diabetes self-care model this includes the web of healthcare players such as providers, healthcare organisations, regulators, payers, suppliers and the patient as well as the clinical, wireless and administrative technologies that support and facilitate healthcare delivery.
Heterogeneous Network: is a network of aligned interests formed by the actors. This is a network of materially heterogeneous actors that is achieved by a great deal of work that both shapes those various social and non-social elements, and 'disciplines' them so that they work together, instead of 'making off on their own' (Latour, 2005).	The wireless technology combined with the specific application is clearly the technology network in this context. However it is important to conceptualise the heterogeneous network not as the technology alone but as the aligning of the actors through the various interactions with the technology so that it is possible to represent all interests and thereby provide the patient with superior healthcare delivery. The key is to carefully align goals so that healthcare delivery is truly patient centric at all times.
Tokens/Quasi Objects: are essentially the success outcomes or functioning of the Actors which are passed onto the other actors within the network. As the token is increasingly transmitted or passed through the network, it becomes increasingly punctualised and also increasingly reified. When the token is decreasingly transmitted, or when an actor fails to transmit the token (e.g., the broadband connection breaks), punctualisation and reification are decreased as well.	In the diabetes self-care model this translates to successful healthcare delivery, such as treating a patient in a remote location by having the capability to access critical information to enable the correct decisions to be made. Conversely, and importantly, if incorrect information is passed throughout the network errors will multiply and propagate quickly hence it is a critical success factor that the integrity of the network is maintained at all times.
Punctualisation: is similar to the concept of abstraction in object oriented programming. A combination of actors can together be viewed as one single actor. These sub actors are hidden from the normal view. This concept is referred to as punctualisation. An incorrect or failure of passage of a token to an actor will result in the breakdown of a network. When the network breaks down, it results in breakdown of punctualisation and the viewers will now be able to view the sub-actors of the actor. This concept is often referred to as depunctualisation.	For example, an automobile is often referred to as a single unit. Only when it breaks down, is it seen as a combination of several machine parts. Or in the diabetes self-care model the uploading task of one key actor, be it a provider or a patient is in reality a consequence of the interaction and co-ordination of several sub-tasks. This only becomes visible when a breakdown at this point occurs and special attention is given to analyse why and how the problem resulted and hence all sub-tasks must be examined carefully.
Obligatory Passage Point (OOP): broadly refers to a situation that has to occur in order for all the actors to satisfy the interests that have been attributed to them by the focal actor. The focal actor defines the OPP through which the other actors must pass through and by which the focal actor becomes indispensable (Callon, 1986).	In the diabetes self-care model we can illustrate this by examining the occurrence of the diabetes or secondary complications that have resulted because of the primary disease. Such incidents form the catalyst for developing shared goals and united focus of effort so necessary to affect superior healthcare delivery.
Irreversibility: Callon (1986) states that the degree of irreversibility depends on (i) the extent to which it is subsequently impossible to go back to a point where that translation was only one amongst others and (ii) the extent to which it shapes and determines subsequent translations.	Given the very complex nature of healthcare operations (von Lubitz and Wickramasinghe, 2006b-e) irreversibility is generally not likely to occur. However it is vital that chains of events are continuously analysed in order that future events can be addressed as effectively and efficiently as possible.

Human actors, such as medical practitioners, can 'negotiate' with non-human actors such as X-Ray or dialysis machines by seeing what these machines can do for them, how easy they are to use, what they cost to use, and how flexible they are in performing the tasks required. If negotiations are successfully completed then an association between the medical practitioner and the machine is created and the machine is used to advantage – the network has become durable. If the negotiations are unsuccessful then the machine is either not used at all, or at least not used to full advantage.

Once this is developed it is then important to apply the techniques of ANT to map the flows of pertinent information and germane knowledge throughout this network and thereby not only enhancing the metacognition of the system but also the ability to rapidly extract and utilise the critical knowledge to support prudent decision making. In this way, it will be possible to be at all times in a state of being prepared and ready (Wickramasinghe and von Lubitz, 2007; von Lubitz and Wickramasinghe, 2006 a,f).

One of the main advantages of an ANT approach in considering the diabetes self-care model is in being able to identify and explore the real complexity involved. Other approaches to technological innovation, Innovation Diffusion for example, put much stress on the properties of the technology or organisation themselves, at the expense of looking at how these interact. Unfortunately in doing this they often tend to oversimplify very complex situations and so miss out on a real understanding. The ANT approach of investigating networks and associations provides a useful means to identify and explain these complexities as well as to track germane knowledge and pertinent information. This is paramount if the doctrine of network-centric healthcare is to be successfully realised.

Borrowing ideas from innovation translation in actor-network theory (Latour, 1996; 1986; 1999; Law, 1991) we will argue that, rather than just the technology, *people* are very important, as they may either accept an innovation in its present form, modify it to a form where it becomes acceptable, or reject it completely. "If we know one thing about innovation and reform, it is that it cannot be done successfully *to* others" (Fullan, 1991 pp xiv). An innovation translation approach has been shown to be useful in considering ICT (information and communications technology) innovation in small business (Tatnall and Davey, 2002) and in education (Tatnall, 2000; Tatnall and Davey, 2001; Bigum, 1998; Busch, 1997).

The innovation translation approach to innovation originates in actor-network theory (ANT) and draws on its sociology of translations. In ANT, translation (Law, 1992) can be defined as: "... the means by which one entity gives a role to others" (Singleton and Michael, 1993 pp. 229). Using an innovation translation approach to consider how the adoption of mobile medical technology occurs, it is necessary to examine the interactions of all the actors involved (Tatnall and Davey, 2003b).

DISCUSION

The longitudinal exploratory case study that we embarked upon started with an examination of the technology solution to facilitate communication of blood sugars between patient and provider, to an assessment of the delivery framework and business model developed by INET to an analysis of the data from trials and pilot studies conducted. Based on rigorous thematic analysis of interview data, gathered following the techniques prescribed by Yin (1994), Kavale (1996) and Boyatsis (1998), triangulated with data from internal documents, reports and medical records as well as our own observations not only was proof of concept attained; ie using the pervasive technology solution did facilitate better self-care and resulted in a decrease in blood sugar, but all patients in the study reported only positive comments about the solution as the following quote highlights:

I loved the solution. It was like an answer to a pray. I could control my sugars more easily and have confidence and peace of mind to enjoy my life.

This serves to underscore the benefits of such an approach.

In the current context, healthcare delivery, especially in the US, is in need of fundamental re-design (Porter and Tiesberg, 2006). The focus on cost containment also necessitates a shift to prevention rather than cure. This is particularly important in the case of chronic diseases such as diabetes thereby making any solution that has the potential to enable cost effective quality care a strategic necessity.

Diabetes is the fifth-deadliest disease in the United States. Since 1987, the death rate due to diabetes has increased by 45 percent, while the death rates due to heart disease, stroke, and cancer have declined. The total annual economic cost of diabetes in 2002 was estimated to be $132 billion. Direct medical expenditures totalled $92 billion and comprised $23.2 billion for diabetes care, $24.6 billion for chronic diabetes-related complications, and $44.1 billion for excess prevalence of general medical conditions. Indirect costs resulting from lost workdays, restricted activity days, mortality, and permanent disability due to diabetes totalled $40.8 billion. The per capita annual costs of health care for people with diabetes rose from $10,071 in 1997 to $13,243 in 2002, an increase of more than 30%. Further, in 2010 alone the total cost of diabetes was $174 billion with approximately $116 billion direct medical costs and at least $58 billion in indirect costs (Wickramasinghe, 2010).

Without a doubt the costs in the United States are alarming, however it should also be noted that evidence also shows that diabetes and its complications incur significant costs for the health system in Australia and elsewhere. These costs include costs incurred by carers, government, and the entire health system (DiabCostAustralia, 2002). For instance, in 2004-05 direct healthcare expenditure on diabetes was A$907 million which constituted approximately 2% of the allocatable recurrent health expenditure in that year (AIHW, 2008). In addition, further costs include societal costs that represent productivity losses for both patients and their carers (DiabCostAustralia, 2002).

CONCLUSION

Technological developments in ubiquitous and mobile computing offer possibilities to many aspects of healthcare and, due to their non-reliance on traditional communications infrastructure they also offer emerging countries an opportunity to jump ahead. Exactly how these technologies should best be used in diabetes treatment, however, is not completely clear at this time (Morel et al., 2005).

There has been a considerable amount of research into the impact and use of technology in healthcare delivery, but not much into explaining the uptake of this technology. We have argued that for research of this type, it is useful to consider this technology in terms of technological innovation and to make use of an approach based on innovation theory. Actor-network theory provides a perspective that can resolve the dilemma of how to handle both the human and non-human contributions to technological innovation, and also provides a useful explanatory system for doing so (Tatnall, 2009).

In the specific case of diabetes discussed in the proceeding sections, clearly stemming the cost pressure is important but of greater importance is providing sufferers of diabetes with a possibility to experience a better quality of life. Where it is possible to provide both a low cost as well as a superior patient-centric solution this would appear to be a very compelling case for the large scale adoption of pervasive wireless solutions to facilitate superior care not only for diabetes patients but for patients with all chronic diseases. Our analysis of this solution drawing upon Actor-network theory demonstrates that there are very important aspects that are being addressed by adopting a pervasive wireless solution and these need further investigation. We close by not only calling for more research which particularly focus on how to move from idea to realisation as rapidly as possible but also for further exploration into the incorporation of pervasive technology solutions in general for supporting superior patient-centric healthcare delivery and the embracement of ANT as an appropriate lens of analysis.

AUTHORS' NOTE

An earlier version of this paper appeared in the proceedings of the 2011 Pacific Asia Conference on Information Systems (PACIS) July 2011 Brisbane, Australia.

REFERENCES

AIHW. (2007). *National Indicators for Monitoring Diabetes: Report of the Diabetes Indicators Review Subcommittee of the National Diabetes Data Working*. AIHW.

AIHW. (2008). *Diabetes: Australian Facts 2008*. Canberra, Australia: Australian Institute of Health and Welfare.

Bigum, C. (1998). Solutions in Search of Educational Problems: Speaking for Computers in Schools. *Educational Policy, 12*(5), 586–601. doi:10.1177/0895904898012005007

Boyatzis, R. (1998). *Transforming Qualitative Information Thematic analysis And Code Development*. Sage Publications.

Britt, H., Miller, G. C., Charles, J., Pan, Y., Valenti, L., Henderson, J., & Knox, S. (2007). *General Practice Activity in Australia 2005-06 (Cat. no. GEP 16)*. Canberra, Australia: AIHW.

Busch, K. V. (1997). Applying Actor Network Theory to Curricula Change in Medical Schools: Policy Strategies for Initiating and Sustaining Change. In *Proceedings of Midwest Research-to-Practice Conference in Adult, Continuing and Community Education Conference*. Michigan State University.

DA. (2007). *Diabetes Facts*. New South Wales, Australia: Diabetes Australia.

Diabcost Australia. (2002). *Assessing the Burden of Type 2 Diabetes in Australia*. Adelaide, Australia: DiabCost Australia.

Diabetes Australia. (2008). *Diabetes in Australia*. Author.

Frost and Sullivan Country Industry Forecast – European Union Healthcare Industry. (2004, May 11). Retrieved from http://www.news-medical.net/print_article.asp?id=1405

Fullan, M. G., & Stiegelbauer, S. (1991). *The New Meaning of Educational Change* (2nd ed.). New York: Teachers College Press.

Geisler, E., & Wickramasinghe, N. (2009). *The Role and Use of Wireless Technology in the Management and Monitoring of Chronic Disease IBM Report*. Retrieved from www.businessofgovernment.org

Goldberg, S., et al. (2002a). Building the Evidence For A Standardized Mobile Internet (wireless) Environment In Ontario, Canada, January Update, internal INET documentation. Academic Press.

Goldberg, S. et al. (2002b). *HTA Presentational Selection and Aggregation Component Summary, internal documentation.* Academic Press.

Goldberg, S. et al. (2002c). *Wireless POC Device Component Summary, internal INET documentation.* Academic Press.

Goldberg, S. et al. (2002d). *HTA Presentation Rendering Component Summary, internal INET documentation.* Academic Press.

Goldberg, S. et al. (2002e). *HTA Quality Assurance Component Summary, internal INET documentation.* Academic Press.

Kavale, S. (1996). Interviews, an introduction to Qualitative Research. *Sage (Atlanta, Ga.).*

Kulkarni, R., & Nathanson, L. A. (2005). *Medical Informatics in medicine, E-Medicine.* Retrieved from http://www.emedicine.com/emerg/topic879.htm

Lacroix, A. (1999). International concerted action on collaboration in telemedicine: G8sub-project 4, Sted. *Health Technol. Inform., 64,* 12–19.

Latour, B. (1986). The Powers of Association. In *Power, Action and Belief: A new sociology of knowledge?* Routledge & Kegan Paul.

Latour, B. (1992). On Recalling ANT. In *Actor Network Theory and After.* Blackwell Publishers.

Latour, B. (1996). *Aramis or the Love of Technology.* Cambridge, MA: Harvard University Press.

Law, J. (Ed.). (1991). *A Sociology of Monsters: Essays on power, technology and domination.* London: Routledge.

Morel, R., Tatnall, A., Ketamo, H., Lainema, T., Koivisto, J., & Tatnall, B. (2005). *Mobility and Education. E-Training Practices for Professional Organizations.* Kluwer Academic Publishers.

OECD. (2010a). *Growing health spending puts pressure on government budgets.* Retrieved January 31st, 2011 from http://www.oecd.org/document/11/0,3343,

en_2649_34631_45549771_1_1_1_37407,00.html

OECD. (2010b). *OECD-Gesundheitsdaten 2010: Deutschland im Vergleich.* Retrieved January 31st, 2011 from http://www.oecd.org/dataoecd/15/1/39001235.pdf

OECD. (2010c). *OECD health data 2010.* Retrieved January 31st, 2011 from http://stats.oecd.org/Index.aspx?DatasetCode=HEALTH

Porter, M., & Tiesberg, E. (2006). *Re-defining health care delivery.* Harvard Business Press.

Rachlis, M. (2006). *Key to sustainable healthcare system.* Retrieved from http:www.improveingchroniccare.org

Singleton, V., & Michael, M. (1993). Actor-Networks and Ambivalence: General Practitioners in the UK Cervical Screening Programme. *Social Studies of Science, 23,* 227–264. doi:10.1177/030631293023002001

Tatnall, A. (2009). Web Portal Research Issues. In *Encyclopedia of Information Science and Technology* (2nd ed.). Hershey, PA: Idea Group Reference.

Tatnall, A. (20002). *Innovation and Change in the Information Systems Curriculum of an Australian University: A Socio-Technical Perspective.* (PhD thesis). Central Queensland University, Rockhampton, Australia.

Tatnall, A., & Davey, B. (2001). How Visual Basic Entered the Curriculum at an Australian University: An Account Informed by Innovation Translation. In *Challenges to Informing Clients: A Transdisciplinary Approach.* Krakow, Poland: Academic Press.

Tatnall, A., & Davey, B. (2002). Understanding the Process of Information Systems and ICT Curriculum Development: Three Models. In *Human Choice and Computers: Issues of Choice and Quality of Life in the Information Society.* Kluwer Academic Publishers. doi:10.1007/978-0-387-35609-9_23

Tatnall, A., & Davey, B. (2003b). Modelling the Adoption of Web-Based Mobile Learning - An Innovation Translation Approach. In *Advances in Web-Based Learning.* Springer Verlag. doi:10.1007/978-3-540-45200-3_40

Wickramasinghe, N. (2007). Fostering knowledge assets in healthcare with the KMI model. *International Journal of Management and Enterprise Development, 4*(1), 52–65. doi:10.1504/IJMED.2007.011455

Wickramasinghe, N. (2010). *Christmas Carol for IBM.* Paper presented at IBM Healthcare Executives Dinner. New York, NY.

Wickramasinghe, N., & Bali, R. (2009). The S'ANT imperative for realizing the vision of healthcare network centric operations. *International Journal of Actor-Network Theory and Technological Innovation, 1*(1), 45–58. doi:10.4018/jantti.2009010103

Wickramasinghe, N., & Goldberg, S. (2003). The Wireless Panacea for Healthcare. In *Proceedings of the 36th Hawaii International Conference on System Sciences* (HICSS-35). IEEE.

Wickramasinghe, N., & Goldberg, S. (2004). How M=EC2 in Healthcare. *International Journal of Mobile Communications, 2*(2), 140–156. doi:10.1504/IJMC.2004.004664

Wickramasinghe, N., & Goldberg, S. (2007a). Adaptive mapping to realisation methodology (AMR) to facilitate mobile initiatives in healthcare. *International Journal of Mobile Communications, 5*(3), 300–318. doi:10.1504/IJMC.2007.012396

Wickramasinghe, N., Goldberg, S., & Bali, R. (2008). Enabling superior m-health project success: A tri-country validation. *International Journal of Services and Standards, 4*(1), 97–117. doi:10.1504/IJSS.2008.016087

Wickramasinghe, N., & Mills, G. (2001). MARS: The Electronic Medical Record System The Core of the Kaiser Galaxy. *International Journal of Healthcare Technology and Management, 3*(5/6), 406–423. doi:10.1504/IJHTM.2001.001119

Wickramasinghe, N., & Misra, S. (2004). A Wireless Trust Model for Healthcare. *International Journal of Electronic Healthcare, 1*(1), 60–77. doi:10.1504/IJEH.2004.004658 PMID:18048204

Wickramasinghe, N., & Schaffer, J. (2006). Creating Knowledge Driven Healthcare Processes With The Intelligence *Continuum*. *International Journal of Electronic Healthcare*, *2*(2), 164–174. PMID:18048242

Wickramasinghe, N., Schaffer, J., & Geisler, E. (2005). Assessing e-health. In T. Spil, & R. Schuring (Eds.), *E-Health Systems Diffusion and Use: The Innovation, The User and the User IT Model*. Hershey, PA: Idea Group Publishing. doi:10.4018/978-1-59140-423-1.ch017

Wickramasinghe, N., & Silvers, J. B. (2003). IS/IT The Prescription To Enable Medical Group Practices To Manage Managed Care. *Health Care Management Science*, *6*, 75–86. doi:10.1023/A:1023376801767 PMID:12733611

Wild, S., Roglic, G., Green, A., Sicree, R., & King, H. (2004). Global prevalence of diabetes: Estimates for the year 2000 and projections for 2030. *Diabetes Care*, *27*(5), 1047–1053. doi:10.2337/diacare.27.5.1047 PMID:15111519

Yin, R. (1994). *Case Study Research: Design and Methods* (2nd ed.). Sage Publications.

KEY TERMS AND DEFINITIONS

Chronic Disease: A disease that has no cure hence once an individually contracts this disease they typically have it for the rest of their life.

Diabetes Management: This involves a combination of both medical and non-medical approaches with the overall goal for the patient to enjoy a life which is as normal as possible. Most focus is placed on keeping blood sugar at appropriate levels.

Diabetes: A chronic disease that is due to incorrect amounts of blood sugar in the body. There are 3 major types of diabetes Type 1, Type 2 and gestational.

Epidemic: A wide spreading a widespread occurrence of a disease in a community at a particular time.

Self-Care: Enabling the patient to take ownership of their healthcare maintenance regimen and consult doctors and care teams as and when required.

Chapter 6
Information and Communication Technology Projects and the Associated Risks

Tiko Iyamu
Namibia University of Science and Technology, Namibia

Petronnell Sehlola
Tshwane University of Technology, South Africa

ABSTRACT

Organisational reliance on Information and Communication Technology (ICT) continues to increase. This is informed and triggered by the premise that ICT will help them to yield solutions that will fulfil or exceed their expectations, thereby making the organisation realise the required return on investment. In order to realise return on ICT investment, many organisations deploy ICT solutions through projects. However, not all ICT projects realise their goals and objectives, due to associated risks. Unfortunately, risks are never easy to identify or managed. This chapter explores and examines the risks factors in the deployment of ICT projects in organisations. Using the case study method, the research employs actor-network theory in the analysis of the data to understand the factors that manifest themselves into risks during the deployment of ICT projects in organisations. The study reveals that factors, such as knowledge base, performance contract, and communicative structure, are used to enable and support and at the same time to constrain the deployment of ICT projects in organisations.

INTRODUCTION

There is an increasing demand for Information and Communication Technology (ICT) solutions by organisations, including government administrations, financial institutions,

higher institutions of learning, and insurance companies (Clancy, 2004). Some of these organisations process high volumes of data at high rates. The effectiveness and efficiency of their operations depend on the capability and capacity of the ICT solutions. The ICT

DOI: 10.4018/978-1-4666-6126-4.ch006

solutions get obsoletes faster as a result of changes in the business environment and rapid development of ICT artefacts. The changes are driven by the business objectives and strategies (Lee & Xia, 2003). This suggests that there is a need to deliver the solutions early while they are still relevant and value-adding to the business. Many of the ICT solutions are employed through ICT projects.

ICT projects are aimed at enabling business processes and activities in order to deliver business benefits and competitive advantage. ICT projects follow project management principles like any other project in other disciplines and professions. They must be delivered on time, on schedule, with expected performance functionality and they must add value to the business. Such ICT projects have inherent risk like any other projects (PMBOK, 2004). What makes them different from other projects is that they deliver intangible products and solutions.

ICT projects have an element of uncertainty and therefore carry inherent risk. According to Ferguson (2004), risk is defined as an uncertain event that could cause an uncertain impact on project schedule, cost, or quality. Many ICT projects are prone to fail (Labuschagne et al, 2008) and this suggests that these are high-risk projects that need project risk management that can effectively improve the project outcome. Risk management is a crucial practice in attaining the successful delivery of IT projects (Tuman&Remenyi, 1999).

Risk management is a systematic process to identify, evaluate and address risks on a continuous basis to prevent such risks from having a negative impact on the institution's service-delivery capacity (Baccarini et al, 2004). Risk management is necessary in the deployment and management of projects and activities.

ICT PROJECTS AND RISK

Information and communication technology (ICT) solutions are often deployed through projects, for organisational processes and activities. According to Morris (2004), projects are a means to yield solutions that include infrastructure, networks, applications, databases or a combination of these. Each deployment is guided by sets of requirements, of both technical and non-technical (business). The requirements are intended to provide opportunities and advantages. However, each requirement does carry potential risk, some visible, and other could be hidden. To manage risk, it must first be identified. According to Taylor (2006), risks are difficult to manage in IT projects if they are not identified when requirements are gathered.

Unfortunately, some project managers seem to focus more on technology artefacts. Kutsch (2008) argued that avoidance, ignorance, delay of risk response actions, and denial of uncertainty by ICT project managers have some influence on the effectiveness of project risk management. However, it is difficult to manage the unknown. Hence identification of risk is foremost, and critical. According to Hillson & Murray-Webster (2004), the attitude of individuals and organisations has a significant influence on whether risk management delivers what it promises. In Ferguson (2004), risk identification and classification are prerequisites for effective risk management.

Identification of risks which are associated with the ICT projects is a major challenge for managers. This can be attributed to the fact that there are numerous ways in which they can be described and categorised. Tesch, Kloppenborg & Frolick (2007) conducted a study to gain better understanding of risk factors, in which they found 92 different factors that could be associated to IT projects in organisations. One of the major ways of identifying and categorising risk is to know the actors and the network that they belong.

Actors are both human and non-human (such as tools or instruments, or technology (Mitev, 2008). Wernick et al, (2008) stated that irrespective of whether the actor is human or non-human, they are both weighed equally as they offer the same contribution to the formation of the network. There are no networks that consist of only humans or only technological components; all networks contain elements of both (Urry, 2003). Actor network theory undertakes the challenge of revealing what constitute an actor, how network are form, and how actor transformed themselves to make a difference in a heterogeneous network (Iyamu & Roode, 2010).

RESEARCH APPROACH

The case study research approach was employed in the study. Lester (1997) defines case studies as a formal report based on the examination of a prearranged subject. Case study is an empirical inquiry that investigates a contemporary phenomenon within its real-life context, especial when the boundaries between phenomenon and context are not clearly evident and in which multiple sources of evidence are used (Yin, 2002). Case study research is an ideal methodology when a holistic, in depth investigation is needed. It is designed to bring out the details from the viewpoint of the participant by using multiple sources of data (Feagin, Orum&Sjoberg, 1999).

According to Blumberg, Cooper &Schindelr (2008), the most benefit of case studies is that they allow the combination of different source of evidence such as Interview, documentation and observation.

Data was collected using Semi-structured interview to allow the researcher to refine the research questions during the interviews to converge issues that arose during the interviewing process. Documents were also used to add to the information received from respondents. Data were analysed using underpinning theory. Organisation used for case study was financial Institution who provides services to its customer.

The study is underpinned by Actor-Network Theory (ANT). The actor-network theory has been widely accepted and used in many studies. Such of the studies include Hanseth and Monteiro, 1997; Walsham and Sahay, 1999; and Mitev, 2008. Wernick et al. (2008) in their studies focused more on human related aspects and afterwards also covered the participation of non-human entities which illustrate the difference of both human and non-human actors.

An actor-network is the act linked together with all of its influencing factors in building a network (Suchman, 1987; Hanseth and Monteiro, 1998). Actor-network Theory (ANT) can be defined as a theory that integrates both human and non-human actors to form or create a network (Macome, 2008). The translation of an actor or actors into a network is achieved through a series of four moments of translation (Callon, 1986). Macome (2008) defined the four stages of moments of translation as follows:

1. **Problematisation:** Key actors attempt to define the problem and roles of other actors to fit the proposed solution, which was made by the key actors.

2. **Interresment:** Processes that attempt to impose the identities and roles defined in problematisation on other actors.

3. **Enrolment:** A process where one set of actors (key actors) imposes their will on others. The other actors will be persuaded

to follow the identities and roles defined by the key actors. This will then lead to the establishment of a stable network of alliances.

4. **Mobilisation:** This is where the proposed solutions gain wider acceptance. The network would grow larger with the involvement of other parties that were not involved previously. This growth is due to the influence of actors.

An ANT perspective was applied in the analysis of the study in order to gain understanding of the sphere of influence that the actors have in risks identification and management of risks in IT projects.

ANALYSIS OF THE EMPIRICAL DATA

The Actor-Network Theory (ANT) was being applied in the data analysis. These include actors and network as defined by ANT. The four moments of translation from the perspective of ANT was used in the data analysis to understand the roles of actors in the network and how they locate themselves.

Actor

From the perspective of ANT, actors are both humans and non-humans. In the organisation, many employees such as business analysts, project managers, IT managers, application developers, and business managers are involved in risk identification and management of IT projects in the organisation. One of the test analysts commented that *the project manager involves business analysts, testers and developers when identifying risks at the beginning of the projects.*

The non-human actors which are involved in risk identification and management of IT projects in the organisation are technology artefacts. The technologies are both hardware and software, which include technological networks, software, and hardware (such as servers and desktops). According to one of the systems analysts, *during risk identification we need to have projectors and computers to be able to see what has been captured and the priority of a specific risk.*

Neither human actors (employees) nor technological actors have the capacity or ability to act alone in a network. As such, the identification, monitoring, mitigation and management of risk in IT projects in the organisation are carried out by both human and non-human actors within a network.

Network

The processes and activities of identification, monitoring, mitigation and management of risk in IT projects in the organisation involve employees and technologies. The employees act in the form of groups (teams and committees). Some of the groups include a project team, a software development team and a process governance committee (PGC). According to a project manager, *at the beginning of the projects, project teams are formed to identify risks and some to manage risks that will harm the projects.*

Some of the employees belong to more than one of these groups. This is as a result of the roles and responsibilities of the individuals, including the organisational structure. For example, the Chief information officer (CIO) is a member of the PGC and Exco. The PGC is headed by the CIO. Another example is cited by the one of the employees: *the business analyst is a member of the PGC and also acts as a coordinator between business units and information technology.*

Another team member of the IT project, aimed to assist with identification and management of risks in IT project is the test team.

One of the test analysts commented as follows: *we test software to assess risk or potential, based on the business requirements which we received from the business analyst.*

Actors act within a certain framework within a given period. These acts are within moments of translation, which consist of four stages. The identification and management of risks in IT projects as analysed using the moments of translation from the perspective of actor-network theory

Moments of Translation: Problematisation

IT projects have elements of uncertainty, risks and potential risks. Some of these risks are difficult to identified, monitor and manage. Hence responsibility and accountability on risks management is critical.

The PGC is accountable for all risk-related matters on IT projects in the organisation. This is in accordance with the organisational structure and mandate mainly because the PGC makes the final decision on all IT project-related matters in the organisation. According to one of the project managers, *the IT projects must have been approved by PGC for me to be able to participate in it.*

Business analysts in conjunction with project managers are responsible for the identification of risks which are associated with IT projects in the organisation. This is based on the roles and responsibilities of the business analysts and project managers as defined by the organisation. However, the identification is never as easy as proclaimed. It requires knowledge more than the reliance on roles and responsibilities. Hence, it is critical to include more stakeholders in the identification and management of risks in IT projects.

According to one of the project managers, *we need to involve all the possible stakeholders to identify the positive and the negative risk in IT projects.*

Knowledge of the environment, as well as of the particular IT project, is essentially important in risks identification and management. This is because the risks come in different forms, and at different stages of IT projects. It is difficult to apply a single formula in the identification and management of risks in IT projects. The identification of risks in IT projects is often based on the knowledge of the actors (such as project managers and business analysts). One of the business analysts explained that, *in most cases I use the knowledge that I have in IT and in the organisation together with my experience that I gained from previous IT projects to identify risks in the projects.*

In IT projects in the organisation, there are many stakeholders, at individual and group levels. This includes the project managers, business analysts, PGC and project teams. Unfortunately, the interests of the stakeholders are not always the same. The interests are influenced by various factors.

Moments of Translation: Interessement

Risks in IT projects in the organisation draw on a variety of interests, which helps to constitute the stakeholders. The stakeholders include individuals and groups such as software developers, business analysts, project managers, PGC and project teams. These stakeholders as defined by the organisation have different roles and responsibilities in identifying, monitoring and managing risk in IT projects. A manager explained that *during brainstorm-*

ing sessions, roles and responsibilities are assigned to individuals and the individuals can negotiate on swapping it with each other if they so wish.

The stakeholders have different interests in the identification, monitoring and management of risk in IT projects in the organisation. The interests are shaped and influenced by many factors. Some of the factors are the financial implications, technical know-how and resources distribution. An interviewee briefly explained as follows: *I am interested in the monitoring and controlling of risk in a project because I do not want project managers to go over the budgeted amount.*

The different interests as upheld and shared by the stakeholders are a challenge to the organisations as they attempt to achieve a common goal and objectives in the identification, monitoring and management of risk in IT projects. This challenge is of primary importance to the PGC, mainly because it is the custodian of all IT projects in the organisation. One of the interviewees opined that *the PGC is interested in the projects as they will add value to the organisation; this is because they will enable employees to execute their work efficiently.*

The employees are also interested in risk in IT projects, because of their required skill sets. Some of these employees take this (solutions for risk in IT projects) as an opportunity to exhibit and demonstrate their wealth of knowledge and expertise. This sometimes leads to conflicts as different employees have different views and understanding relating to risks in IT projects. According to one of the project coordinators, *employees differ in their views and understanding about the risks in IT projects. Some of them will say it is an issue not a risk, while others will say it is a risk in IT projects.*

Not all employees and stakeholders who are interested in particular IT projects and their associated risks, participate in the identification and management of risk in the organisation.

Moments of Translation: Enrolment

Participation in the identification and management of risks in IT projects are expressed through both negotiations and mandatory agreements between managers and subordinates in the organisation. The negotiations and mandatory agreements are subject to terms and conditions of employment in the organisation. The project manager and the PGC are responsible and accountable, respectively, for risks in IT projects. Workshops are held to inform PGC and employees about the risks which had been identified and how the risks would be managed. According to one of the System analysts, *the business analyst notifies everybody involved in the projects about the risks and how these risks will be solved. She does this through meetings and mini workshops.*

Participation in the identification and management of risks at the executive and senior manager's level is not a challenge. This is because they have a better view of the organisation, and how managing risks in IT projects benefits the business of the organisation. Some managers participate in the identification and management of risks in IT projects for reasons such as to be recognised by their line manager. According to one interviewee, *Executive managers participate in most of the IT projects because they want to be recognised and some because they have the best interest of the organisation at heart.*

The responsibilities of the project manager on risk-related issues are shared or allocated to the team members in the form of tasks. The allocation of tasks is done through processes of negotiation and is mandatory between the project managers and the subordinates (team members). One of the interviewees expressed herself as follows: *Roles and responsibilities are assigned to individuals to monitor risks, but the project manager together with my boss sometimes allows us to negotiate roles and responsibilities allocated to us.*

The tasks relating to risks in IT projects which are allocated to employees can be negotiated through one-on-one meetings between the project manager and project team members, which include business analysts and software developers. The meetings create opportunities for both parties to discuss and negotiate roles and responsibilities allocated to individuals. According to one of the senior managers, *Individual employees are more than welcome to schedule one-on-one meetings to discuss or to negotiate the tasks which are allocated to them.*

The project managers ensure that the enrolled employees are aware of the roles and responsibilities which are allocated to them in risks identification and management within the organisation. Meetings and brainstorming sessions are used to allocate tasks to individuals. One of the project managers expressed that *Sometimes the project manager assigns roles and responsibilities to individual team members during project meetings to identify risks in certain areas.*

The project manager and others who are involved in the identification, monitoring and management of risks in IT projects are expected to be aware and conversant with customer requirements. The rapidly and constantly changing nature of business requirements make it difficult for some employees to participation in risk-related tasks. According to one of the senior managers, *there are*

challenges in managing risks in IT projects because of continuous changes to requirements by the client.

The project team members often have different interests in their participation in the identification, monitoring and management of risks in IT projects in the organisation. Due to a variety of interests, criteria are employed in the allocation of tasks. These are based on skills, knowledge and experience. One senior manager expressed his view as follows: *I deploy experience resource in risk management in IT projects, to be able to manage risks efficiently and to deliver a quality product.*

However, the allocation of tasks relating to the identification, monitoring and management of risks sometimes deviate from the factors mentioned above. Personal interests such as relationships and incentives are other factors that influence employees to participate in the tasks regarding risk-related matters in the organisation. According to one of the system analysts: *Most employees want to be involved in risk identification and management in the IT projects because they know that managers give bonuses or vouchers if the project is successful or risk free.*

Executive managers, project managers and business analysts are sometimes used as spokespersons for the focal actor to be able to engage individual employees in risk management.

Moments of Translation: Mobilisation

Many of the employees who show an interest and participate in the identification and management of risks in IT projects in the organisation do so based on their individual understanding. The understanding is based on communication and persuasion by other employees, who for various reasons act on behalf of the focal actor (initiator) in the identification and management of risks in IT projects in the organisation.

One of the reasons is the mandate bestowed on individuals and groups in accordance with the organisational structure. For example, the PGC acts and speaks on behalf of the organisation on the identification and management of risks in IT projects in the organisation. The PGC within the organisation ensures that the facilitator (spokesperson) acts according to the agreement and does not betray the initiator's interest. Initiators seek to continue support for the underlying ideas from the enrolled employees on risk identification and management within IT projects. One of the interviewees briefly stated that: *The CIO facilitates the PGC, but he is not supposed to change anything from the project owner's request.*

There are different forms of mobilisation during the identification and management of risks in IT projects in the organisation. Business analysts mobilise the IT projects team, business units and information technology unit to ensure that they participate in the identification and management of risks in IT projects. One of the business analysts commented as follows: *I also act as a coordinator between the business unit and the IT unit. In short, my job is to facilitate in a projects, making sure that information is shared.*

Consciously and unconsciously, spokespersons lure and get other employees interested, as well as to participate in matters related to risks on IT projects in the organisation. Some of the spokespersons were able to influence and persuade others mainly because of their positions and because they belong to many networks in the organisation. For example, an employee represents the project team to the stakeholders and to the project owner. This process strengthens and stabilises the roles or responsibilities, as well as the employees' formation of networks in their involvement in the IT projects in the organisation.

FINDINGS AND DISCUSSION

From the analysis of the data, five factors were found to have a critical influence on the identification and management of risks on IT projects in the organisation. The factors include knowledge base, performance contract, personalisation (personal relationship), formalisation of task allocation and communicative structure. Figure 1 depicts the factors and how they relate to each other. The factors are discussed below:

Knowledge-Based Centre

The knowledge-based centre is a centralised repository for information. It is used to optimise information and knowledge within the organisation. From the knowledge-base centre, knowledge and information can be shared and used to carry out tasks in IT projects.

The knowledge-base concept organisation needs to implement knowledge based in the organisation to increase the speed of response to IT projects and associated risk. Employees can be self-motivated in IT projects through knowledge base. Knowledge base will reduce time spent trying to find the impact of risk in IT projects and can help organisations save money.

The knowledge-based concept would ease the negotiation that takes place during tasks allocation. This is mainly owing to the existence of a repository of knowledge and information to draw from.

Figure 1. Factors influencing risk management

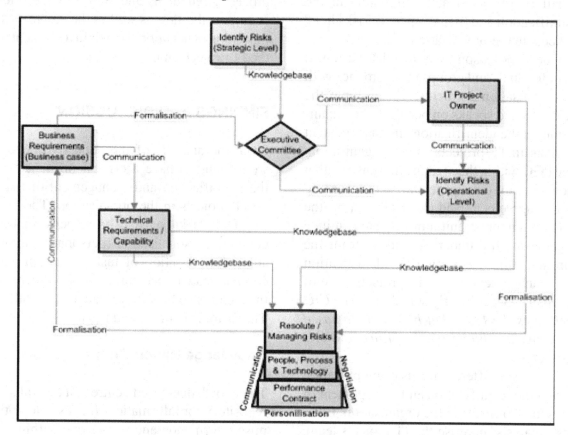

Formalisation of Task Allocation

Tasks allocation among employees is critical to the identification and management of risks in IT projects in the organisation. This helps to define roles and responsibilities, as well as trace the tasks and activities of employees in the identification and management of risks in IT projects.

Organisations need to have task allocation processes in place to guide managers in allocating tasks to the employees. A lot of time is spent deciding which task should be allocated to which individual employee and this has a negative impact on managing risks in IT projects. This is because the impact and effect of risks happens too fast. The guidelines which define criteria can assist managers in the allocation of tasks to the employees, based on skills, experiences and proficiencies. Unfortunately, not all employees who have skills and experience in dealing with impact of risk in IT projects within the organisation are allocated relevant tasks.

For the organisation to successfully assess the impact of risk in IT projects, tasks should be allocated to individuals or groups of employees based on skills, experience and knowledge. However, in attempts to manage the negotiation that takes place between employees and managers, the organisation adopted the performance agreement system.

Performance Contract

Performance management is the systematic process by which an organisation involves its staff, as individuals and members of a group, in improving organisational effectiveness in the accomplishment of its mandate and goals. Performance contracts define individual employee performance tasks and expectations. The contract plays an important role in determining employees' performance ratings and bonuses.

Organisations have performance contracts between employees and managers on behalf of the employer that forces employees to participate in the identification and management of risks in IT projects. Without performance contracts, the feelings and attitudes of the employees are challenged which has an impact on the objectives of the organisation. Under such circumstances, some employees are de-motivated and others are motivated to achieve the tasks of identification, monitoring and management of risks in IT projects.

Moreover, performance contracts in the organisation assist in building relationships between employees and employers, because the rights and obligations of both parties involved in the performance contracts are covered.

Personalisation

In the organisation, many employees personalise matters of organisational interest. Personalisation of matters of organisational interest has a huge impact on the roles and responsibilities. As a result, it is of vital importance for employees to have relationships at a personal level when and where possible.

In an organisation, personal relationships with managers and colleagues are important. However, personal relationships are not always easy between employees. The relationship with a manager is of fundamental importance for employees' ability to perform and develop well within their roles. A good personal relationship between employees and their supervisors or their managers is an important aspect of a good working environment.

Employees and managers must know that a successful personal relationship is based on communication. Employees who have worked together for years are sometimes like friends or they have built a personal relationship in the sense that they have developed a "shorthand" way of communicating because they understand each other.

Communication Structure

Communication is a key to project success, and as such, emphasis should be placed on effective communication. A two-way approach to communication between employers and the employees is preferred. The basic function of communication during projects is the coordination of actions and activities. The organisation therefore needs to have a spokesperson who speaks on behalf of business units.

All teams need to have a clearly defined goal, objective or vision defined by the project manager. The project manager uses project team meetings and mini-workshops, including one-on-one meetings to communicate to all project team members and stakeholders during the processes, activities, and updates of the projects. The goal should be communicated to all team members and referred back to during the projects to ensure that all the teams work towards the same objectives. Workshops are held to communicate information such as who is committed to a specific task in an IT project and what are the challenges they are likely to face.

INTERPRETATION OF THE FINDINGS

Some of the factors which influence identification and management of risks in IT projects in the organisation are revealed in the analysis, using the moments of translation. The factors, which include documents, incentives and a semi-structured environment, are interpreted in this section.

Documentation

All processes and activities which include communication, information, skill-set, and negotiation procedures and guidelines are vital to documentation. The documentation of these processes and activities would help create a knowledge base on which the organisation can rely for continuity and as source of stock knowledge. IT is even more important how the documentation is carried out. If not done properly it will become another form of challenge regarding how the information is retrieved, used and managed.

Information Retrieval Usage and Management: These refer to the organisational ability to document information and the ability to access information needed for IT projects and associated risks, as well as the ability to share the knowledge. It is also an address for sharing information and expertise among employees who are on the same job level without hiding any information that they think could lead to threats.

Employees might trust transferring skills to their colleague, which means the organisation at large can benefit from it. They should also trust that the information contributed to the knowledge-base centre cannot be used against them in future. Thus, the organisation should assure security for employees to contribute freely.

There is pressure to deliver a product on time and within budget. Project managers must ensure that lessons learnt by individual employees and groups are captured onto a knowledge-base centre for sharing.

Criteria for Task Allocation: The organisation applies general rules to employees, which stipulates that the more experienced and knowledgeable they are, the more they will be involved in IT projects. Some projects in the organisation sometimes have to be kept on hold due to lack of experienced employees in the IT projects.

Organisation need to document criteria to use for tasks allocation to employees. Such criteria will balance the tasks allocated to each employee. It will also minimise employees accepting tasks that they are not knowledgeable on and it will speed up the process of completing the project on time.

Policy Guiding: Policies are the strategic link between the organisation's vision and its day-to-day operations. It is difficult to justify any act in the organisation without documented policy. Policies also provide the organisation with significant protection against legal action. Organisational policies guide employees to reduce some conflict between employees and to carry out their task effectively.

Communication Content and Channel: Effective communication is crucial to the success of IT projects in the organisation. The organisation needs to reach out to fulfil its goals or mission. A proper communication channel must be established within the organisation. Communication channels used in organisations differ, but may include websites, workshops, letters, newsletters, email, telephone conversations, video conferences and face-to-face meetings. Effective communication relies on selecting an appropriate communication channel for the particular message. Selecting the incorrect communication channel causes communication obstacles including information overload and inadequate feedback.

A documented procedure on how employees must communicate in the organisation improves productivity. Thus, communication channels should be made known to employees because bad communication will cause confusion among employees about the messages that are sent out. This will result in employees being less likely to do what is expected from them.

Incentives

The processes and activities that take place in the identification and management of risks in IT projects in the organisation are carried out either voluntarily or as mandated by the organisation, based on personal or organisational interest. Irrespective of interest and enforcement of tasks, no organisation has total control of employees' actions. As a result, the organisation needs to encourage and motivate its employees through incentive. Only then collective contributions from individual knowledge, as well as healthy enforcement of the performance contract could be achieved.

Stock of Knowledge: Managers often transfers knowledge and skills. In the organisation, a lot depends on teamwork and collective knowledge. However, it is only a handful of people who have knowledge that their colleagues do not have thus giving them power over their colleagues. Some experts have been in the organisation for many years and have built up a stock of knowledge, which they use as a source of influence over other employees. For the experts to share their knowledge with other employees, the organisation needs to guarantee security and that their jobs will not be at risk if they share their information and expertise with others.

There is a need to provide incentive bonuses to those who have stocks of knowledge in the organisation to encourage them to contribute to the development of a knowledge-base centre.

Performance Contract: This refers to organisational standards to evaluate employees' performance and provide incentives where it is due. Another aspect is the use of aligning rewards to support appropriate behaviour, efficiency and productivities of employees. Performance contracts build relationships between employees and employers and as a result it improves productivity in the organisation.

Semi-Structured Environment

As revealed in the analysis, the organisational environment is either too structured (formal) or too informal (semi-structured). This can have serious implications regarding the way individuals and groups relate and communicate with each other. Some of the implications include lack of cooperation, working in silence and misinterpretation of communicated content. The manifestation of the implications has an impact on the identification and management of risks in IT projects in the organisation.

Personalisation: This refers to organisations allowing employees to have freedom to express themselves in the working environment. Personalisation in the organisation allows employees to build or to develop personal relationships. Personal relationships in the working place improve productivity and open up channels to communicate at all levels.

Communication: Communication in the organisation encourages employees to participate in activities aimed at discussion, protection and management of the environment. The organisation realises the importance of communicating about risk in IT projects. Sometimes communication is overlooked, but the ability to communicate effectively is necessary to communicate the ideas and visions of an organisation to the employees.

When communication is too structured in the organisation, employees tend to hold back their thoughts, opinions and ideas. They do not share their thoughts because of the formality in the organisation. Spending time discussing risks in IT projects and how risk will be managed enhance communication in the organisation.

Formalisation: One of the most important things in the organisation is to allow employees to take ownership of what they are doing. Formalisation allows employees to operate in the way they feel comfortable or the way they want to do their work. This helps employees to become flexible and open up in the working environment, which assists in managing risks in IT projects.

CONCLUSION

Identification of risks is significant to the deployment of ICT projects in organisations. However, the identification is never easy. Hence monitoring and management of risks during ICT projects poses challenges to the organisations that deploy them. The empirical nature of this study makes it useful and reliable, thereby instil confidence in ICT projects managers and sponsors in organisation. The uses of ANT help to understand why certain factors are categorised as risks or potential, and why those factors are deemed critical during ICT projects deployment. to the use of It has been recognized in this paper that risk has effects on ICT projects.

The findings from the study could be used to formulate policy in organisations, to help addressing risks on ICT projects, through identification, monitoring, and management.

The paper also revealed that for an organisation to be successful in their ICT project deployment, the interactions that happen amongst actors should be guided within heterogenous network that they operates, and channelled towards organisational objectives.

REFERENCES

Baccarini, D., Salm, G., & Love, P. E. D. (2004). Management of risk in information technology projects. *Industrial Management & Data Systems*, *104*(4), 286–295. doi:10.1108/02635570410530702

Callon, M. (1986). Some elements of the sociology of translation: Domestication of the scallops and the fisherman of St Brieuc Bay. In J. Law (Ed.), *A New Sociology of Knowledge, power, action and belief* (pp. 196–233). London: Routledge.

Clancy, T. (2004). *IT development projects.* Retrieved from www.standishgroup.com/msie.htm

Cooper, D. R., & Schindler, P. S. (2003). *Business Research Methods* (8th ed.). McGraw-Hill.

Feagin, J., Orum, A., & Sjoberg, G. (Eds.). (1991). *A case for case study.* Chapel Hill, NC: University of North Carolina Press.

Ferguson, R. W. (2004). A Project Risk Metric. *CrossTalk: The J. Defense Software Eng.*, *17*(4), 12–15.

Hanseth, O., Aanestad, M., & Berg, M. (2004). Guest editors' introduction actor-network theory and information systems. *Information Technology & People*, *17*(2), 116–123. doi:10.1108/09593840410542466

Hanseth, O., & Monteiro, E. (1996). Developing Information Infrastructure: The tension between standardization and flexibility. *Science, Technology & Human Values, 21*(4), 407–426. doi:10.1177/016224399602100402

Hillson, D., & Murray-Webster, R. (2004). Understanding and managing risk attitude. In *Proceedings of 7th Annual Risk Conference.* London, UK: Academic Press.

Iyamu, T., & Roode, D. (2010). The use of Structuration and Actor Network Theory for analysis: A case study of a financial institution in South Africa. *International Journal of Actor-Network Theory and Technological Innovation, 2*(1), 1–26. doi:10.4018/jantti.2010071601

Kutsch, E. (2008). The effect of intervening conditions on the management of project risk. *International Journal of Managing Projects in Business, 1*(4), 602–610. doi:10.1108/17538370810906282

Labuschagne, L., Marnewick, C., & Jakovljevic, M. (2008). IT project management maturity: A South African perspective. In *Proceedings on the PMSA Conference: From Strategy to Reality.* PMSA.

Lee, G., & Xia, W. (2003). *An Empirical Study on the Relationships between the Flexibility, Complexity and Performance of Information Systems Development Projects.* University of Minnesota.

Lester, V. M. (1997). *A Program for Research on Management Information.* Academic Press.

Macome, E. (2008). On Implementation of an information system in the Mozambican context: The EDM case viewed through ANT lenses. *Information Technology for Development, 14*(2), 154–170. doi:10.1002/itdj.20063

Mitev, N. (2008). In and out of actor-network theory: a necessary but insufficient journey. *Information Technology & People, 22*(1), 9–25. doi:10.1108/09593840910937463

Morris, P. W. G. (2004). *The irrelevance of project management as a professional discipline.* INDECO Management Solutions.

PMBOK. (2004). *A Guide to the Project Management Body of Knowledge (PMBOK® Guide)* (3rd ed.). Newtown Square, PA: Project Management Institute.

Suchman, L. A. (1987). *Plans and situated actions: The problem of human-machine communications.* Cambridge, UK: Cambridge University Press.

Taylor, H. (2006). Risk Management and Problem Resolution Strategies for IT Projects. *Project Management Journal, 37*(5), 49–63.

Tesch, D., Kloppenborg, T., & Frolick, M. N. (2007). IT project risk factors: The project management professional's perspective. *Journal of Computer Information Systems, 47*(4), 61–69.

Tuman, J. (1993). *Project management decision-making and risk management in a changing corporate environment.* Paper presented at the Project Management Institute 24th Annual Seminar/Symposium. Vancouver, Canada.

Urry, J. (2003). *Global Complexities.* Polity Press.

Walsham, G., & Sahay, S. (1999). GIS for district-level administration in India: Problems and opportunities. *Management Information Systems Quarterly, 23*(1), 39–66. doi:10.2307/249409

Wernick, P., Hall, T., & Nehaniv, C. L. (2008). Software evolutionary dynamics modelled as the activity of an actor-network. *The Institution of Engineering and Technology*, 2(4), 321–336.

Yin, R. K. (2002). *Case Study Research*. Design and Methods.

KEY TERMS AND DEFINITIONS

Actor-Network Theory: Actor network theory is a sociotechnical theory that integrates both human and non-human actors to form or create networks.

ICT Projects: Projects are a means to yield solutions through ICT artefacts, such as infrastructure (networks included), applications, databases or a combination of these.

Information and Communication Technology (ICT): Consist of hardware and software, including database, and communication tools.

Risk Identification: This is to differentiate the factors that could potentially cause danger or have negative impact on ICT projects.

Risk Management: This is control of the factors, known or unknown that could possibly cause danger to ICT projects.

Risk: Risk is an uncertain factor that could cause an uncertainty or negative impact on ICT projects, through schedule, cost, or quality.

Chapter 7
Actor–Network Theory:
A Bureaucratic View of Public Service Innovation

Noel Carroll
University of Limerick, Ireland

ABSTRACT

Public sector institutions continue to significantly invest in Information and Communication Technology (ICT) as a solution for many of their service provision challenges, for example, greater efficiency and quality of services. However, what has come to light is that there is a lack of research on understanding the contributory value or "success" of technological innovations. This chapter introduces a socio-technical view of public service innovation. The aim of this research is to extend on the notion of bureaucracy, which is traditionally focused on the politics of office environments. This socio-technical view extends this traditional view to include the politics of service networks, particularly within IT-enabled public service innovation. The chapter focuses on how service innovation is exploited to align specific interests through the process of translation and shifts the focus from value co-creation to value co-enactment. In essence, this chapter explains how public service technological innovations act as an agent of bureaucracy that alters the relational dynamics of power, risk, responsibility, and accountability. For demonstrative purposes, this chapter describes a case study that examines IT-enabled service innovation with an academic service environment.

1. INTRODUCTION

Service comprise of socio-technical (human and technological) factors which exchange various resources and competencies. Service network environments remain one of the most significant, yet 'invisible' infrastructures within the modern business era (Carroll et al., 2010). Service networks become increasingly complex when technology is implemented to execute specific service processes. This ultimately adds to the complexity of a service environment, making it one of the most difficult environments to examine and manage. In addition, although the emerging paradigm of 'Service Science' calls for more theoretical focus on understanding complex service systems, few efforts have surfaced which ap-

DOI: 10.4018/978-1-4666-6126-4.ch007

ply a new theoretical lens on understanding the underlying trajectories of socio-technical dynamics within a service system (Spohrer et al. 2007). Often, researchers are tasked with defending a particular theory to focus their research, but as Walsham (1997; p. 478) suggests:

There is not, and never will be, a best theory. Theory is our chronologically inadequate attempt to come to terms with the infinite complexity of the real world. Our quest should be for improved theory, not best theory, and for theory that is relevant to the issues of our time.

While it may seem contradictory after reading Walsham's quote, this chapter sets out to advance theoretical developments to extend our understanding of the traditional view of bureaucracy to include the politics of service networks, particularly within IT-enables public service innovation. To achieve this, the chapter proposes the need to examine the socio-technical impact of technology on public service network dynamics. This empirical research explores an academic service network, with particular attention paid towards a critical end-to end exam grading process. A single case study (Yin, 2009) is employed to examine the introduction of a Web-based system on a traditionally bureaucratic public service system and its transformation from a paper-based system to an automated system. The research adopts actor-network theory (ANT) as a research lens. ANT offers a rich vocabulary to describe the interplay of socio-technical dynamics which influence the service system reconfiguration. Thus, this chapter also offers a discussion on how ANT may be employed to examine the complexity of service systems and service innovation and builds on the efforts of Carroll et al. (2012). More specifically, this chapter examines the 'translation' process of IT-enabled public

service network innovation. In essence, it explains how public service technological innovations commands control over public sector behaviour and therefore acts as an agent of bureaucracy which alters the relational dynamics of power, risk, responsibility, and accountability. The descriptions resulting from ANT assists us to uncover the difficulties of service innovation diffusion by various actors and structural forces. This highlights the importance of 'translation' in service provision. This chapter argues that the concept of service co-enactment replaces co-creation within public sector networks since interaction is inscribed and governed through service regulation.

2. SOCIETY AND TECHNOLOGY

There have been numerous conceptualisations of the relationship which exists between technology and society and many studies highlight the important factor in which information technology (IT) plays to enable and increase the transformations of organisations (Orlikowski, 1991; Demirkan et al. 2008). However, it is difficult for Service Science practitioners to accept a presumptuous attitude towards the promise of technology, and suggest that these assumptions regarding the affordance of technology are becoming a cliché (for example, Demirkan et al. 2008).

In the past, there have been two differing schools of thought on the relationship of IT and social factors. One school of thought focused on technological determinism (Winner, 1977), which suggests that technology follows it own logic and patterns of usage. Alternatively, there was considerable support for constructionism which suggests that society develops the technology and society determines technology's role (Woolgar, 1991). These schools of thought were much debated throughout literature over in the 1970's and 1980's. But, in recent years, researchers began to examine the role in which both arguments played simultaneously to advance our understanding of the embedded relationship of IT and the organisation. Continued interest focused towards the question of how IT and the organisational roles interplay and how they come into 'being', suggesting the need to pay more attention to the characteristics and properties which support their co-existence (Kling, 1991; Orlikowski, 1992). Nowadays, we acknowledge that there is a mid-point between the two schools of thought which offers us a 'truer' picture of technologies ability to 'enable' and 'restrict' transformations. There have been increasing efforts to propose suitable models to explain the socio-technical factors of organisations. One approach in particular which is gaining more research ground across diverse research fields is ANT, which offers a radical vocabulary to examine the socio-technical building blocks on the nature of service networks.

A service system comprises of socio-technical systems which stabilise a service network through the exchange of resources and competencies which generate value (Carroll, 2012). The transformation of system thinking began during the 1960's and people began to view the organisation as an 'open system' made up of socio-technical factors. Within this school of thought, Emery and Trist (1960) examine how a system maintains quasi-stationary equilibrium despite changes in the environment. A socio-technical view of organisations incorporates the need to examine the hybrid nature of social (i.e. people) and the technical (i.e. things) in order to understand how actions are executed and the factors which influence the actions' outcomes (Carroll, 2012). Although technical factors are often concerned with machinery, it also includes methods and procedures to explore

how work is organised as a process. Nowadays, technology (i.e. service systems) plays a critical role in supporting critical organisational functions which highlight the importance of understanding how socio-technical systems impact of service relational structures. Thus, ANT is a fitting research approach to gain insight of socio-technical systems particularly in the field of Service Science.

2.1 The Emergence of Service Science

In response to the growing importance placed on understanding service complexities (technical, social, economic, etc.) the field of 'Service Science' has emerged to guide the effective design, implementation, and management of service systems. However, although Service Science calls for more theoretical focus on understanding complex service systems, few efforts have surfaced which apply a new theoretical lens on understanding the underlying trajectories of socio-technical dynamics within a service system. This chapter argues that ANT provides a suitable theoretical lens to examine and explain the underlying trajectories of socio-technical dynamics within a service system.

The design, management and delivery of complex service systems suggest that we need to develop a scientific understanding regarding the configuration of resources to deliver service excellence. In order to extend our understanding on service delivery, there is a need to establish alternative methods to examine service formation and the value propositions which connects them. Within the service-dominant environment (Normann, 2001; Vargo and Lusch, 2008), organisations are faced with increasing challenges to develop their capabilities in complex service models (Vargo et al. 2008). The emergence of Service Science as a discipline in recent years confirms the fundamental change which continues to alter the nature and application of technology within business environments. Service Science is an attempt to understand the complex nature of service systems and acts as an interdisciplinary umbrella which incorporates widely diverse disciplines to construct, manage, analyse and evolve service systems (Spohrer et al, 2007). This suggests that we need a more systematic, analytical, and overarching approach (Carroll, 2012) to examine service co-production operations to generate knowledge regarding the overlap between the social, business, and technology factors within a service environment (i.e. bridging service management and service computing). As services become more "open", collaborative, flexible, agile, and adaptive, there are greater pressures on business to reconfigure and meet change through strategic realignments (Carroll et al., 2010; Carroll 2012). In doing so, managers should develop an understanding as to how this impacts the 'value' of the service system. A service system comprises of a provider(s) and a client(s) who collaborate to deliver (i.e. co-create) and benefit from a service (Vargo et al. 2008). A service system may be defined as (IfM and IBM, 2007; p. 5):

...a dynamic value co-creating configuration or resources, including people, technology, organisations and shared information (language, laws, measures and methods), all connected internally and externally by value propositions, with the aim to consistently and profitably meet the customer's needs better than competing alternatives.

The environment in which the configuration of resources is achieved is described as a service network. A service network comprises of clear linkages which define the service structure and interactions in which it co-ordinates its tasks to achieve a certain

business objective. Since it accounts for the collective effort of all service interactions to generate and realise value, co-creation is an important concept within a service network. This suggests that ANT can assist Service Science researchers in their quest to understand the complexity of service systems.

3. INTRODUCING 'PUBLIC' SERVICE SCIENCE

As discussed earlier in this chapter, the concept of a service had received much interest over the last decade, in particular the applications of technology to extend services. As the business environment continues to shift from a goods-dominant environment towards a service-dominant one (Normann, 2001), service networks play a central role towards supporting and delivering services across the global economy (Fitzsimmons and Fitzsimmons, 2004). However, although, practitioners continue to make substantial investments on services and supporting service infrastructure, one of the main problems is that little is known about the socio-technical infrastructure to deliver a service and understand the factors which influence service 'value'.

Service science introduces a new paradigm shift to deal with new service realities. A paradigm shift may be described as "a broad model, a framework, a way of thinking, or a scheme for understanding reality" (Tapscott and Caston, 1993; p. xii). The influence of Internet technologies and other Web services has altered our understanding of the modern service environment. For example, in business, the marketplace has become a volatile and 'flattened' (Friedman, 2006) global stage. In addition, business is considered to be a 'networked effort' rather than an individualistic effort (Carr, 2003; Carr, 2004), while customers co-create value within a business (Normann,

2001), and technology continues to be the dominant force in the service sector. Thus, Service Science acts as an interdisciplinary umbrella which incorporates widely diverse disciplines to construct, manage, analyse and evolve service systems. This suggests that we need a more systematic approach (Spohrer et al. 2007; Spohrer and Maglio 2008) to examine service co-production operations to generate new knowledge regarding the overlap between business and technology (i.e. service management and service computing).

The delivery of services is facilitated through a service system which is supported by technological innovations and organisational networks, connected through value propositions and information exchanges. In today's service-dominant business environment (Normann, 2001; Spohrer et al., 2007), harnessing innovative applications of technology is considered one of the critical factors towards organisational sustainability (for example, Weill et al., 2002) and achieving greater efficiencies. Consequently, the application of technology to support services has altered our traditional understanding of the 'organisation', making it more difficult to conceptualise the paradigm of services. Understanding the socio-technical factors is critical for several phases of service network development, including; conceptualisation, design, delivering, monitoring, optimisation, planning, and analysis. What has come to light is that there is a lack of theoretical developments on what these intertwining dynamics are and how they are influenced by technology. This is particularly true within public sector service when one considers the influence of bureaucracy in service change and actor-network reconfiguration, hence the need to shed more light on Public Service Science.

3.1 Need for Public Service Science

Public services typically refer to direct or indirect services which are provided to citizens by a government through the provision of finance to support fundamental services. Examples of public services include health, education, transport, infrastructure, and security and defence. Public services tend to be heavily regulated to define service composition and orchestration. However, it becomes evident that although much research focuses on public administration, few efforts exist on exploring how public administration bodies incorporates the concept of 'service networks' and there is a lack of understanding regarding the socio-technical implications of process patterns on administrative tasks (O'Toole, 1997a; O'Toole, 1997b; O'Toole, 1997c; Bannister, 2001; Berry et al. 2004; Rethemeyer, 2005).

Over the last decade, there has been a dramatic shift in the delivery of public services which are often characterised by political agendas, legislation-driven rather than market-driven goals, and deficiencies in financial and human resources (Rusaw 2007; Feller and Finnegan, 2008). Digitising public services appears to be an emerging focus in public services to improve efficiency through the affordance of IT (Pedersen et al. 2006), which often impacts service management issues (Jansen, 2005). Feller and Finnegan (2008) discuss how early adopters of technology in the public sector such as across Scandinavian countries, the UK, and Canada are considered to avail of a more developed technical infrastructure to deliver of e-services. However, Feller and Finnegan (2008) caution that placing substantial efforts on IT issues often allows organisation issues to become increasingly neglected and therefore a certain balance is required to address both technical and organisation factors. This places greater emphasises on the need to examine socio-technical factors within the public service.

3.2 Traditional View of Public Services

Public services play a critical role in the 'public good' and economic development. The provision of public services is considered to be a universal guarantee within governmental developments to support human rights, for example, health care and education. Many of these service efforts may be difficult to supply through individual efforts (for example, rail transport) and may be difficult to determine quality metrics (for example, health care). Thus, it is inevitable that bureaucracy plays a central role on public services. This is also evident from Weber's classical description of bureaucracy which may be characterised as (Weber, 1919):

- A dictation of labour based on functional specialisation;
- A hierarchy of authority;
- A system of rules which limit discretion;
- Impersonality;
- A career structure based on technical competence;
- Written records of activities.

Bureaucracies exists across many organisational structures and cultures although there have been efforts in the past to minimise or remove bureaucratic procedures, for example, business process re-engineering. Such organisational change initiatives are significant catalysts for change, although it suggests that there is a clash with the nature of bureaucracy and innovative technological culture. Rethemeyer (2009) provides a discussion on the challenges which exist in public services.

He examines how Terry Moe once described the politics–administration dichotomy as the "dichotomy that won't die" (1994, p. 17). This dichotomy is also embedded within the modern public service sector. However, Thompson and Alvesson (2005) argue that that due to the changing nature of the business and social environment, the bureaucratic model is obsolete. Thompson and Alvesson (2005) list two main assumptions throughout bureaucracy literature. Firstly, although bureaucracy is the dominant organising logic of modernity, "it produces degrees of inefficiency, dehumanisation, and ritualism" (p. 90). Secondly, although bureaucracy may be the desired functionality within a particular environment, it does not fit all environments, particularly where there is a high degree of unpredictability, instability, innovation, and adaptiveness.

The role of IT within the public sector continues to come under scrutiny. McIvor et al. (2002) suggest that there is a lack of transparency in public sector service although technologies may provide greater insight on service behaviour (for example, Internet technologies). There are many factors which contribute to the lack of transparency within public services. Throughout the 1980's there was considerable promise of IT to deliver more cost effective public services although McIvor et al., (2002) explain that there were many limitations to realising IT effectiveness since IT developments affected the "*public purse*". Other problems surrounded the confusion of IT potential and the lack of know-how at management level, lack of IT strategy developments, poor project management, lack of training and staffing, and difficulties arising from inter-departmental politics. Willcocks (1994) explains that as a result of these significant hurdles, many public services began to outsource IT consultants at an extremely high cost although they rarely benefited from their skills. Considering the cost and often publicity of technological efforts, IT implementation was often prioritised although there was a sense of low confidence attached to public sector IT implementation. McIvor et al., (2002) explains that one of the most significant problems facing organisations is their lack of ability to align organisational structures and peoples' roles with technological innovation and is often hindered by the authority figures. As a result, this began to separate policy from administration which effects how information was sanctioned through bureaucratic structures and lacking in strategy which slowly became embedded in the public service culture. Bichard (2000) suggests that public service culture focuses on risk aversion, whereby bureaucratic structures control service operations rather than developing structures to 'connect' information silos through the service. McIvor et al., (2002) suggests that there is a need for cultural change and viewing technology as a tool to share information to enhance public service performance and remove the perception of *"power perceived to accompany the possession of information and where managers often act as controlling pretty tyrants, jealously guarding their own patch of turf"* (p. 175). This is not often the case within the private sector. For example, Bannister (2001) examines how IT-orientated change in public administration has lagged behind that in the private sector. He discusses how the conservative and bureaucratic nature of public administration often prevents change. However, nowadays, particularly within the service sector, there are growing pressures to secure increased 'value' from technological investments which is ripe with challenges in the past. Many of the challenges experienced include cultural, structural, resource, technical, or isolated developments which do not interrelate (Bannister, 2001). One way to overcome some of these factors, as suggested

by Lynn (2005) is to 'divorce' politics from management within the public sector. In order to do so, this chapter proposes the need to bring the public sector into the Service Science focus to examine the trajectories of public service networks. Therefore, it is important to examine these boundaries and explore how technologies support relational structures and 'meaning' through a socio-technical lens such as that provided by ANT.

4. ACTOR-NETWORK THEORY: AN OVERVIEW

To put it very simply: A good ANT account is a narrative or a description or a proposition where all the actors do something and don't just sit there. – Latour, (2005, p. 128)

ANT continues to make a significant contribution to science and technology studies. ANT is often described as a systematic approach to explore the infrastructure which supports the 'scientific and technological achievements' within a network making it a more profound approach to researching and understanding service networks (Carroll et al. 2012; Carroll, 2012). ANT suggests that the world is made up of intertwining networks which are comprised of many complex interactions (locally and globally) which constantly reconfigure itself on a regular basis. This systematic approach focuses on the infrastructure which supports socio-technical developments and their interactions. ANT also provides us with a lens to examine the links between the so-called social and the technical and suggests that actors can be enrolled to stabilise the network. Steps may involve identifying stakeholders and their interactions; the development of an actor-network model; the identification of irreversible technologies, enablers and inhibitors of specific processes and activities which are socially embedded in a service network. ANT breaks away from the social science school of thought as it does not fix itself upon any set theory per se, but rather enjoys the radical uncertainty of human behaviour in which actions are not predetermined. ANT provides an approach to understand how both social action shapes technology and how technological innovations shape social action. Thus, ANT acts as a toolkit to explore how human and non-human actors interact with one another to make sense of their world (Latour, 2005). Law (2007; p.2) provides an account of ANT and explains that:

Theories usually try to explain why something happens, but Actor-Network Theory is descriptive rather than foundational in explanatory terms, which means that it is a disappointment for those seeking strong accounts. Instead it tells stories about 'how' relations assemble or don't. As a form, one of several, of material semiotics, it is better understood as a toolkit for telling interesting stories about, and interfering in, those relations. More profoundly, it is a sensibility to the messy practices of relationality and materiality of the world. Along with this sensibility comes a wariness of the large-scale claims common in social theory: these usually seem too simple.

This supports what the author refers to as Walsham's (1997) argument at the beginning of the chapter that *'our quest should be to improve theory'*. ANT provides a vocabulary to examine how powerful networks emerge and pays particular attention to assemblage and the influence of objects and people. Therefore, it establishes networks and determines particular actions or behaviour. Although there are many aspects to ANT, the process of 'translation' is fruitful in examining the implementation of service innovation to describe how technology impacts on service

network dynamics and impacts the structure of a service network. To appreciate the value of ANT, it is also important to understand the background of ANT.

4.1 Actor-Network Theory: A Brief Background

The fundamental aim of ANT is to explore how networks are built or assembled and maintained to achieve a specific objective. Identities (networks and actants) are established by their represented or delegated interactions which acknowledge the importance of the inseparable socio-technical factors (Carroll et al. 2012). ANT rejects "any sundering of human and non-human, social and technical elements" (Hassard et al., 1999) since ANT adopts socio-technical symmetry to explore actants' (human and non-human) participation within heterogonous network assemblages through negotiation and translation.

ANT provides the ability to uncover the chain of actions or influences from various actors which are carried out to deliver a specific action and outcome. Therefore, it breaks away from the social science school of thought since it does not fixate upon any set theory per se, but rather enjoys the uncertainty of human behaviour in which actions are not predetermined. Latour (2005) explains that the ANT approach rejects a social dimension, social order, a social force, frame of reference, actors are not embedded in a social context, and suggest that actors know what they are doing and are connected to many other elements. In this alternative view, 'social' is not some glue that could fix everything: it is what is glued together by many other types of connectors (Latour, 2005; p. 5) and the specific associations provided which are of importance. This draws the author's attention towards the linkage, relations, assemblages, or interactions of service networks. During the interactions, one of the key factors which emerge from the negotiations is the concept of translation (Callon, 1986a). Translation is a complex view of interactions which suggest that actors:

- Assemble similar definitions and meanings;
- Define network representatives;
- Encourage one another towards the pursuit of self-interest and collective objectives.

After negotiation with certain states of power relations, actants eventually conceive what they want and what they can achieve. Actants have the ability to (re)construct a network which their interactions to stabilise the system. Of course, the reverse is also true, i.e. the lack of interactions can destabilise the network until it eventually dissolves. In addition, ANT identifies objects as boundary objects which foster interconnections (Star and Griesemer, 1989). They describe boundary objects as being adaptable to different viewpoints and robust enough to maintain identity across them and identify four types of boundary objects (Star and Griesemer, 1989):

- Repositories;
- Ideal types;
- Coincident boundaries;
- Standardised forms.

These boundary objects relate to how information may be interpreted by different communities but with enough fixed content to maintain its reliability. They also discuss how problems from conflicting views are often managed from a variety of ways including (list extracted from Star and Griesemer, 1989; p. 404):

- Via a 'lowest common denominator' which satisfies the minimal demands of

each world by capturing properties that fall within the minimum acceptable range of all concerned worlds;

- Via the use of versatile, plastic, reconfigurable (programmable) objects that each world can mould to its purposes locally;
- Via storing a complex of objects from which things necessary for each world can be physically extracted and configured for local purposes, as from a library;
- Each participating world can abstract or simplify the object to suit its demands; that is, 'extraneous' properties can be deleted or ignored;
- Work in the worlds can proceed in parallel except for limited exchanges of standardised sorts; or
- Work can be staged so that stages are relatively autonomous.

The list above places emphasis on actant configuration and their properties which may be interpreted to facilitate the exchange of resources and competencies across a service network. In addition, this list acts as a platform upon which we can develop a socio-technical view of a service network. Berg and Timmermans (2000) explain that ANT does not assume that order can hold totalitarian control but rather, order is a co-produced achievement. This is an interesting concept which links ANT to Service Science logic while both schools of thought are focused on examining the intertwining nature of co-creation and co-production interactions which will be outlined in the case study. One of the main differences between actors and actants is that actors have the ability to circulate actants within a system. Latour (2005) denies that sociology can never attain an objective viewpoint and look beyond its participants (i.e. a meta-language). Actants influence one another. Law (2008) refutes that technology is transferable since it does not originate from a fixed point and instead suggests that technology is passed and changed to a point that it becomes 'less and less recognisable'. Within a network, actors tend to present one another with a version of their necessities, and from that other actors understand the strategies they attribute to each other (Latour, 2005; p.163). This often allows them to create their own society, sociology, language, and meta-language. ANT suggests that there is no single theory of action (Walsham, 1997; Latour, 2005), i.e. it denies a fixed frame of reference as indicated from a relativistic sociology (which examines deviant phenomena through a fixed theory), and instead embrace a fluctuating reference approach (*follow the actors*). Due to the complex and intertwining nature of actants within service networks, ANT presents a significant contribution towards Service Science research undertakings. It has significant potential to provide a contribution towards the emerging paradigm of Service Science, for example service formation, service evolution and service innovation. Thus, one can examine the formation of service systems through a radical and rich vocabulary offer through ANT.

4.2 Actor-Network Theory: Key Concepts and Vocabulary

While exploring the underlying mechanics of a service network, ANT presents us with a 'vocabulary' to examine and discuss, for example, how the introduction of an IT system impacts the structure of a service network. Latour, Callon and Law are among the most cited scholars whom introduce a vocabulary which is used to distinguish between objects and subjects and explore particular network phenomena, i.e. the objective and the subjective. Many ANT studies examine 'success' and 'failure' and examine the concept of 'power' which established actor-networks and impos-

ing 'order' on actants to meet specific interests (for example, Berg and Timmermans, 2000). Additional studies began to examine multiplicity and difference of multiple 'orders' (Gad & Jensen, 2010) which act almost automatically and simultaneously. ANT suggests that 'reality' is dependent, contextual, and emergent and refutes the notion that there may be a 'fixed point' of analysis (Carroll, 2012). Rather than suggest that factors such as culture or globalisation impact a certain phenomena, ANT suggests that these factors need explanation and sets out to describe how environments (i.e. networks) come into being. These studies adopt ANT to incorporate a different language and viewpoint to describe the network's operations. This is also suggested by Latour (1992), as he explains that ANT overcomes the need to discuss knowledge and objects using a one dimensional language and instead adopts a dualism as a second dimensional approach. He suggests that, "instead of being opposite causes of our knowledge, the two poles are a single consequence of a common practice that is now the single focus of our analysis" (p. 281). There are a number of key concepts which one has to become familiar with while adopting ANT (see Table 1).

Although Table 1 lists the key vocabulary used throughout ANT studies, Hassard et al. (1999; p. 392) explain that the success of ANT is with the "*habit of failing to forge its own internal and external boundaries*", which presents us with a large degree of exploration freedom. Law (1999) suggests that ANT has become a strategy which has an "*obligatory passage point....with a more or less fixed location*" (p. 2). Latour (2005) provides what he describes as the 'intellectual architecture' in his account of the social explanations of social phenomenon. He explains that the word 'social' cannot be conceptualised as a 'kind of material or domain' which can be discussed using a 'social explanation' (p. 1). ANT is

Table 1. ANT main concepts (Carroll et al., 2012)

Concept	Explanation
Actant	"Any element which bends space around itself, makes other elements dependent upon itself and translates their will into the language of its own" (Callon and Latour, 1981; p. 286).
Actor-network	A heterogeneous network of aligned interests formed through translation of interests (Walsham and Sahay, 1999).
Assemblages	Built out of social ties rather than physical and explores what is the social made of, e.g., how we act, or who else is acting.
Associations	Non-social ties which can be used to trace associations and does not designate a thing among other things.
Black box	A snapshot of the network which illustrates its irreversible properties.
Translation	The creation process of an actor-network through four main phases (Callon, 1986a): 1. Problematisation, 2. Interessement, 3. Enrolment, 4. Mobilisation.
Problematisation	Defines identities and interests of other actors which align with its own interests (i.e. *obligatory passage point*).
Obligatory Passage Point	A situation that has to occur in order for all the actors to satisfy the interests (Callon, 1986a).
Interessement	Convince other actors to agree on and accept the definition of the focal actor (Callon, 1986a).
Enrolment	An actor accepts the interests defined by the focal actor and sets out to achieve them through actant allies which align with the actor-network (Callon, 1986a).
Mobilisation	Ensuring actors represent actors interests (Callon, 1986a).
Inscription	Creating technical objects which ensure an actor's interests are protected, e.g. a particular piece of software or regulations to meet organisational objectives (Latour, 1992).
Performativity	"Entities achieve their form as a consequence of the relations in which they are located…they are *performed* in, by and through those relations" (Law, 1999).
Irreversibility	The point to which it is impossible to return to a point where alternative opportunities may exist (Walsham and Sahay, 1999).
Immutable mobile	Strong properties within a network which establishes it irreversibility, e.g. software standards (Walsham, 1997)
Speaker/delegate/ representative	An actor that speaks on behalf of (or stands in for) other actors (Callon, 1986a; Sarker et al., 2006).
Betrayal	A situation where actors do not abide by the agreements arising from the enrolment of their representatives (Callon, 1986a; Sarker et al., 2006).

often referred to as the sociology of translation (Callon, 1986a) which suggests that one must identify the meaning of 'assemblages' through ANT (Latour, 2005). ANT examines the "*motivations and actions of groups of actors who form elements, linked by associations, of heterogeneous networks of aligned interests*" (Walsham, 1997; p. 468).

There are some subtle differences between the social science literature and ANT. For ex-

ample, an actor may be considered as anything which compromises of a process or a number of processes to execute a certain task, i.e., a person, group, department, organisation, or an information system. In ANT literature an actant (human and non-human) is more than what social science would describe as an actor, since an actant is often 'enrolled' in a certain position to strengthen it. For example, software (actant) executes code (action) to perform an action to meet a business objective (the network) or, an elevator (the actant) strengthens the accessibility (the action) of the floors (the network) within a building. In addition, ANT promotes that humans are not the only beings of agency, and that we should consider machines, animals, and as demonstrated in other studies, matter (Latour, 1998) and thus can be considered an actant "*if it performs, or might perform [agency]*" (Callon and Law, 1995; p. 491). Actants may be considered as human and non-human stakeholders whom are focused on interests that influence technological applications (Monteiro and Hanseth, 1996; Walsham, 1997; Hanseth et al., 2004; Sarker et al., 2006; Carroll, 2012). In the pursuit of specific interests, networks are formed and aligned through technological innovations. As actors continue to translate (align interests) and enrol additional actors, the network becomes increasingly more stable. Succeeding in alignment is particularly important. This is achieved through inscriptions. Inscriptions are common procedures such as managerial practice, employee contracts, standards, regulations or software requirements documentation (i.e., indicates how the network should operate). Latour (2005), discusses the notion of the neologism "valorimeter" which refers to a measurement of a network's ability to meet actor's requirements are being addressed, and is of particular interest from an IS requirements perspective. Inscriptions also support the translation process through the design of the network and determine who will participate, how they will participate, and the impact on their roles. For example, once business processes have been established and automated, the software which supports business processes adopts the inscription role which often becomes fixed and irreversible, i.e. making it impossible to start the process again or explore alternative opportunities. The

actors which participate in the network and operate the technology form the actor-network which creates an embedded black-box model of the system of what appears to be the optimum system operations.

4.2.1 Materiality, Inscription and Translation

ANT suggests that objects have agency to establish relations and translate interests. For example, Latour (1992) discusses how a hydraulic door system is considered more reliable than a human operator or, how a car seatbelt imposes morality on humans. Although it is often considered controversial, ANT practitioners insist that researchers must refuse to distinguish between human and non-human as prior categories and is considered one of the main contributions to this research approach. Callon and Muniesa's (2005) provide an interesting account of materiality and they caution that we should not confuse materiality with physicality. Instead, we should examine how properties are supported through specific process.

The concepts of inscription (Akrich 1992; Akrich and Latour 1992) and translation (Callon 1991; Latour 1987) are of particular relevance within Service Science (Carroll et al. 2012; Carroll, 2012). Translation treats actants within an actor-network as a heterogeneous unit of analysis with particular on network formation. Translation examines the various meanings which actors provide about a specific phenomenon which actors discuss the interessement process of various interests. The ultimate aim of the interessement is to enrol actors to support a set of defined interests and stabilise a network. Translation suggests that the nature of power plays a significant role in actor-network formation. For example, Callon (1986a; p. 223) explains that:

To translate is to displace... to express in one's own language what others say and want, why they act in the way they do and how they associate with each other: it is to establish oneself as a spokesman.

Translation is a very complex task which undergoes four main phases (Callon, 1986a):

1. **Problematisation:** Defines the problem or opportunity with which an actor proposes a solution. Defining the proposed solution acts as the obligatory passage point;
2. **Interessement:** Attracts other actors in the proposed solution to favour a new opportunity which confirms the problematisation phase.
3. **Enrolment:** Is a negotiation process to exhibits how the interessement meets the actors' interests and needs and persuades them to accept the new actor-network.
4. **Mobilisation:** Is an important process which ensures that actors represent other actors' interests.

Another key concept in ANT is inscription. Inscription refers to what may be described as the patterns of use but is relatively flexible in the nature of use, for example, computer applications. In order to stabilise a network and establish social order, actors engage in continuous negotiation to align particular interests (interpretation, representation, or self-interests) to mobilise support as part of the translation process. Inscription translates specific interests within technical objects, for example, text, software, user requirements, or regulations, which typically impacts on actors' roles. This process varies substantially as there are many factors which impact on its success, for example:

1. What is the desired outcome from the inscription?
2. What medium is utilised for the inscription process?
3. Which actors inscribe the particular interests?
4. How strong are the inscriptions (what level of resistance could oppose the inscriptions)?

Consequently, the design of the translation process is realised to align with users needs to provide a particular solution. The solution is then translated to complete a task, while actions are translated to specific outcomes. Inscriptions are typically provided with more concrete content to record actors' interests within a material which vary in their flexibility, for example, policy and regulations. Therefore, the strength of the inscription may be determined by the possibility of irreversibility. Translation and inscription play a fundamental role in the formation approach of an actor-network. In addition to understanding the theory of ANT, one can also adopt ANT as an approach to examine service systems.

5. ACTOR-NETWORK THEORY: THE APPROACH

Technology is impacted and consequently shaped by a number of factors including, for example, social interests, existing networks or network formation, power structures within a network, influence structures, political nature of the network, and attitudes (Carroll, 2012). In order to understand how social actions shapes technology and technological innovations shape social action there are a number of phases one can adopt as a roadmap which is significant to the research methodology. The phases listed in Table 2 (adapted from McBride, 2000) outlines the research methodology which may be adopted by the Actor-Network theorist to apply in Service Science research.

According to Latour (1993), ANT's theoretical ability rests in its refusal to reduce network explanations to natural, social, or discursive categorisations although identifying the importance of each one (p. 91). In addition, to support this logic, Law (1990; p. 113) suggests that "the stability and form of artifacts should be seen as a function of the interaction of heterogeneous elements as these are shaped and assimilated into a network". The phases listed in Table 2 provide a roadmap on how to employ ANT to explore the nature of a service network. This is important as Latour (2005) denies that sociology can ever attain an objective viewpoint and look beyond its participants (i.e. to develop a meta-language). In addition, Mitchell (2002) suggests that with the continuous pursuit social abstraction, there is a growing division of our social world "*into image and object, representation and reality*" (p 93).

Social abstraction also takes into account the performance of the actor-network. For example, Knox et al. (2007) discuss performances of 'calculability' (or the 'effects' of calculability) which accounts for social practices in terms of the abstract workings of a 'locationless logic'. In addition, Knox et al. (2007) draws on Callon and Law (2005; p. 25) to suggests that 'calculation' is, "*a process in which entities are, so to speak, released from local entanglements and detached from specific contexts so that they can be 'reworked, displayed, related, manipulated, transformed and summed up in a single space*". But the question remains: how is this accomplished?

From the extensive literature review on ANT, Figure 1 depicts the authors conceptualisation of how an actor-network is formed and stabilised, starting with the centre of the diagram which motivates network formation through 'interests'. Figure 1 provides an overview of ANT and illustrates what one can conceptualise how the main concepts operate and intertwine with one another. Networks are created through aligned interests into which actants enrol. When they enrol, they accept

Table 2. Phases of adopting the ANT approach (adopted from McBride, 2000)

Phase	Description
Identify the stakeholders	Comprise of human or non-human actors which influence or becomes influenced by other actor's policies and practices.
Investigate the stakeholders	Understanding the character of the stakeholders through, surveys, or interviews with network representatives, accessing documentation, understanding their attitudes, interactions, interests, etc.
Identify stakeholder interactions	Tracing interactions between stakeholders to explore the level of influence between stakeholders (e.g. trust and control).
Construct an actor-network model	Construct an actor-network model to determine for example, the networks complexity, cohesion, strength, and influence.
Examine irreversibility	Determine to what degree it is difficult to make a change, e.g. through understanding the culture and the nature of acceptance in the network.
Source of inhibitors and enablers	Determine who enables and inhibits actions to shape technology and the network under investigation, e.g. technology, attitudes, resistance, or network infrastructure.
Tracing actions	Identify what activities led to the alignment of the actor-network, for example, training.
Reporting on the actor-network	Report on the overall nature of the network and explain how social actions shapes technology and technological innovations shape social action within the network.

Figure 1. ANT overview

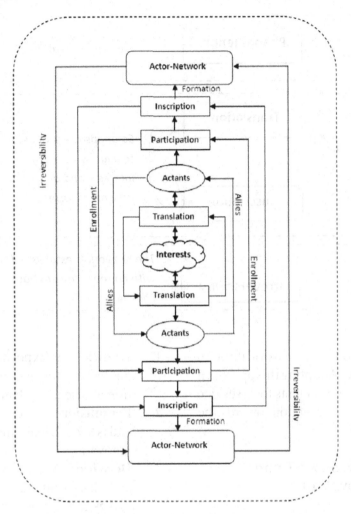

allies' interests through a process of translation which effectively states their agreement with their participation and efforts to stabilise the network. Thus, an agreement is established on what the network represents, how it shall operate, and the rolls various actants will adopt to stabilise the network. These processes form the network into what becomes known as the actor-network. The actor-network becomes irreversible and cannot explore alternative opportunities at this point (i.e. the actor-network becomes a black box). The process of translation and inscription are illustrated as follows (Figure 2).

Figure 2 depicts the relationship between translation and inscription to address a phenomenon (the formation of an actor-network) through various interests and to establish an irreversible network. Traditionally, organisations would implement technology to mediate complex or laborious tasks. This would essentially disentangle knowledge from one actant (e.g. a department) and transfer the knowledge and repackage it in various other locations for other people to benefit from the records. Therefore, one should consider how technology often 'replaces' methods of process execution and relocates knowledge

Figure 2. Process of translation and inscription

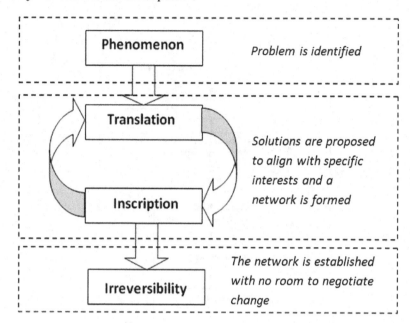

which alters the socio-technical world through a representative view. Adopting this approach places emphasis on 'how' actants form service networks which directs attention towards the application of ANT.

5.1 Overlap between ANT and Diffusion of Innovation

Carroll (2012) identifies a significant overlap between the ANT concepts listed above and the theoretical works provided on 'Diffusion of Innovation' (Rogers, 1962; Rogers and Shoemaker, 1971; Rogers, 1976; Rogers, 2010). The theoretical developments on the diffusion of innovation examine the characteristics of an innovation and its context that correlate with its diffusion (Rogers, 1962). The process examines how innovation is communicated among various interested parties through various channels within a social system. The success of an innovation is largely dependent on decisions made within the social system as they adopt five steps:

1. **Knowledge:** Exposure to an innovation but lacks information about the innovation and seeks to learn more.
2. **Persuasion:** Interested in the innovation and is keen to learn more detail about the innovation.
3. **Decision:** Weights up the advantages and disadvantages of innovation and makes a decision whether to adopt the innovations.
4. **Implementation:** Employs and examines the usefulness of innovations.
5. **Confirmation:** Finalise decision to continue using the innovation.

Individuals have different levels of enthusiasm towards the adoption of innovations and often the amount of time required depends on certain characteristics of the person or social system. These are innovators, early adopters, early majority, late majority, or laggards. Although this theory provides a useful insight of the implementation of innovation, it fails to address the socio-technical factors which this study sets out to examine. To investigate this, the author has explored the relationship between service IT-enabled innovation and Rogers (1962) diffusion of innovation. The findings present a significant overlap between the diffusion of innovation and ANT (Table 3).

Table 3 highlights that public service IT-enabled innovation presents a significant overlap between the concept of diffusion of innovation and actor-network translation. The promise of innovation is only realised if management adopt a successful interessement process to convince other actors to learn and agree on a specific innovative solution to support public service operations. Thus, it is important to gain an understanding the problematic relationship in IS and organisational development (van den Hooff and Winter, 2011). This is critical when, for example, it is applied within the public sector considering the unsatisfactory application and culture of IS in public administration and the effect of the diffusion of innovation. Latour (2005) discusses the "*division of labour*" which suggests that one can only create sub-projects after a project succeeds which is often determined by whether it is of a continuous compromise nature. Interestingly, Latour (2005, p.126) suggest that the more a technological project advances, the more likely the impact

of technology diminishes in relative terms. ANT prescribes two main methodological approaches:

1. Follow the actor (i.e. using interviews and ethnographic research);
2. Examine inscriptions (i.e. text sources which are also central to credibility, e.g. the strategy of enrolling others).

Although traditional research approaches have guided researchers to gain insights on various demographics of the social world, there appears to be a void in our ability to truly understand how technology continues to shape our world. Latour's expression, "*follow the actors*", suggests that we can examine what actors do, why they do it, and their interests or beliefs in doing so through their interactions which support their existence. The focus of the theory is to trace and explain where stabilising networks are the result of aligned interests, or in some cases, fail to establish themselves (Walsham, 1997). This, it is suggested, provides us with insight as to what shapes network infrastructures which is significant when applied to the public sector (Ali and Green, 2007; Feller and Finnegan, 2008; Cordella and Iannaccin, 2010; Davis, 2010). Within an IS perspective, there are several

Table 3. Overlap between the diffusion of innovation and ANT

Diffusion of Innovation	ANT	Description
Knowledge	Problematisation	Identifies how innovation will support interests and align others interests.
Persuasion	Interessement	Convince other actors to learn and agree on a specific innovation to support operations.
Decision	Enrolment	Weights up the advantages and disadvantages of innovation and decides to adopt the defined interests represented via the innovation.
Implementation	Inscription	Creating technical objects such a piece of software to demonstrate the usefulness of the innovation.
Confirmation	Irreversibility	Making a final decision to implement or continue using an innovation to a point to which it is impossible to return to another point where alternative opportunities may exist.

key studies which develop ANT concepts, in particular the IS-related studies.

6. CASE STUDY

The author examined the introduction of a web-based grading system which replaced a paper-based process within a university service environment. The exam administration service department (EASD) initiated the need for change in order to improve operational efficiency. Central to this research study was the need to develop an understanding of the changeover process and its impact on the service network dynamics. This was important as Latour (2005), suggests that one must ask how the 'system' has complied through various associations. The purpose of this case study was therefore to explore the socio-technical dynamics of the translation process during the implementing of IT innovation within a public service network. Thus, through the implementation and evolution of the service network, the author could examine the shift in norms of a bureaucratic nature. In doing so, the author had to embrace the unexpected conduits of information which appear on the discovery for the truth, which often raised more questions as to the motivations of public service innovation. The evidence suggests

that there were some considerable tensions between actors interests within the service network. Some of the initial findings indicate that academics' had no insights as to the impact of this on EASD operations which is illustrated in Figure 3.

As Figure 3 illustrates, there are many factors which influence service actor behaviour which impacts on service delivery quality and efficiencies. For example, the concept of 'responsibility' appears to be a transferred property from academics' to EASD and vice-versa as EASD prompt academics and academics' submit grading sheets for the grading deadline. The conceptual design supports the author's enquiry and provides a model for some level of structure or consistency in answering the research questions and reach new conclusions. However, it is worth noting that a certain level of flexibility is necessary in order for the research to comply with the underlying anthropological/ethnographic strategy or philosophy which this research supports. Therefore the conceptual design plays a central role in framing the research since theory building relies on the general construct of concepts which emerge from the findings. Miles and Huberman (1984) discuss how they identify terms and store them in labelled bins which come from theory, experience and the objectives of the study. The bins contain many discrete

Figure 3. Issues with the manual paper-based grading process

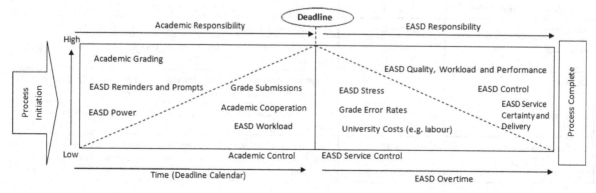

events or behaviours although one may not fully understand at the initial stage how they relate to one another. Organising the bins in a specific layout supported my quest to develop a conceptual framework to clarify their interrelationships. The conceptual framework (Figure 4) explains, either graphically or in narrative forms, the main dimensions which are studied, i.e., the key factors or variables, and the presumed relations among them (Miles and Huberman, 1984). The conceptual model acted as the initial framework which "...*explains, either graphically or in narrative form, the main things to be studied—the key factors, construct or variables, and the presumed relationships among them*" (Miles and Huberman, 1994; p. 19).

Figure 4 depicts the initial framework which supported framing this research study. The framework above illustrates and specifies which concepts were identified in the initial literature review are used to examine service networks and supported the decisions of the initial inductive process. For example, the model considers:

- A service network is comprised of both social and technical factors;
  ○ ANT is an appropriate research lens to explore the socio-technical nature of service network.

- Both the service micro and macro environments should be compared for their socio-technical components and irregularities;
  ○ The questions highlight the need to examine 'how' social-enabled and technical-enabled factors influence the service network.
  ○ The model also suggests the need to explore 'why' the enablers and inhibitors of socio-technical entities influence a service network.
  ○ It is also important to examine how relational structures influence service dynamics.

The framework assumes some relationship as indicated by the arrows which are illustrated for logical reasons, for example, how the social and technical factors influence the service network interactions. These topics are discussed in the following sections.

6.1 From Service Co-Creation to Co-Enactment

The case study revealed much insight on the translation process which was motivated by the need to control service interactions and behaviour to meet specific interests. The concept of inscription came in many forms, for example, job descriptions, grading sheets,

Figure 4. Conceptual model of the research

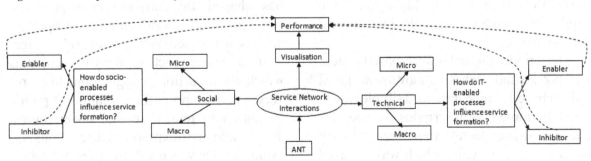

software, and policy and regulation. Thus, it became clear the notion of 'creation' was not as obvious within the public sector, but rather behaviour in a forced action which suggested that actions were enacted to comply with service regulations. The literature suggests that services are co-created which implies a sense of voluntary enrolment in a service network. This is applicable within the private sector since there is a reward to 'co-create'. However, within the public sector, services are stabilised through actions governed by regulations. This suggests that enrolment is not voluntary per se, but rather, actions are mandatory to comply with service governance to stabilise service networks.

The inscription of an online grading system was mobilised to an external actor-network (i.e. outsourced). In doing so, it also served specific self-interests and acted as a representative to understand the university service logic. This led to the formation of some tensions within the university actor-network and was viewed as a sense of betrayal of the problematisation phase which highlighted key requirements and concerns through a pilot test. Betrayal in this context refers to a situation where actors do not abide by the agreements arising from the enrolment of their representatives (Callon, 1986; Sarker et al., 2006). This sense of betrayal was protected through inscriptions of service regulations. In addition, additional senior actors were enrolled to support the fast implementation of the IS which was mobilised through email inscriptions highlighting the need to reassemble a technological-enabled actor-network.

The introduction of the Web-based system was not a single cohesive movement, i.e. it did not have a simple start nor a simple finish. The requirement for inscripting the need for change and discussion of the problematisation dates back for 30 years which was reported

by several actors, yet it was not considered important to entertain this notion. The motivation for change highlights other 'interests'. Thus, the evolution of a service network stems from the evolution of interests which generates the need for innovation and innovative associations and assemblages. For example, when one considers the core ideals of service science, it is not necessarily a historical process but rather an evolving process that shaped the emergence and translation of service values, norms, culture which (re)framed an information system reputation through a hybrid of socio-technical evolutionary developments. Thus, service is an evolving entity. For example, although ANT provided the research lens to examine service networks, the notion of service evolution challenges the concept of actor-network irreversibility. Therefore, the author would argue that in order to have a state of irreversibility, one must conclude that a service network is viewed as a fixed state, i.e. a static service but rather an entity which reacts and is influences by the relational structure which stabalises and alters it defining structure, behaviour, and composition. As this research demonstrates and will tie back to the literature, in order for a service network to function and stabilise, it must remain dynamic through continuous interaction (i.e. associations and assemblages'). This suggests that service actions are complex and evolve with service eco-system factors and demands. In this regard, the research delves into to what has 'shaped' the collective concept of the 'service' actor-network and the 'acceptable' actions which deliver the 'modern' service. Thus, a service reacts to many uncertainties which allows action or regulations to be reversed and supports the evolution of policy formation to reduce service uncertainties. This enables us to explore specific actions and examine what was involved in service provision. The findings also explore how service norms are shaped to support the academic service network considering the challenges they continue to face.

What is of considerable interest throughout the findings lies within the translation process which reconfigures an actor-network and how the service 'architects' enrol all participants. Technological innovation in this context altered a service network and highlighted some concerns regarding 'backward innovation' – depending upon whose interests were best served. The reach or accessibility of virtual methods to port service information may be embraced as an innovative asset, or rejected as an intrusive extension of working practices outside of the working day. For example, email implies that actors are accessible anywhere, at any-time through PC, laptop, or increasingly through mobile phones. Understanding this view, allows us to see services from a new perspective as the notion of service has no clear boundaries. While developing theoretical insights on service networks, the also examined concepts such as bureaucracy, politics, power, influence, and methods of control which is formed into a theory of *servicracy* (i.e. the bureaucracy of IT-enabled service networks).

Insights on human experiences provided true 'explanations' on what constitutes as a service and the actions which deliver a service. Service behaviour may be viewed as the socio-technical interaction of actants within a service environment often dictated by decision makers who act as actor-network representatives. In this sense, it is the actual interaction (social, technical, or socio-technical) which generated the network linkage that we refer to as 'service'. Therefore, it is more natural to view a service as being connected to the social and technical world, rather than separate autonomous entity which the traditional view of services once adopted. The research argues that the 21st century service network

supports self-interest, self-referencing, self-aware, self-validating, socially embedded model of service autonomy, whose interactions or associations recognises not only service strengths and opportunities, but also its limitations, weaknesses, and even threats. The findings support the notion that public service networks are protected by the inscription of self-made service policy and regulation by which quality acts as the 'trust' obligatory passage point which actors must comply with. The findings suggest that one must pay particular attention towards the 'collective' ability to shape a service network. It is through the reflection of the findings, that we can understand how actors generate certain behaviour to facilitate service requirements or how behaviour becomes controlled. To elaborate, actors must therefore have direct relations with their actions rather than be captive of them, which suggest relationships have both a direct and in-direct impact on service environments (Brynjolfsson, 1993). In this regard, there is an obvious need for greater process ownership, rather than offloading process which fosters a defensive service environment (Normann, 2001). Considering the nature of IT-enabled service structures, accountability or responsibility appears to have become 'liquidised' and in some cases 'unbundled' within a virtual environment. It is worth noting that the literature suggests that the bureaucratic model is the best approach to get work complete since "it deals with size, complexity, and the need for accountability" (Thompson and Alvesson, 2005; p. 91). Within the bureaucratic nature of the academic service network, it introduces the notion of virtual bureaucracy which will be discussed further in this chapter (summarised in Table 4).

Table 4 summarises some of the initial findings which steered the research theoritical developments towards the emergence of virtual bureaucracy. While software plays

Table 4. Summary of the emergence of virtual bureaucracy

	Main Theme
1	Need for a socio-technical view of public service networks and IT-enabled innovation
2	Shift from value co-creation to value co-enactment
3	Need to visualise service relational structures
4	Unpredictable nature of public service innovation
5	The role of 'language' in service network evolution
6	Need to explain the material-semiotice relationality of service networks
7	Refocus attention on the need to examine service foundations

an increasingly critical supportive role on service provisions, the concept of service co-creation or co-production appears to be less evident in the implementation of IS in the public service network. For example, co-creation appears to become evident after the Web-based implementation phase and only adopts a role which participates in predefined actions, i.e. less evident in decision-making tasks and policy formation. This creates tensions in various interest groups, arguably reflecting a failure by senior actors to balance the imperatives of service science and the ideals of service management and computing. Tensions also arise from the emphatic capacity of the groups most affected by the service network change. Service management evolves from their transition from obligatory passage points through enrolled agreement of IS short-to-medium term strategy and departmental interests. However, the links or associations which lead into interpersonal, departmental decisions on a micro and macro scale are considerably complex. To protect these decisions, management consume quiet a lot of time shaping regulations through the inscription process and entering discussion to complete the translation process while consequently shaping 'acceptable' service behaviour.

Although much focus nowadays is placed upon innovation and how this alludes to 'progress', the author would argue that fostering collaborative service capacity is just as important to achieving a service environment. It would provide an individualistic sense of participation, worth, ownership, whose requirements must be discussed, debated, and decided upon while removing the unhealthy sense of service disparagement through the inscription of software and authoritarian language. Adding software to the socio-technical layers of *servicracy* increases the complexity of service while removing social factors in the strive towards the more 'modern' technical factors. Actors should be mindful as to "what is progress" (Postman, 1992) which raises substantial ethical and managerial questions which an actor network must be more willing to examine and debate before entering the process of translation. The idea that progress should be theoretically designed to increase human happiness (or self-interests) but appears to be turned into the assumption that pursuing progress is the same as improving actor-network welfare. This research demonstrates how this is not the case and examines the implications of this and the so-called 'general good'. This supports the concept of bureaucracy and its tendency to put the rationality of rules above the rationality of ends, therefore restricting behaviour and remaining open to self-promoting opportunities. Adopting such a view only adds to the increasingly fragmented efforts of public service networks making it extremely difficult to integrate a united service network infrastructure. In addition, this introduces the concept of 'missed opportunity' in public service innovation.

7. CONCLUSION AND FUTURE RESEARCH

Service comprise of socio-technical (human and technological) factors which exchange various resources and competencies. Service networks are used to transfer resources and competencies, yet they remain an underexplored and 'invisible' infrastructure. Service networks become increasingly complex when technology is implemented to execute specific service processes. This ultimately adds to the complexity of a service environment, making it one of the most difficult environments to examine and manage. In addition, although the emerging paradigm of 'Service Science' calls for more theoretical focus on understanding complex service systems, few efforts have surfaced which apply a new theoretical lens on understanding the underlying trajectories of socio-technical dynamics within a service system. Despite the burgeoning number of studies of public service sector information systems, none of these research efforts focus on the dynamic relationship between technology and the impact of the assemblage-like configuration of service relational structures. This empirical research explores an academic service network, with particular attention paid towards a critical end-to end exam grading process. The author employs a single embedded case study to examine the impact of a Web-based system on a traditionally bureaucratic public service system and its transformation from a paper-based system to an automated system. The research adopts ANT as a research lens which offers a rich vocabulary to describe the interplay of socio-technical dynamics which influence the service system reconfiguration with particular attention paid towards the shift from service co-creation to service co-enactment.

Public sector institutions continue to significantly invest in ICT as a solution towards many of their service provision challenges, for example, greater efficiency and quality of services. However, what has come to light is that there is a lack of research on understanding the contributory value or 'success' of technological innovations. This chapter introduces a socio-technical view of public service innovation. The aim of this research is to extend on the notion of bureaucracy which is traditional focused on the politics of office environments. This socio-technical view extends this traditional view to include the politics of service networks, particularly within IT-enables public service innovation. The chapter focuses on how service innovation is exploited to align specific interests through the process of translation and shifts the focus from value co-creation to value co-enactment. In essence, this chapter explains how public service technological innovations act as an agent of bureaucracy which alters the relational dynamics of power, risk, responsibility, and accountability. For demonstrative purposes, this chapter describes a case study which examines IT-enabled service innovation with an academic service environment. While developing theoretical insights on service networks, the author also examined concepts such as bureaucracy, politics, power, influence, and methods of control which is formed into a theory of *servicracy* (i.e. the bureaucracy of IT-enabled service networks). Servicracy stems from the merging of 'service' and '-cracy' the Greek for rule or political power. Servicracy develops our understanding of bureaucracy with service network theoretical developments and examines the implementation process of technology in a public service network which is a significant contribution to service science. As part of the future research work, the theory of servicracy will be tested across a number of research fields to test its validity. [Note: While this chapter provides a brief overview of the case study, a more detailed account of this research is available from the Carroll (2012) resource].

REFERENCES

Akrich, M. (1992). The De-Scription of Technical Objects. In W. Bijker, & J. Law (Eds.), *Shaping Technology, Building Society: Studies in Sociotechnical Change*. Cambridge, MA: MIT Press.

Akrich, M., & Latour, B. (1992). A summary of a convenient vocabulary for the semiotics of human and nonhuman assemblies. In W. Bijker, & J. Law (Eds.), *Shaping Technology, Building Society: Studies in Sociotechnical Change*. Cambridge, MA: MIT Press.

Ali, S., & Green, P. (2007). IT governance mechanisms in public sector organisations: An Australian context. *Journal of Global Information Management*, *15*(4), 41–63. doi:10.4018/jgim.2007100103

Bannister, F. (2001). Dismantling the silos: Extracting new value from IT investments in public administration. *Information Systems Journal*, *11*(1), 65–84. doi:10.1046/j.1365-2575.2001.00094.x

Berg, M., & Timmermans, S. (2000). Orders and their others: On the constitution of universalities in medical work. *Configurations*, *8*, 31–61. doi:10.1353/con.2000.0001

Berry, F. S., Choi, S. O., Goa, W. X., Jang, H., Kwon, M., & Word, J. (2004). Three traditions of network research: What the public management research agenda can learn from other research communities. *Public Administration Review*, *64*(5), 539–552. doi:10.1111/j.1540-6210.2004.00402.x

Bichard, M. (2000). Creativity, leadership and change. *Public Money and Management*, *20*(2), 41–46. doi:10.1111/1467-9302.00210

Bowker, G. C., & Star, S. L. (2000). *Sorting things out – Classification and its consequences*. Cambridge, MA: MIT Press.

Braa, J., Monteiro, E., & Sahay, S. (2004). Networks of action: Sustainable health information systems across developing countries. *Management Information Systems Quarterly*, *28*(3), 337–362.

Brynjolfsson, E. (1993). The productivity paradox of information technology. *Communications of the ACM*, *36*, 66–77. doi:10.1145/163298.163309

Callon, M. (1986a). The sociology of an actor-network: The case of the electric vehicle. In M. Callon, J. Law, & A. Rips (Eds.), *Mapping the dynamics of science and technology* (pp. 19–34). Basingstoke, UK: Macmillan.

Callon, M. (1986b). Some elements of a sociology of translation: Domestication of the scallops and the fishermen of St Brieuc Bay. In J. Law (Ed.), *Power, Action and Belief: A New Sociology of Knowledge*. London: Routledge & Kegan Paul.

Callon, M. (1991). Techno-economic Networks and Irreversibility. In J. Law (Ed.), *A Sociology of Monsters? Essays on Power, Technology and Domination*. London: Routledge.

Callon, M., & Latour, B. (1981). Unscrewing the Big Leviathan: How actors macrostructure reality and how sociologists help them to do so. In K. D. Knorr-Cetina, & A. V. Cicoure (Eds.), *Advances in Social Theory and Methodology: Toward an Integration of Micro- and Macro-Sociologies*. Routledge and Kegan Paul.

Callon, M., & Latour, B. (1992). Don't throw the baby out with the Bath School! A reply to Collins and Yearley. In A. Pickering (Ed.), *Science as Practice and Culture*. Chicago: Chicago University Press.

Callon, M., & Law, J. (1995). Agency and the Hybrid Collectif. *The South Atlantic Quarterly*, *94*, 481–507.

Callon, M., & Law, J. (2005). On Qualculation, Agency and Otherness. *Society and Space*, *23*, 717–733.

Callon, M., & Muniesa, F. (2005). Economic markets as calculative collective devices. *Organization Studies*, *26*(8), 1229–1250. doi:10.1177/0170840605056393

Carr, N. (2003). IT Doesn't Matter. *Harvard Business Review*, *81*(5), 41–49. PMID:12747161

Carr, N. (2004). *Does IT matter? Information technology and the corrosion of competitive advantage*. Harvard Business School Press.

Carroll, N. (2012). *Service science: An empirical study on the socio-technical dynamics of public sector service network innovation*. (PhD Thesis). University of Limerick. Retrieved from https://www.academia.edu/2259268/Service_Science_An_Empirical_Study_on_the_Socio-Technical_Dynamics_of_Public_Sector_Service_Network_Innovation

Carroll, N., Whelan, E., & Richardson, I. (2010). Applying Social Network Analysis to Discover Service Innovation within Agile Service Networks. *Journal of Service Science*, *2*(4), 225–244. doi:10.1287/serv.2.4.225

Carroll, N., Whelan, E., & Richardson, I. (2012). Service Science – An Actor Network Theory Approach. *International Journal of Actor-Network Theory and Technological Innovation*, *4*(3), 51–69. doi:10.4018/jantti.2012070105

Cordella, A., & Iannacci, F. (2010). Information systems in the public sector: The e-Government enactment framework. *The Journal of Strategic Information Systems*, *19*(1), 52–66. doi:10.1016/j.jsis.2010.01.001

Cordella, A., & Shaikh, M. (2006). *From Epistemology to Ontology: Challenging the Constructed Truth of ANT* (Working Paper). London: Department of Information Systems, London School of Economics.

Cussins, C. (1998). Ontological Choreography Agency for Women Patients in an Infertility Clinic. In M. Berg, & A. Mol (Eds.), *Differences in Medicine: Unravelling Practices, Techniques and Bodies*. Durham, NC: Duke University Press.

Darking, M. L., & Whitley, E. A. (2007). Towards an Understanding of Floss: Infrastructures, Materiality and the Digital Business Ecosystem. *Science Studies, 20*(2), 13–33.

Davis, P. (2010). *Development of a Framework for the Assessment of the Role and Impact of Technology on the Public Procurement Process: An Irish Health Sector Study.* (Doctoral Thesis). Dublin Institute of Technology, Dublin, Ireland. Retrieved on 28/10/2011 from website: http://arrow.dit.ie/cgi/viewcontent.cgi?article=1026&context=engdoc&sei-redir=1#search=%22ireland%20must%20examine%20technology%20public%20sector%22

Demirkan, H., Kauffman, R. J., Vayghan, J. A., Fill, H. G., Karagiannis, D., & Maglio, P. P. (2008). Service-oriented technology and management: Perspectives on research and practice for the coming decade. *Perspectives on the Technology of Service Operations. Electronic Commerce Research and Applications,* (4): 356–376. doi:10.1016/j.elerap.2008.07.002

Emery, F. E., & Trist, E. L. (1960). Socio-Technical Systems. In *Management sciences, models and technique.* London: Pergamon.

Feller, J., Finnegan, P., & Nilsson, O. (2008). We have everything to win: Collaboration and open innovation in public administration. In *Proceedings of ICIS.* Retrieved on 19/07/2011 from http://aisel.aisnet.org/icis2008/214

Fitzsimmons, J. A., & Fitzsimmons, M. J. (2004). *Service Management – Operations, Strategy, and Information Technology* (4th ed.). McGraw-Hill.

Friedman, T. L. (2006). *The world is flat.* New York: Penguin Books.

Gad, C., & Jensen, C. B. (2010). On the Consequences of Post-ANT. *Science, Technology & Human Values, 35*(55). PMID:20526462

Gao, P. (2005). Using Actor-Network Theory to Analyse Strategy Formulation. *Information Systems Journal, 15,* 255–275. doi:10.1111/j.1365-2575.2005.00197.x

Hanseth, O., Aanestad, M., & Berg, M. (2004). Guest editors' introduction: Actor-Network Theory and information systems: What's so special? *Information Technology & People, 17*(2), 116–123. doi:10.1108/09593840410542466

Hanseth, O., & Monteiro, E. (1998). *Understanding information infrastructure.* Retrieved on 19th July 2011 from http://heim.ifi.uio.no/oleha/Publications/bok.pdf

Hassard, J., Law, J., & Lee, N. (1999). Actor network theory and managerialism. *Organization, 6*(3), 387–390. doi:10.1177/135050849963001

IfM and IBM. (2007). *IfM and IBM, Succeeding through Service Innovation: A Discussion Paper.* Cambridge, UK: University of Cambridge Institute for Manufacturing.

Jansen, A. (2005). Assessing E-government progress – Why and what. In *Proceeding of Norsk konferanse for organisasjoners bruk av IT (Nokobit 2005).* Bergen, Norway: Academic Press.

Kling, R. (1991). Computerization and social transformations. *Science, Technology & Human Values, 16*(3), 342–367. doi:10.1177/016224399101600304

Knox, H., O'Doherty, D., Vurdubakis, T., & Westrup, C. (2007). Transformative capacity, information technology, and the making of business 'experts'. *The Sociological Review, 55*(1), 22–41. doi:10.1111/j.1467-954X.2007.00679.x

Latour, B. (1987). *Science in Action.* Cambridge, MA: Harvard University Press.

Latour, B. (1991). Technology is Society Made Durable. In J. Law (Ed.), *A Sociology of Monsters? Essays on Power, Technology and Domination*. London: Routledge.

Latour, B. (1992). Where are the missing masses? The sociology of a few mundane artifacts. In W. E. Bijker, & J. Law (Eds.), *Shaping technology/building society: Studies in sociotechnical change*. Cambridge, MA: MIT Press.

Latour, B. (1993). *We have never been modern*. Brighton, UK: Harvester Wheatsheaf.

Latour, B. (1996). *Aramis, or the love of technology*. Cambridge, MA: MIT Press.

Latour, B. (2005). *Reassembling the social: An introduction to Actor-Network-Theory*. new york: oxford university press.

Law, J. (1986). On methods of long-distance control: Vessels, navigation and the Portuguese route to India. In J. Law (Ed.), *Power, Action and Belief* (pp. 196–233). Routledge & Kegan Paul.

Law, J. (1990). Technology and heterogeneous engineering: The case of Portuguese expansion. In *The social construction of technological systems: New directions in the sociology and history of technology*. Cambridge, MA: MIT Press.

Law, J. (1999). After ANT: Topology, naming and complexity. In J. Law, & J. Hassard (Eds.), *Actor Network Theory and After*. Oxford, UK: Blackwell.

Law, J. (2007). *Actor-Network Theory and Material Semiotics*. Retrieved from http://www.heterogeneities.net/publications/Law2007ANTandMaterialSemiotics.pdf

Law, J. (2008). On Sociology and STS. *The Sociological Review*, *56*(4), 623–649. doi:10.1111/j.1467-954X.2008.00808.x

Levy, M., & Powell, P. (2005). *Strategies for Growth in SMEs: The Role of Information and Information Systems*. Amsterdam: Elsevier Butterworth-Heinemann.

Linderoth, H. C. J., & Pellegrino, G. (2005). Frames and Inscriptions: Tracing a Way to Understand IT-Dependent Change Projects. *International Journal of Project Management*, *23*(5), 415–420. doi:10.1016/j.ijproman.2005.01.005

Lynn, L. E. (2005). Public management: A concise history of the field. In E. Ferlie, L. E. Lynn, & C. Pollitt (Eds.), *The Oxford handbook of public management* (pp. 27–50). Oxford, UK: Oxford University Press.

McBride, N. (2000). *Using actor-network theory to predict the organizational success of a communications network*. Department of Information Systems, De Montfort University. Retrieved from http://www.cse.dmu.ac.uk/~nkm/WTCPAP.html

McIvor, R., McHugh, M., & Cadden, C. (2002). Internet technologies: supporting transparency in the public sector. *International Journal of Public Sector Management*, *15*(3), 170–187. doi:10.1108/09513550210423352

Miles, M. B., & Huberman, A. M. (1984). *Qualitative Data Analysis*. Newbury Park, CA: Sage.

Mitchell, T. (2002). *Rule of Experts: Egypt, Techno-politics Modernity*. Berkeley, CA: University of California Press.

Mitev, N. (2009). In and out of Actor-Network Theory: a necessary but insufficient journey. *Information Technology & People*, 22(1), 9–25. doi:10.1108/09593840910937463

Monteiro, E., & Hanseth, O. (1995). Social shaping of information infrastructure: On being specific about the technology. In W. Orlikowski, G. Walsham, M. R. Jones, & J. I. DeGross (Eds.), *Information technology and changes in organisational work* (pp. 325–343). Chapman & Hall.

Mutch, A. (2002). Actors and networks or agents and structures: Towards a realist view of information systems. *Organization*, 9(3), 477–496. doi:10.1177/135050840293013

Navarra, D. D., & Cornford, T. (2009). Globalization, networks, and governance: Researching global ICT programs. *Government Information Quarterly*, 26(1), 35–41. doi:10.1016/j.giq.2008.08.003

Ngosi, T., Helfert, M., Carcary, M., & Whelan, E. (2011). *Design science and Actor-Network Theory Nexus: A perspective on content development of a critical process for enterprise architecture management.* Paper presented at the 13th International Conference on Enterprise Information Systems (ICES). Beijing, China.

Normann, R. (2001). *Reframing business: When the map changes the landscape.* Chichester, UK: Wiley.

O'Toole, L. J. (1997a). Implementing public innovations in network settings. *Administration & Society*, 29(2), 115–138. doi:10.1177/009539979702900201

O'Toole, L. J. (1997b). The implications for democracy in a networked bureaucratic world. *Journal of Public Administration: Research and Theory*, 7(3), 443–459. doi:10.1093/oxfordjournals.jpart.a024358

O'Toole, L. J. (1997c). Treating networks seriously: Practical and research-based agendas in public administration. *Public Administration Review*, 57(1), 45–52. doi:10.2307/976691

Orlikowski, W. J. (1992). The duality of technology: Rethinking the concept of technology in organizations. *Organization Science*, 3(3), 398–427. doi:10.1287/orsc.3.3.398

Orlikowski, W. J., & Baroudi, J. (1991). Studying information technology in organizations: Research approaches and assumptions. *Information Systems Research*, 2(1), 1–28. doi:10.1287/isre.2.1.1

Pedersen, M. K., Fountain, J., & Loukis, E. (2006). Preface to the focus theme section: electronic markets and e-government. *Electronic Markets*, 16(4), 263–273. doi:10.1080/10196780600999601

Postman, N. (1992). *Technopoly: The surrender of culture to technology.* New York: Vintage Press.

Ramiller, N. C., & Wagner, E. L. (2009). The element of surprise: Appreciating the unexpected in (and through) actor-networks. *Information Technology & People*, 22(1), 36–50. doi:10.1108/09593840910937481

Rethemeyer, R. K. (2005). Conceptualizing and measuring collaborative networks. *Public Administration Review*, 65(1), 117–121. doi:10.1111/j.1540-6210.2005.00436.x

Rethemeyer, R. K. (2009). Making sense of collaboration and governance issues and challenges. *Public Performance & Management Review*, 32(4), 565–573. doi:10.2753/PMR1530-9576320405

Rogers, E. M. (1962). *Diffusion of Innovations.* New York: The Free Press.

Rogers, E. M. (1976). New product adoption and diffusion. *The Journal of Consumer Research*, 290–301. doi:10.1086/208642

Rogers, E. M. (2010). *Diffusion of innovations*. New York: Simon and Schuster.

Rogers, E. M., & Shoemaker, F. F. (1971). *Communication of Innovations: A Cross-Cultural Approach*. Academic Press.

Rusaw, C. A. (2007). Changing public organizations: Four approaches. *International Journal of Public Administration*, 30, 347–361. doi:10.1080/01900690601117853

Sarker, S., Sarker, S., & Sidorova, A. (2006). Understanding Business Process Change Failure: An Actor-Network Perspective. *Journal of Management Information Systems*, 23(1), 51–86. doi:10.2753/MIS0742-1222230102

Spohrer, J., & Maglio, P. P. (2008). The emergence of service science: Toward systematic service innovations to accelerate co-creation of value. *Production and Operations Management*, 17(3), 238–246. doi:10.3401/poms.1080.0027

Spohrer, J., Maglio, P. P., Bailey, J., & Gruhl, D. (2007). Steps Toward a Science of Service Systems. *IEEE Computer*, 40(1), 71–77. doi:10.1109/MC.2007.33

Star, S. L., & Griesemer, J. R. (1989). Institutional ecology, 'translations' and boundary objects: Amateurs and professionals in Berkeley's Museum of Vertebrate Zoology, 1907-39. *Social Studies of Science*, 19(3), 387–420. doi:10.1177/030631289019003001

Star, S. L., & Ruhleder, K. (1996). Steps Toward an Ecology of Infrastructure: Design and Access for Large Information Spaces. *Information Systems Research*, 7, 111–133. doi:10.1287/isre.7.1.111

Tapscott, D., & Caston, A. (1993). *Paradigm shift: The new promise of information technology*. New York: McGraw-Hill, Inc.

Tatnall, A. (2005). Actor-Network Theory in Information Systems. In *Encyclopaedia of Information Science and Technology*. Hershey, PA: Idea Group Reference.

Tatnall, A., & Burgess, S. (2006). Innovation Translation and E-Commerce in SMEs. In *Encyclopaedia of Information Science and Technology*. Hershey, PA: Idea Group Reference.

Tatnall, A., & Gilding, A. (1999). Actor-Network Theory and Information Systems Research. In *Proceedings of the 10th Australasian Conference on Information Systems*. Wellington, New Zealand: Academic Press.

Thompson, P., & Alvesson, M. (2005). Bureaucracy at Work: Misunderstandings and Mixed Blessings. In P. Du Gay (Ed.), *The Values of Bureaucracy*. Oxford University Press.

Van den Hooff, B., & de Winter, M. (2011). Us and them: A social capital perspective on the relationship between business and IT departments. *European Journal of Information Systems*, 20(3), 255–266. doi:10.1057/ejis.2011.4

Vargo, S. L., & Lusch, R. F. (2008). From Goods to Service(s), Divergences and Convergences of Logics. *Industrial Marketing Management*, 37, 254–259. doi:10.1016/j.indmarman.2007.07.004

Walsham, G. (1997). Actor-Network Theory and IS research: Current status and future prospects. In *Information systems and qualitative research*. London: Chapman and Hall. doi:10.1007/978-0-387-35309-8_23

Walsham, G., & Sahay, S. (1999). GIS for District-Level Administration in India: Problems and Opportunities. *Management Information Systems Quarterly*, *23*(1), 39–66. doi:10.2307/249409

Weber, M. (1919). Bureaucracy. In *From Max Weber: Essays in sociology*. London: Routledge.

Weill, P., Subramani, M., & Broadbent, M. (2002). Building IT infrastructure for strategic agility. *Sloan Management Review*, *44*(1), 57–65.

Willcocks, L. (1994). Managing information system in the UK public administration: issues and prospects. *Public Administration*, *72*, 13–32. doi:10.1111/j.1467-9299.1994.tb00997.x

Winner, L. (1977). *Autonomous Technology*. Cambridge, MA: MIT Press.

Winner, L. (1993). Upon opening the black box and finding it empty: Social constructivism and the philosophy of technology. *Science, Technology & Human Values*, *18*(3), 362–378. doi:10.1177/016224399301800306

Woolgar, S. (1991). The turn to technology in social studies of science. *Science, Technology & Human Values*, *16*(1), 20–50. doi:10.1177/016224399101600102

KEY TERMS AND DEFINITIONS

Actor-Network Theory: Provides an approach to understand how both social action shapes technology and how technological innovations shape social action. ANT acts as a toolkit to explore how human and non-human actors interact with one another to make sense of their world.

Conceptual Model: Explains (graphically or in narrative form) the main focus of study, i.e. the key factors, construct or variables, and the presumed relationships among them.

Service Science: Acts as an interdisciplinary umbrella which incorporates widely diverse disciplines to construct, manage, analyse and evolve service systems. The explosive growth in Service Science is motivated by the need to develop more systematic, analytical, and overarching approaches to understanding the complexity of services.

Service System: comprises of socio-technical systems which stabilise a service network through the exchange of resources and competencies which generate value.

Service: Comprise of socio-technical (human and technological) factors which exchange various resources and competencies.

Servicracy: Extends the notion of bureaucracy of IT-enabled service networks and develops our understanding of how bureaucracy shapes service networks during the implementation process of technology in a public service network.

Chapter 8
E-Health in Australia and Germany

Manuel Zwicker
RMIT University, Australia

Juergen Seitz
DHBW Heidenheim, Germany

Nilmini Wickramasinghe
Epworth Healthcare, Australia & RMIT University, Australia

ABSTRACT

This chapter focuses on two specific e-health solutions, the PCEHR in Australia and the German EHC. National e-health solutions are being developed by most if not all OECD countries, but few studies compare and contrast these solutions to uncover the true benefits and critical success criteria. The chapter provides an assessment of these two solutions, the possibility for any lessons learnt with regard to designing and implementing successful and appropriate e-health solutions, as well as understanding the major barriers and facilitators that must be addressed. Finally, ANT is used to provide a rich lens to investigate the key issues in these respective e-health solutions.

BACKGROUND

Healthcare industries continue to be at the forefront of agendas globally. Between 1970 and 1997 the average percentage of Gross Domestic Product (GDP) on healthcare by members of the Organization for Economic Cooperation and Development (OECD) countries rose from about 5% to roughly 8% (Huber, 1999). Since 2000, total spending on healthcare in these countries has been rising faster than economic growth, which resulted in an average ratio of health spending to GDP of 9.0% in 2008. Challenges including the technological change, longer life expectancy and population ageing will serve to push health spending up further in the future. Hence, this growing health spending creates a significant cost pressure for several countries (OECD, 2010a).

Reducing these expenditures as well as offering effective and efficient quality healthcare

DOI: 10.4018/978-1-4666-6126-4.ch008

treatment is a priority worldwide. Technology and automation in general have the potential to reduce these costs (Ghani et al., 2010). Moreover, the use of information and communication technologies (ICT) in e-health solutions in particular appears to be the key to respond to these challenges (Wickramasinghe & Schaffer, 2010).

In addition, several countries are changing their thinking about healthcare, because they know that the current situation is no longer feasible. Therefore, we are witnessing new healthcare reforms. Based on this fact, countries like Australia, Finland, Germany, UK and U.S., to name but a few countries, have started to change their healthcare system because they have recognized that with the use of information and communication technologies (ICT) in general and specifically e-health, healthcare costs can be reduced, while the quality of healthcare delivery can be improved (Wickramasinghe & Schaffer, 2010). This means that e-health is becoming an essential part of modern healthcare delivery, which in turn means it is now an essential one to fully understand.

In their enthusiasm to develop e-health solutions, it appears to us that countries are focusing efforts to only address the internal issues. However, healthcare delivery is a global phenomenon and in an age where global business operations are prevalent, it is essential for e-health to also have a global, network centric perspective, including being able to support healthcare information exchange between different countries. This is the central thesis behind the doctrine of network centric healthcare (Wickramasinghe et al., 2007). Thus, a current problem is that in an attempt to address escalating healthcare costs countries are turning to developing e-health solutions, but because these solutions are not being designed with a global perspective rather than provide effective and efficient quality healthcare solutions they are likely to exacerbate the current situation and create more costly, poor quality healthcare solutions. One way to develop an e-health solution, which has a global perspective, is to investigate the possibility of transferring an e-health solution across countries. In this way, it will be possible to support healthcare information exchange between countries. This is essentially the strategy that the banking industry adopted. Hence, this paper serves to explore the research question "How can we transfer e-health solutions?"

In looking at the possibility of transferring different e-health solutions it is also necessary to have a rich theoretical lens to facilitate in depth analysis of all critical issues as well as their interactions. We therefore adopt ANT (Actor Network Theory) as such a lens.

LITERATURE REVIEW

In order to investigate the proposed research question it is necessary first to examine key issues of NCHO (network centric healthcare operations). In addition, it is important to discuss the different healthcare systems and to define e-health. Finally, ANT is briefly presented.

Network Centric Healthcare

The doctrine of network centric healthcare operations (NCHO) has been defined as "unhindered networking operations within and among the physical, information, and cognitive domains that govern all activities conducted in healthcare space based on free, multidirectional flow and exchange of information without regard to the involved platforms or platform-systems, and utilizing all available means of ICCTs to facilitate such operations" (von Lubitz & Wickramasinghe,

2006, p. 334; Jamshidi, 2009). The abbreviation ICCT stands for information, computer and communication technologies.

The confluence of three domains is critical to the success of network centric healthcare operations (Wickramasinghe & Schaffer, 2010):

1. **Information Domain:** Contains all elements, which are required for generation, storage, dissemination/sharing, manipulation of information and in addition its transformation and dissemination/sharing as knowledge in all its forms.

2. **Physical Domain:** Encompasses the structure of the entire environment healthcare operations intend to influence indirectly or directly – political environment, fiscal operations, patient and personnel education, etc.

3. **Cognitive Domain:** Relates to all human factors, which affect operations – education, training, experience, motivation and intuition of individuals involved in the relevant activities.

Based on this information, it is important to look at a Worldwide Healthcare Information Grid (WHIG), which is shown in Figure 1. This WHIG allows a full and hindrance-free sharing of information among the individual domains and their constituents as well as among constituents across the domains. The WHIG must consist of an interconnected matrix of ICT systems and capabilities (communication platforms, data collection, etc.), associated processes like knowledge and information storage, people (healthcare providers, etc.) and agencies (governmental and non-governmental organizations) at a local, national or international level, so as to achieve such a function (Wickramasinghe et al., 2007).

This complex network structure is required to support NCHO and is represented through the interconnected grids. The ellipses all over the network are entry points. The cubes represent nodes, where intelligence capabilities further refine and process the data and information, which reside in the network.

As the literature mentions, it is essential to position the healthcare organization and the

Figure 1. Network centric information/knowledge grid system (adapted from Wickramasinghe et al., 2007)

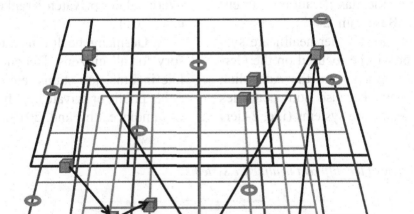

e-health initiative in view of external pressures and present trends. In order to achieving this, it is fundamental to take a network centric perspective and to design an e-health solution to support network centric healthcare operations. The solution here is not the technology rather the macro understanding of the need to design suitable solutions (Wickramasinghe & Schaffer, 2010).

Different Healthcare Systems

Figure 2 provides a schematic which gives an overview of the major types of healthcare systems. On the one side we have a healthcare system that is primarily predicated on private healthcare (e. g. the United States healthcare system) while on the other side, a healthcare system that is mainly based on a public healthcare system (e. g. the United Kingdom healthcare system), and then finally in the middle a 2-tier healthcare system is shown.

This study defines a 2-tier healthcare system as a system, where, in addition to a guaranteed public health insurance, a private healthcare system exists, which can have a substitutive or complementary function, and different factors like income and job status which influence the classification of an enrollee (ArticlesBase.com, 2009).

As examples for a 2-tier healthcare system, this paper will be focused on the German healthcare system and the Australian healthcare system, because both countries have similar healthcare systems (i. e. 2-tier)

and both are focused on e-health solutions to improve their respective healthcare delivery systems.

The German Healthcare System

In 2008, Germany had a total expenditure on health (% GDP) of 10.5%, which was 1.5% higher than the average ratio of the OECD countries. Concurrently, Germany's total expenditure on health per capita (US$) was $3,737, whereas the OECD countries spent on average $3,060 per capita (OECD, 2010b).

The healthcare actors in Germany are divided into enrollees, service providers (medical doctors, pharmacists, hospitals) and cost units (health insurance companies). Germany has around 82.14 million inhabitants, where around 70.23 million people have public health insurance and around 8.62 million people use a private health insurance. Furthermore, Germany has 319,697 medical doctors, 2,087 hospitals and 21,602 pharmacies, where 48,030 pharmacists work (BMG, 2009) In addition, the 200 health insurance companies are divided into 153 public health insurance companies (GKV-Spitzenverband, 2011) and 47 private health insurance companies (Verband der privaten Krankenversicherung e. V., 2011).

In Germany, health insurance is compulsory for all citizens. Depending on factors like income, job status, etc. enrollees have either public or private health insurance (The Commonwealth Fund, 2010).

Figure 2. Overview of the different healthcare systems

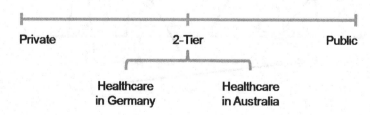

Public health insurance companies are autonomous, non-profit oriented and non-governmental bodies, which are regulated by law. This system is financed by compulsory premiums charged as a percentage of the gross wages up to a defined threshold. Based on the facts of July 2009, the employee contributes 7.9% of their gross wage, while the employer adds 7.0% on top of the gross wage, which is in total a premium of 14.9% of each individual's gross wage. In addition, dependents like kids and spouses without income are also included (The Commonwealth Fund, 2010).

Private health insurance is taking mainly a substitutive function in Germany. This private health insurance scheme is covering two groups, who are mostly exempt from public health insurance (The Commonwealth Fund, 2010):

- Civil servants, who get partly a refund of their healthcare costs by their employers as well as the self-employed people.
- People with high incomes, who decide to opt out of the public health insurance scheme.

The Australian Healthcare System

In 2007 (last data available), Australia had a total expenditure on health (% GDP) of 8.5%. In addition, Australia's total expenditure on health per capita (US$) was $3,353 (OECD, 2010c).

The healthcare system in Australia consists of both a public and private component. The key feature is public health insurance under Medicare, which is funded by taxation. Here enrollees have the possibility to use subsidized medical services and pharmaceuticals as well as free of charge hospital treatment according to their status as a public health enrollee. Besides Medicare, Australian patients have the possibility to use, in addition, a private health insurance, which gives for example patients access to dental services and hospital treatment as a private patient (The Commonwealth Fund, 2010).

Australia's healthcare system is financed by (The Commonwealth Fund, 2010):

- **National Health Insurance:** Medicare is a national health insurance, which is compulsory and administered by the government; Medicare is funded in large part by general revenues and partly by a 1.5% levy based on taxable income; depending on the income of an individual or family the amount of levy can change; in 2007 till 2008, the governments funded 69% of all health expenditures, while 43% came from the Australian government and 26% from State or Territory governments.
- **Private Insurance:** 7.6% of Australian's total health spending can be contributed to the private health insurance; since 1999, the Australian government has supported private health insurance by giving enrollees a rebate of 30% of private health insurance premiums; the government's rebate increases for elderly; in mid of 2009, 44.6% of Australia's population had a private hospital insurance; private health insurance is in Australia community-rated, which means that "everyone pays the same premium for their health insurance" (Australian Government, 2011); in Australia there exists non-profit as well as for-profit insurers.
- **Out-of-Pocket Spending:** In 2007 till 2008, 16.8% of Australia's health expenditures were out-of-pocket spending; examples are dental services and copayments on medical fees.

E-Health

E-health is defined by the WHO as "a new term used to describe the combined use of electronic communication and information technology in the health sector" or as "the use, in the health sector, of digital data-transmitted, stored and retrieved electronically-for clinical, educational and administrative purposes, both at the local site and at a distance" (WHO, 2005).

For the purposes of this paper the term "e-health" will be used as defined by WHO.

As the literature defines, e-health is the main engine for three significant changes within the healthcare environment (Maheu et al., 2001):

- Patients are becoming better informed.
- Patients are becoming more active in their healthcare.
- Healthcare is becoming more efficient.

E-Health in Germany

Currently, the German e-health sector is especially focused on a new e-health card concept, which is at this time in the test stage and should be established within the next years. The e-health card (eHC) will change healthcare delivery in Germany, because this concept allows several new functions and will lead to a more connected healthcare system. The following paragraphs will describe the four implementation steps of the German eHC and will highlight the benefits of these new functions.

Generally, the functions of the e-health card are divided into two category groups. Firstly, there is the area of administrative functions, which are compulsory for all card owners. Secondly, there is the area of medical functions, which are optional for the card holders. Both category groups consist of two steps (Barmer GEK Krankenkasse, 2011).

The first implementation step of the German eHC begins with the implementation of the administrative functions. Within this

first implementation step information about the insurance agreement and the necessity of additional payments will be stored. The data will be stored on the e-health card respectively on a server and can be updated for example during every consultation of a doctor through an online process. Furthermore, this first implementation step includes data about the care provider and personal information about the enrollee. In addition, each enrollee will become a lifelong valid insurance number, which will be printed on the e-health card. Moreover, each eHC will be equipped with a personal photo of the enrollee. Through this lifelong valid insurance number and the personal photo the eHC is well protected against misusage (gematik, 2011a).

Furthermore, this first step of the administrative area includes an insurance coverage for enrollees within the European Union. But this issue is based on the requirement that the appropriate countries have a social agreement among each other. The back side of the e-health card is perfect as identity card for this European Health Insurance Card (EHIC) (European Commission, 2011).

The second step of the administrative functions introduces electronic prescriptions (e-prescription), which is also compulsory for all involved healthcare actors in Germany. Based on this electronic solution, it is possible to remove the approximately 600-700 million paper-based prescriptions and to process these transactions electronically. Thereby, the process looks like this: At first, the doctor has to look on the insurance data of the patient. Therefore, the doctor needs the eHC of the patient, his electronic health professional card (HPC) and a special reading devise, which is connected with his computer. Both cards act as a key. When the patient needs some medicine, the doctor can store the data of the decreed medicine in electronically form (e-prescription) on the e-health card or on an e-prescription server. The necessary signature of the doctor will be generated electronically with the aid of the electronic health professional card (HPC) (Die gesetzlichen Krankenkassen, 2007).

When a patient wants to redeem the electronic prescription in a pharmacy then the procedure goes in the reverse direction. Firstly, the pharmacist needs his HPC and also the eHC of the enrollee. Secondly, the pharmacist retrieves the e-prescription from the eHC respectively from the server and checks the validity of the doctor's signature. Thirdly, the pharmacist or another authorized staff member will present the medicine and explain their risks to the enrollee. After the presentation of the medicine through a pharmacist or another authorized staff member, the electronic prescription will become invalid (Die gesetzlichen Krankenkassen, 2007).

The medical functions in step 3 introduce the voluntary part for the enrollees and their e-health card. This means that the enrollees can decide for their own, if they want to use these additional functions or not. The main focus here is on the storage of personal health data about the enrollees. Examples are the documentation of medicine, which an enrollee has used or the storage of an emergency data record of the enrollee in case of emergency. Due to the medicine documentation it is possible to avoid interdependencies between the individual drugs (gematik, 2011b). Furthermore, the emergency data should help the emergency doctor to medicate purposefully and effectively, because patient's allergies and chronic illnesses should be taken into consideration during the therapy (ZTG, 2010).

The fourth step of the eHC implementation includes, among other things, the electronic health record (EHR). With this EHR it is possible to have access to the entire patient's data. For example, the EHR can include information about medications, past medical history, immunizations, laboratory data, progress notes and radiology reports. Thereby it is not important, if the data are stored at one place or at different places, because the patient's data can be accepted, processed and attended centrally (GVG, 2004).

All these functions will lead to a more connected healthcare system in Germany, which is the goal of network centric healthcare. This means that the eHC concept is not just an innovative solution in conjunction with several new functions, it will also result in a better information exchange between all healthcare actors.

E-Health in Australia

The problem with Australia's healthcare system was, and still is, in comparison to its other economy sectors that the usage of ICT is very low (Pearce & Haikerwal, 2010). This was also evidenced through a comparison of health information technologies between developed countries, where Australia's system was ranked in the middle, because its use of modern electronic solutions for communication and information exchange within the health systems was low (Jha et al., 2008).

In Australia, the National E-Health Transition Authority Limited (NEHTA), which was founded by the Australian Commonwealth, State and Territory governments in July 2005, plays an important role in order to achieve a higher level of electronically collecting and securely exchanging healthcare information (NEHTA, 2011a).

NEHTA's strategy includes four priorities (NEHTA, 2011b):

- Urgently develop the essential foundations required to enable e-health.
- Coordinate the progression of the priority e-health solutions and processes.
- Accelerate the adoption of e-health.
- Lead the progression of e-health in Australia.

Based on NEHTA's strategy, one service, which is currently planned, is the personally controlled electronic health record (PCEHR). This PCEHR is an individual electronic health record of a patient, which summarize key medical information of a patient from different systems centrally. Only the patient or authorized healthcare providers will have access to the PCEHR. The information of a PCEHR will support healthcare providers in their decisions and treatments. In addition, it will be also possible for a patient to add own information to his PCEHR (NEHTA, 2011c). The benefits of this solution are for example a higher level of quality of care, time savings through a faster availability of health information and a better communication between the healthcare actors (NEHTA, 2011d).

In order to realize this PCEHR, the Australian government allocated around $466 million over 2 years in the 2010 budget (Pearce & Haikerwal, 2010). In addition, a countrywide quantitative survey has shown that overall Australians will support the idea of an individual electronic health record, but they have concerns about data security and privacy. 80% want that the participation is voluntary and most of the Australians want that the Federal government will manage this implementation (NEHTA, 2008). In this study the terms individual electronic health record and personally controlled electronic health record are used as synonyms.

NEHTA is also working on e-communications in practice (what they also call "E-Health Solutions"). Thereby, this plan includes (NEHTA, 2011e):

- **E-Pathology:** Seamless pathology results.
- **E-Discharge Summaries:** Instant and correct patient records.
- **E-Referrals:** Streamlining the handover of care.
- **E-Medication Management:** Reducing risks included.

In addition, Australia is also planning to displace the paper-based prescriptions through an electronic prescription by developing a national electronic prescription system (NEHTA, 2011f).

But for using e-health services, it is essential to have a Healthcare Identifiers (HI) Service. The HI Service includes identifiers, which is a unique 16-digit reference number, for consumers, healthcare providers and healthcare organization (The Royal Australian College of General Practitioners, 2010):

- Individual Healthcare Identifier (IHI).
- Healthcare Provider Identifier - Individual (HPI-I).
- Healthcare Provider Identifier - Organisation (HPI-O).

These identifiers make it possible to identify a healthcare individual at the point of care uniquely and consistently and associate healthcare information with an individual within a healthcare context (NEHTA, 2010).

In spite of these future e-health plans, which will surely help to improve healthcare delivery in Australia, Australia could be still confronted with the barrier of different e-health services/

solutions for different states, healthcare actors or healthcare domains. For example, the PCEHR will lead to a countrywide solution; however this solution will exist besides the e-prescription infrastructure, which means that for two different healthcare domains there will exist two different e-health infrastructure solutions. This situation will still exist as long as the different healthcare actors and governments will not be focused on one national e-health solution, which includes all states, healthcare actors and all necessary functions in one infrastructure. NEHTA is a first step to guarantee Australian standards and among other things a national PCEHR service, but this currently does not avoid that different e-health services from healthcare providers are planned and will be developed additionally.

KEY CONCEPTS OF ACTOR NETWORK THEORY

Succinctly, ANT embraces the idea of an organizational identity and assumes that organizations, much like humans, possess and exhibit specific traits (Latour, 1996). ANT was developed by British sociologist, John Law and two French social sciences and technology scholars Bruno Latour and Michel Callon (Latour, 1986, 1996, 1999; Law, 1987, 1991; Wickramasinghe & Bali, 2009). It is an interdisciplinary approach that tries to facilitate an understanding of the role of technology in specific settings, including how technology might facilitate, mediate or even negatively impact organizational activities and tasks performed. Hence, ANT is a material-semiotic approach for describing the ordering of scientific, technological, social, and organizational processes or events. Hence, we believe ANT is an appropriate analysis tool in this context.

Central to ANT and relevant for this specific context are the six key concepts as follows:

1. Actor/Actant.
2. Heterogeneous Network.
3. Tokens/Quasi Objects.
4. Punctualization.
5. Obligatory Passage Point (OPP).
6. Irreversibility.

After the presentation of the SWOT analyses of e-health in Germany and Australia we shall return to a discussion and application of these six key concepts in the context of e-health.

SWOT ANALYSES OF E-HEALTH IN GERMANY AND AUSTRALIA

An effective way to compare and contrast the German e-health solution and the Australian e-health solution is a SWOT analysis. This SWOT analysis was developed alter gathering multispectral data from various sources including documents and interview data. A summary of the results of these analyses is provided in Table 1.

From Table 1 we can see that the overall difference between the German and Australian e-health solutions is centered around the different connections between the different stakeholders within the respective countries. While Germany is designing a national e-health solution, which results in a better information exchange, cost and time savings and higher quality of care, Australia is to date focused on a fragmented e-health solution, which is orthogonal to the doctrine of network centric healthcare and can therefore not guarantee smooth seamless information exchange between the different healthcare

Table 1. SWOT analyses about e-health in Germany and Australia

	E-Health in Germany	E-Health in Australia
Strengths	• The German eHC connects all healthcare stakeholders nationally ➡ better information exchange, cost and time savings and higher quality of care (mhplus Krankenkasse, 2011; vdek, 2009).	• First step for the fragmented structure to become a more coordinated system ➡ in the context of its possibilities this approach should lead to a better information exchange, cost and time savings and higher quality of care.
Weaknesses	• High implementation costs of approximately 1.7 billion Euros and 150 million Euros running costs a year (Scheer, 2009). • Time schedule: the implementation of the eHC does not meet the deadline for several times ➡ several decision makers.	• Costs. • Still underlying fragmented system ➡ healthcare information exchange is difficult respectively not possible between different stakeholders. • Time schedule uncertain to attain.
Opportunities	• Contribution to life savings. • Potential to reduce health expenditures. • Extending scope of healthcare delivery. • Possibility for health information exchange with other countries. • First step for a healthcare information exchange between different stakeholders worldwide and is consistent with NCHO.	• Australia's e-health solution is a first step to develop a national e-health solution ➡ doctrine of NCHO. • Possibility for health information exchange with other countries. • First step for a healthcare information exchange between different stakeholders worldwide and is consistent with NCHO.
Threats	• Data security and data protection. • Information overload of a patient ➡ doctor can lose overview and lose time. • Doctors need more time for documentation. • Acceptance by Germans is questionable. • New law changes from the government. • A big range of decision makers.	• Australia's e-health solution has still a complex structure because of different systems ➡ therefore higher risk for succeed superior healthcare delivery. • Australians are concerned of costs. • Several decision makers. • New laws from the government means maybe changes in the e-health solution. • Data security and data protection. • Information overload of a patient ➡ doctor can lose overview and lose time. • Doctors need more time for documentation. • Acceptance by Australians is questionable also steep learning curve for patients.

actors in Australia. Finally, it is important to note that given the similarity between the German and Australian healthcare systems; namely, that they are both 2-tier, it is unsurprising that the respective SWOT analyses in Table 1 also share many similarities on all considerations of strengths, weaknesses, opportunities and threats.

DISCUSSION

This paper set out to explore the research question "How can different e-health solutions be transferred to other countries?"

From the SWOT analyses (Table 1) we can see both similarities and differences between the German and Australian e-health solutions. For example, both solutions try to establish national e-health solutions to cut costs and provide high quality of care. However, the German e-health solution to date appears to integrate all the web of healthcare players, while the Australian e-health solution is struggling to bring together a fragmented healthcare delivery system.

This preliminary analysis provides directional data and initial support to the merits of the transferability of e-health solutions; i. e. transferring the German e-health solution to the Australian context. Since, given there are similarities it should be possible, at least in theory, to transfer such a solution. Clearly, in the transferring of e-health solutions, one must also be mindful of local structures, policies and protocols so that the transferred solution can be successful. Furthermore, we note that much more research is needed in this area; however, based on our initial findings we can also make the following recommendations:

- For Australia, NEHTA should try to bring the different stakeholders together.

- For Germany, the government together with the key stakeholders and organizations need to work together to ensure high acceptance by the citizens.

- There are potential lessons to be learnt from other industries such as banking that can be applied into the healthcare context.

- International bodies such as the United Nations and/or World Health Organization should take the lead in developing global protocols and policies regarding e-health solutions; in this way, increasing the likelihood of appropriate network centric healthcare solutions resulting.

- Given the importance of e-health today as well as its growing importance in the future, it may be prudent for the development of a new international organization that solely focuses on e-health considerations, most importantly the transferability of e-health solutions.

While the SWOT analyses presented in Table 1 serve to provide an initial indication from which we have derived some salient recommendations, we believe in order to truly develop effective and successful e-health solutions, it is now necessary to apply a richer lens of analysis and recommend the use of ANT (Actor Network Theory) as follows. Table 2

maps the findings of the SWOT analyses in terms of the key concepts of ANT.

In particular, what ANT underscores is the importance to consider human and non-human actors as equally important. This is of paramount importance in the development of an appropriate e-health solution. Moreover, as can be seen from Table 2 both the German and Australian e-health solutions benefit from such a rich understanding. In fact, many of the threats and weaknesses identified in the respective SWOT analyses with the application of ANT can be converted into strengths and opportunities. Further, ANT provides us with a rich objective framework to analyze all key issues.

CONCLUSION

All OECD countries are currently being confronted with rising healthcare expenditures. In addition, they are all turning to various e-health solutions. Such e-health solutions, in particular, seem to be the key to respond to challenges like rising health expenditures. Based on this, the proceeding has focused on a deeper examination of Germany and Australia and the critical issues regarding the respective developments of their own e-health solutions. These countries were chosen, because both countries are based on a 2-tier healthcare system and have mainly equal general conditions to develop an e-health solution. However, to date both countries are focused on developing their own unique solution from the ground up. It is interesting indeed to note however that even so there exist many similar strengths, weaknesses, opportunities and threats. The SWOT analyses have demonstrated that the construction of e-health solutions is a challenging endeavor, and thus appropriate facilitators must be identified in order to realize the

Table 2. Key concepts of ANT

Concept	Relevance to Germany	Relevance to Australia
Actor/Actant: Typically, actors are the participants in the network which include both the human and non-human objects and/or subjects. However, in order to avoid the strong bias towards human interpretation of Actor, the neologism Actant is commonly used to refer to both human and non-human actors. Examples include humans, electronic instruments, technical artifacts, or graphical representations.	In the German e-health context this includes the web of healthcare players such as provides, healthcare organizations, regulators, payors, suppliers and the patient, the telematics infrastructure as well as the clinical and administrative technologies that support and facilitate healthcare delivery.	In the Australian e-health context this includes the web of healthcare players such as provides, healthcare organizations, regulators, payors, suppliers and the patient, the telematics infrastructure as well as the clinical and administrative technologies that support and facilitate healthcare delivery.
Heterogeneous Network: Is a network of aligned interests formed by the actors. This is a network of materially heterogeneous actors that is achieved by a great deal of work that both shapes those various social and non-social elements, and "disciplines" them so that they work together, instead of "making off on their own" (Latour, 1996, 2005).	The German e-health card concept will lead to a more connected healthcare system in Germany, which is the goal of network centric healthcare. This means that the eHC concept is not just an innovative solution in conjunction with several new functions, it will also result in a better information exchange between the healthcare actors.	The Australian e-health approach is focused on a fragmented e-health solution to date, which is orthogonal to the doctrine of network centric healthcare and can therefore not guarantee smooth seamless information exchange between the different healthcare actors in Australia.
Tokens/Quasi Objects: Are essentially the success outcomes or functioning of the Actors which are passed onto the other actors within the network. As the token is increasingly transmitted or passed through the network, it becomes increasingly punctualized and also increasingly reified. When the token is decreasingly transmitted, or when an actor fails to transmit the token (e. g., the broadband connection breaks), punctualization and reification are decreased as well.	This translates to successful healthcare delivery, such as treating a patient by having the capability to access critical information to enable the correct decisions to be made. Conversely, and importantly, if incorrect information is passed throughout the network, errors will multiply and propagate quickly; hence, it is a critical success factor that the integrity of the network is maintained at all times.	This translates to successful healthcare delivery, such as treating a patient by having the capability to access critical information to enable the correct decisions to be made. Conversely, and importantly, if incorrect information is passed throughout the network, errors will multiply and propagate quickly; hence, it is a critical success factor that the integrity of the network is maintained at all times.
Punctualization: Is similar to the concept of abstraction in Object Oriented Programming. A combination of actors can together be viewed as one single actor. These sub actors are hidden from the normal view. This concept is referred to as Punctualization. An incorrect or failure of passage of a token to an actor will result in the breakdown of a network. When the network breaks down, it results in breakdown of punctualization and the viewers will now be able to view the sub actors of the actor. This concept is often referred to as depunctualization.	This only becomes visible when a breakdown at this point occurs and special attention is given to analyze why and how the problem resulted and hence all sub tasks must be examined carefully.	This only becomes visible when a breakdown at this point occurs and special attention is given to analyze why and how the problem resulted and hence all sub tasks must be examined carefully.
Obligatory Passage Point (OPP): Broadly refers to a situation that has to occur in order for all the actors to satisfy the interests that have been attributed to them by the focal actor. The focal actor defines the OPP through which the other actors must pass through and by which the focal actor becomes indispensable (Callon, 1986).	With the eHC this includes all the protocols for accessing medical data during a patient consult.	NEHTA is still developing this component.
Irreversibility: Callon (1986) states that the degree of irreversibility depends on (i) the extent to which it is subsequently impossible to go back to a point where that translation was only one amongst others and (ii) the extent to which it shapes and determines subsequent translations.	Given the very complex nature of healthcare operations (von Lubitz & Wickramasinghe, 2006) irreversibility is generally not likely to occur. However, it is vital that chains of events are continuously analyzed in order that future events can be addressed as effectively and efficiently as possible.	Given the very complex nature of healthcare operations (von Lubitz & Wickramasinghe, 2006) irreversibility is generally not likely to occur. However, it is vital that chains of events are continuously analyzed in order that future events can be addressed as effectively and efficiently as possible.

vision and thereby achieve better healthcare outcomes. In addition, these analyses serve to help us understand some of the important considerations regarding the transferability of e-health solutions, especially within similar healthcare systems e. g. Germany and Australia. As noted in the discussion section above, this paper presents initial findings from an ongoing research in progress. Even so, these preliminary findings serve to indicate the importance of investigating this issue of transferability of e-health solutions. In addition, we have highlighted the need for using ANT as an appropriate and rich framework to analyze all germane issues. We close by calling for more research into this vital area.

REFERENCES

ArticlesBase.com – McMillan. (2009). *Two-tier health care*. Retrieved August 26, 2011, from http://www.articlesbase.com/health-articles/twotier-health-care-1020480.html

Australian Government – Department of Health and Ageing. (2011). *Private health insurance*. Retrieved August 26, 2011, from http://www.health.gov.au/internet/main/publishing.nsf/Content/health-privatehealth-consumers-glossary.htm

Barmer, G. E. K. Krankenkasse. (2011). *Elektronische Gesundheitskarte*. Retrieved July 29, 2011, from https://www.barmer-gek.de/barmer/web/Portale/Versicherte/Leistungen-Services/Leistungen-Beitraege/Multilexikon_20Leistungen/Alle_20Eintr_C3_A4ge/Gesundheitskarte.html?w-cm=CenterColumn_tdocid

BMG – Bundesministerium fuer Gesundheit. (2009). *Daten des Gesundheitswesens 2009*. Retrieved August 26, 2011, from http://www.bmg.bund.de/uploads/publications/BMG-G-09030-Daten-des-Gesundheitswe-sens_200907.pdf

Callon, M. (1986). Some Elements of a Sociology of Translation: Domestication of the Scallops and the Fishermen of St Brieuc Bay. In J. Law (Ed.), *Power, Action and Belief: A New Sociology of Knowledge?* (pp. 196–223). London: Routledge & Kegan Paul.

Commonwealth Fund. (2010). *International Profiles of Health Care Systems*. Retrieved August 26, 2011, from http://www.commonwealthfund.org/~/media/Files/%20Publications/Fund%20Report/2010/Jun/1417_Squires_Intl_Profiles_622.pdf

Die gesetzlichen Krankenkassen. (2007). *Das elektronische Rezept*. Retrieved August 26, 2011, from http://www.g-k-v.de/gkv/fileadmin/user_upload/Projekte/Telema-tik_im_Gesundheitswesen/2.1.7_das_elek-tronische_rezept.pdf

European Commission. (2011). *The European Health Insurance Card*. Retrieved August 26, 2011, from http://ec.europa.eu/social/main.jsp?catId=559

gematik – Gesellschaft für Telematikanwend-ungen der Gesundheitskarte mbH. (2011a). *Versichertenstammdaten*. Retrieved August 26, 2011, from http://www.gematik.de/cms/de/egk_2/anwendungen/verfuegbare_an-wendungen/versichertendaten/versicherten-daten_1.jsp gematik – Gesellschaft für Telematikanwendungen der Gesundheitskarte mbH. (2011b). *Anwendungen der eGK*. Retrieved August 26, 2011, from http://www.gematik.de/cms/de/egk_2/anwendungen/anwendungen_1.jsp

Ghani, M., Bali, R., Naguib, R., Marshall, I., & Wickramasinghe, N. (2010). Critical issues for implementing a Lifetime Health Record in the Malaysian Public Health System. [IJHTM]. *International Journal of Healthcare Technology and Management, 11*(1/2), 113–130. doi:10.1504/IJHTM.2010.033279

GKV-Spitzenverband. (2011). *Alle gesetzlichen Krankenkassen*. Retrieved August 26, 2011, from http://www.gkv-spitzenverband.de/ITSGKrankenkassenListe.gkvnet

GVG – Gesellschaft für Versicherungswissenschaft und –gestaltung. (2004). *Managementpaper Elektronische Patientenakte*. Retrieved August 26, 2011, from http://ehealth.gvg-koeln.de//cms/medium/676/MP_ePa_050124.pdf

Huber, M. (1999). Health Expenditure Trends in OECD Countries 1970-1997. *Health Care Financing Review, 21*(2), 99–117. PMID:11481789

Jamshidi, M. (2009). *System of Systems Engineering: Innovations for the 21st Century*. Hoboken, NJ: John Wiley & Sons.

Jha, A. K., Doolan, D., Grandt, D., Scott, T., & Bates, D. W. (2008). The use of health information technology in seven nations. *International Journal of Medical Informatics, 77*, 848–854. doi:10.1016/j.ijmedinf.2008.06.007 PMID:18657471

Latour, B. (1986). The Powers of Association. In J. Law (Ed.), *Power, Action and Belief: A new sociology of knowledge?* (pp. 264–280). London: Routledge & Kegan Paul.

Latour, B. (1996). *Aramis or the Love of Technology*. Cambridge, MA: Harvard University Press.

Latour, B. (1999). On Recalling ANT. In J. Law, & J. Hassard (Eds.), *Actor Network Theory and After* (pp. 15–25). Oxford, UK: Blackwell Publishers.

Latour, B. (2005). *Reassembling the Social: An Introduction to Actor-Network-Theory*. Oxford, UK: Oxford University Press.

Law, J. (1987). Technology and Heterogeneous Engineering: The Case of Portuguese Expansion. In W. E. Bijker, T. P. Hughes, & T. J. Pinch (Eds.), *The Social Construction of Technological Systems: New Directions in the Sociology and History of Technology* (pp. 111–134). Cambridge, MA: MIT Press.

Law, J. (Ed.). (1991). *A Sociology of Monsters: Essays on power, technology and domination*. London: Routledge.

Law, J. (1992). *Notes on the Theory of the Actor Network: Ordering, Strategy, and Heterogeneity*. Retrieved August 26, 2011, from http://www.lancs.ac.uk/fass/sociology/papers/law-notes-on-ant.pdf

Maheu, M. M., Whitten, P., & Allen, A. (2001). *E-Health, Telehealth, and Telemedicine: A Guide to Start-Up and Success*. San Francisco, CA: Jossey-Bass.

mhplus Krankenkasse. (2011). *Infos zu Vorteilen bei der eGK*. Retrieved August 26, 2011, from http://www.mhplus-krankenkasse.de/1228.html

NEHTA – National E-Health Transition Authority Limited. (2008). *Quantitative Survey Report*. Retrieved August 26, 2011, from http://www.nehta.gov.au/component/docman/doc_download/585-public-opinion-poll-iehr

NEHTA – National E-Health Transition Authority Limited. (2010). *Healthcare Identifiers Service: Implementation Approach.* Retrieved August 26, 2011, from http://www.nehta.gov. au/component/docman/doc_download/1139-healthcare-identifiers-implementation-approach

NEHTA – National E-Health Transition Authority Limited. (2011a). *Delivering E-Health's Foundations.* Retrieved August 26, 2011, from http://www.nehta.gov.au/media-centre/nehta-news/416-delivering-e-healths-foundations

NEHTA – National E-Health Transition Authority Limited. (2011b). *The National E-Health Transition Authority Strategic Plan (2009-2012).* Retrieved August 26, 2011, from http://www.nehta.gov.au/about-us/strategy

NEHTA – National E-Health Transition Authority Limited. (2011c). *What is a PCEHR?* Retrieved August 26, 2011, from http://www. nehta.gov.au/ehealth-implementation/what-is-a-pcher

NEHTA – National E-Health Transition Authority Limited. (2011d). *Benefits of a PCEHR.* Retrieved August 26, 2011, from http://www.nehta.gov.au/ehealth-implementation/benefits-of-a-pcehr

NEHTA – National E-Health Transition Authority Limited. (2011e). *e-Communications in Practice.* Retrieved August 26, 2011, from http://www.nehta.gov.au/e-communications-in-practice

NEHTA – National E-Health Transition Authority Limited. (2011f). *Electronic Transfer of Prescription (ETP) Release 1.1 draft.* Retrieved August 26, 2011, from http://www. nehta.gov.au/media-centre/nehta-news/704-etp-11

OECD. (2010a). *Growing health spending puts pressure on government budgets.* Retrieved July 29, 2011, from http://www.oecd. org/document/11/0,3343, en_2649_34631_45549771_1_1_1_37407, 00.html

OECD. (2010b). *Gesundheitsdaten 2010 – Deutschland im Vergleich.* Retrieved August 26, 2011, from http://www.oecd.org/datao-ecd/15/1/39001235.pdf

OECD. (2010c). *OECD health data 2010.* Retrieved June 06, 2011, from http://stats.oecd. org/Index.aspx?DatasetCode=HEALTH

Pearce, C., & Haikerwal, M. C. (2010). E-health in Australia: time to plunge into the 21st Century. *The Medical Journal of Australia, 193*(7), 397–398. PMID:20919969

Royal Australian College of General Practitioners. (2010). *Model Healthcare Community.* Retrieved August 26, 2011, from http://www.racgp.org.au/scriptcontent/ehealth/201002MHC_Handout.pdf

Scheer, A.-W. (2009). *BITCOM-Pressekonferenz E-Health.* Retrieved August 26, 2011, from http://www.bitkom.org/files/documents/BITKOM_Praesentation_E-Health_PK_04_03_2009.pdf

vdek – Verband der Ersatzkassen e. V. (2009). *Fragen und Antworten zur elektronischen Gesundheitskarte (eGK).* Retrieved August 26, 2011, from http://www.vdek.com/presse/Fragen_und_Antworten/egk.pdf

Verband der privaten Krankenversicherung e. V. (2011). *Die elektronische Gesundheitskarte in der privaten Krankenversicherung.* Retrieved August 26, 2011, from http://www. pkv.de/zahlen/

von Lubitz, D. K. J. E., & Wickramasinghe, N. (2006). Healthcare and technology: The doctrine of network centric healthcare. *International Journal of Electronic Healthcare*, 2(4), 322–344. PMID:18048253

WHO – World Health Organization. (2005). *What is e-health?* Retrieved August 26, 2011, from http://www.emro.who.int/his/ehealth/AboutEhealth.htm

Wickramasinghe, N., & Bali, R. K. (2009). The S'ant Imperative For Realizing The Vision Of Healthcare Network Centric Operations. *International Journal of Actor-Network Theory and Technological Innovation*, 1(1), 45–58. doi:10.4018/jantti.2009010103

Wickramasinghe, N., Bali, R. K., & Tatnall, A. (2007). Using actor network theory to understand network centric healthcare operations. *International Journal of Electronic Healthcare*, 3(3), 317–328. doi:10.1504/IJEH.2007.014551 PMID:18048305

Wickramasinghe, N., & Schaffer, J. L. (2010). *Realizing Value Driven e-Health Solutions*. Washington, DC: IBM Center for the Business of Government, Improving Healthcare Series.

ZTG – Zentrum für Telematik im Gesundheitswesen GmbH. (2010). *Notfalldaten*. Retrieved August 26, 2011, from http://www.egesundheit.nrw.de/content/elektronische_gesundheitskarte/e3338/e3380/index_ger.html

KEY TERMS AND DEFINITIONS

Different Healthcare Systems: Healthcare systems can be predominantly public such as the Uk, predominantly private such as Us or two tier where there is a blend of both private and public systems co-existing. Tow tier systems can be either complementary or substitutive.

eHealth Card: This is an e-health system that is card based. Access to electronic health data is initiated via a card.

Healthcare Information Grid: This is a large network structure for electronic dissemination, use and re-use and transfer of medical information of any type in digital format.

Network Centric HealthCare: This is a type of healthcare operations that includes a wide and holistic approach that includes the domains of informational issues, physical issues and cognitive issues in applying healthcare delivery.

PCEHR (Personally Controlled Electronic Health Record): This is the national e-health solution chosen and developed in Australia. It is based on a web portal architecture.

Chapter 9
Human Interactions in Software Deployment:
A Case of a South African Telecommunication

Tefo Sekgweleo
Tshwane University of Technology, South Africa

Tiko Iyamu
Namibia University of Science and Technology, Namibia

ABSTRACT

Software is intended to enable and support organisations, in order for them to function effectively and efficiently, towards achieving their strategic objectives. Hence, its deployment is critically vital to the focal actors (managers and sponsors). Software deployment involves two primary components, technology and non-technology actors. Both actors offer vital contributions to software deployment from different perspectives. Unfortunately, there has been more focus on the technological actors over the years. This could be attributed to why the same types of challenges, such as disjoint between development and implementation, persist. This chapter holistically examines the roles of non-technology actors in the deployment of software in organisations. The Actor-Network Theory (ANT) provides a lens in the analysis of the empirical data.

INTRODUCTION

Computer software can be defined as a combination of both technical and non-technical resources adopted within the organisational requirement in support of specific needs of business (Bistričić, 2006; Chen et al., 2010). The main components of computer software include people, processes and technology infrastructure. Computer software has been shown to play a very important role in organisations.It plays a vital role in improving the capability of organisations to conduct business and develop new opportunities, as well as enabling them to remain profitable and competitive (Bergeron and Raymond, 1992;

DOI: 10.4018/978-1-4666-6126-4.ch009

Tetteh and Snaith, 2006).Organisations make a substantial investment in the implementation of systems (software) to help run their businesses effectively so that they achieve their goals, improve efficiency, provide value added service to customers and provide both employee and customer satisfaction (Avison and Fitzgerald, 2006).

Organisations such as financial institutions and retail stores strive to simplify their customers' needs to save time. For example, they provide them with conveniences such as automatic teller machines (ATM), internet banking, mobile banking and online shopping. It is through the use of communication networks that organisations are able to provide customers with products and services from any location, at any time (Požgaj et al., 2007). There is no longer a need for customers to wait in long banking and shopping queues when they can simply do it online. According to Bielski (2008), ATMs and internet banking enable customers to make banking transactions at any time without having to physically present themselves at the bank. Goswami and Raghavendran (2009) argue that with mobile banking, banks are able to offer their customers more convenience, as a result of the provision of banking transactions through cellular phones, at any given time of the day.This allows organisations to carry out their business transactions uninterrupted, twenty four (24) hours a day, seven (7) days a week.Avison and Fitzgerald (2006) posit that trading via the internet (electronic commerce) has made it easier for customers to access worldwide markets and to expand trading hours to twenty four hours a day, seven days a week.Increased competition amongst businesses forces organisations to take advantage of the new technology to provide value added service to fulfil their customers' needs (Pappa and Stergioulas, 2008).

The purpose of the research was to investigate, from non-technical perspective, the disjoint between the development and implementation of computer software in the organisation. The study took cognisance that those who develop computer software are not necessarily the ones that implement it.

RESEARCH APPROACH

A case study is one of the research methods commonly used in social science or other related fields (Yin, 2009). According to Noor (2008) a case study focuses on conducting an in-depth investigation into one or a few cases in order to gain a holistic insight about the phenomenon. Parè (2004:233) defines a case study as "an empirical enquiry that investigates a contemporary phenomenon within its real-life context, especially when the boundaries between phenomenon and context are not clearly evident". The case study method was employed for this study mainly because of the nature of the study as well as the features of the case study method.

Letsatsi is a Telecommunication company based in South Africa. The organisation was used in the study mainly for the two reasons: of the eight organisations that were approached, it was only the only one that agreed to participate in the study; and there was a *prima facie* evidence that they had challenges in the development of software in the company.

The organisation had about 625 employees in its Computing department. The computing department was further divided into units such as Infrastructure Services, Customer Relations, Software development, and Project Management.

The semi-structured approach was used in the collection of data. There were two main questions to the study: "What were the processes involved in the software development and implementation?" and "What were the factors influencing the disjoint between software development and implementation in both organisations?" A total of fifteen employees from different units and at various levels were interviewed.

The moments of translation was employed in the data analysis. This is mainly because of strength to focus on negotiation and interactions of actors within networks. Iyamu and Tatnall (2009) posit that actors are allowed to make decisions in the creation of the networks in which they choose to participate. The moments consist of problemitisation, interessement, enrolment and mobilisation:

1. **Problematisation:** An Actor-network can be formed to solve a problem or take advantage of a new opportunity. At this stage the focal actor problematizes an issue (Greenhalgh and Stones, 2010).
2. **Interessement:** At the interessment stage, actors show interest in the problematized issue. This could be through negotiations with the other actors.
3. **Enrolment:** Actors partake in the tasks that are assigned to them through negotiations. According to Luoma-aho and Paloviita (2010), communication plays a vital role in the enrolment stage because it clarifies what is expected from the actors and what actions need to be executed.
4. **Mobilisation:** At this stage spokespersons are identified to represent other actors within the network (Macome, 2008). Good representation helps to strengthen the network.

The obligatory passage point (OPP) is an entity responsible for representing other actors in a way that suits their significance and actions in the world of translation (Gonçalves and Figueiredo, 2010). Actors require a representative who will act on their behalf to address problems and communicate issues. It is the responsibility of the focal actor to define the OPP as it is the channel of communication.

DATA ANALYSIS: ANT VIEW

Actor-network Theory incorporates the principles that integrate both human and non-human actors into the same network of common interest. However, non-human actors do not reason like human actors but they are able to perform what they are directed to do. Both human and non-human actors complement each other in a way that completes the network.

Actor

In actor-network theory, actors are both human and non-human. The human actor in software deployment (development and implementation) included employees such as project managers, software developers, software testers and operations and billing specialists are involved in the software development and implementation in the organisation.

However, in the software deployment, the employees who were involved in the development were not necessarily the ones who were involved in the implementation. Some of the employees who were involved in the software development in the organisation included project managers, software developers and software testers.

During the software development and implementation in the organisation, different technologies which are the non-human actors are involved. Through direct or indirect connectivity or interface, the technologies depend on each other to enable processes and activities of systems (software). Some of the technologies included computers, servers, networks, modelling tools, and programming languages (e.g. Visual Basic, Java and Dephi).

The employees, as well as the technologies operate within heterogeneous network in the software deployment (development and implementation) in the organisation.

Network

Actor-network Theory assumes that all the elements involved in a network, irrespective of whether they are human or non-human should be described in the same terms. During software deployment in the organisation, all actors (employees) within the network were directly or indirectly connected (related) through individual or group roles and responsibilities. The groups included the project team, the Integrated Change Control Forum (ICCF), and the Marketing team.

In the organisation, the marketing team were the key stakeholders in the initiation of the project. The ICCF was the deciding body within the organisation. The project team was one of the main stakeholders from the perspectives of software development and implementation. The ICCF determines which projects should be allocated to which departments. Once that process is completed the projects are allocated to the senior manager who then engages the line managers and their subordinates in the software development and implementation in the organisation. One of the software developers explained that: *"all the projects requested by business are approved by the ICCF"*.

The engagement of actors, both human and non-human becomes a network which contributes equally to the software development and implementation. To be able to deploy software using actors within a network, a process of negotiation is required. From the perspective of ANT, the process of negotiation follows the moments of translation.

Moments of Translation

The four stages of the moments of translation were used in the analysis of the data. This was done in order to understand the factors which influence the disjoint between the development and implementation of software in the organisation. These stages include problematisation, interessement, enrolment and mobilisation:

Moments of Translation: Problematisation

The success of software development and implementation is never guaranteed. Anything can happen during the systems deployment life cycle. Hence, responsibility and accountability for the development and implementation of software are vital in organisations.

The marketing team were the initiators of IT projects in the organisation, and also regarded as the internal customers. Due to the stiff competition which occurs between competitors, the marketing team is obliged to introduce new products in the market or enhance the existing ones. A business analyst briefly explained: *"Marketing logs a change request with the ICCF if they want to introduce a new product or service to the market"*.

On the other hand, the integrated change control forum (ICCF) was the deciding body within the organisation. This committee (ICCF) was responsible for determining, prioritising and allocating software projects to different domains and departments. One of the interviewees explained as follows: *"The integrated change control forum (ICCF) determines which domain is impacted by the change request and according to the urgency of the change request, allocations are made to the impacted domains".*

The project team is responsible for the software development and implementation in the organisation. The project managers, together with the business analyst, are responsible for the software development and implementation in the organisation. This was based on the roles and responsibilities of the project managers and business analysts. However, software deployment is not as easy as proclaimed. It requires knowledge as well as personnel who have the roles, responsibilities and accountabilities. Therefore it is vital to include relevant stakeholders in the development of software and its implementation. The business analyst team manager explained that: *"Once the change requests are routed to our domain, the project manager and business analyst gets assigned to the project".*

The software development and implementation occurs differently at various stages of the development life cycle. However, it is not obvious that the same resources that were involved in the previous project are part of the new project because for each project a new project team is formed. The systems development was often based on the knowledge of actors (project manager, business analysts, software developers, software testers and support specialists).

Planning plays a vital role in the software development and implementation. However some teams are involved from the initiation of the software development project while others are involved later in the project. According to some of the interviewees, who were from different business units: *"Once the change request is assigned to our domain, the project manager and the business analyst get assigned to the project". Operations get involved closer to the implementation phase".*

Software deployment was problematized in accordance to the organisational structure and mandates. The project manager and business analyst were primarily responsible and accountable for software development and implementation in the organisation. According to one of the interviewees, *"the project manager is assigned a project and the business analyst is assigned a project by the business analysis team leader and line manager".*

In the organisation, the stakeholders' interests in the deployments of software were not always the same. The next section explores the various interests of the stakeholders in the software development and implementation in the organisation.

Moments of Translation: Interessement

However, the team members had diverse interests in software development and its implementation within the organisation. Their interests were motivated by numerous factors. Some wished to experience new challenges, others wanted to be recognised for their efforts and a few wanted to be part of the winning team. A personal view by one of the interviewees was expressed as follows: *"My experience in the ISP environment is good because we work on a lot of new projects, there is a lot of interesting challenges and we learn on every project, you learn something new so that is what I enjoy the ISP environment".*

Some of the stakeholders who were involved in software deployment have adequate skills. Some stakeholders were unhappy about how things were done during the software development and implementation within the organisation. These, including manipulation of rules and processes, were manifestation of interests other than the objectives of the organisation. One of the employees vocally stated: *"Unfortunately the system documenta-*

tion only applies to big projects while small projects are not documented and they end up growing to be big projects that do not have documentation".

It is vital to be considerate of other stakeholders during the development and implementation of software. Unfortunately, considerations were sometimes overshadowed by personal interests. Personal interests begin to affect some of the team members understanding of users' requirements.Other employees, such as IT technologists, were interested in the development and implementation of software because they want to introduce new approaches in the organisation.

The interests of individuals and groups resulted in certain actions, which were opposed to the organisation's objectives.For example, some important aspects were overlooked by some stakeholders during the software deployment. An employee sympathetically explained: *"Software developers often do not take security as well as configuration files into account because usernames and passwords are hardcoded, which results to problem during implementation".*

Other employees are interested in the development and implementation of software because they are bound by the performance contract they signed with their line managers on behalf of the organisation. Therefore the employees were expected to perform their duties as required for the development and implementation of software within the organisation. One of the interviewees briefly explained the contract binding as follows: *"During performance evaluation, what counts is the number of change requests you worked on, irrespective of whether they have been implemented or not".*

Task allocation by various managers during the development and implementation of software was one of the factors that interested some employees. Some employees were assigned tasks they were comfortable doing while others did not like what was assigned to them. This had a negative impact on productivity during the development and implementation of software.

Unfair treatment during the development and implementation of software in the organisation caused tensions and jealousy among the employees. As a result, the highly skilled employees did not accept responsibility for assisting the less skilled when they required assistance. One of the interviewee further explained as follows: *"At times I have seen the highly skilled become jealous of the fact that the lesser skilled get rewarded in the same fashion. This caused lack of cooperation between employees in the development and implementation teams".*

While the management made several efforts to build interest for the proposed solution in order to rectify the problem, the building of interest among the employees could not be regarded a success. The enrolment of employees was not fully successful as some employees unwillingly accepted the tasks allocated to them and their roles and responsibilities in the development and implementation of software.

Moments of Translation: Enrolment

Enrolment was partially communicated through negotiations and agreements about the development and implementation of software within the organisation. Meetings are held to update the employees about the progress in the software development and implementation. The update influences some employees to participate in the processes and activities relating to the development and implementation of software in the organisation.

Line managers used the mandate bestowed on them by the organisation to get employees to participate and to accept the roles and responsibilities allocated to them in the development and implementation of software. Line managers also linked the tasks allocated to individuals to the organisation's compulsory performance evaluation and as such other employees were forced to participate. The performance evaluation gave little or no room for negotiation during tasks allocation in the software deployment. According to one of the developers, "*the* managers used *their discretions to assign resources to projects*".

Processes and activities involved in software development and implementation require a team effort. Hence, a personal relationship among employees in the organisation was vital. Line managers used their personal relationship with employees, especially those who fall into the same language group, to get employees to participate in the development and implementation of software in the organisation. However, other employees participated in the system's (software) development because they shared the same vision as the organisation. One of the interviewees, in her personal view stated that the: "*the organisation requires people who are able to make sacrifices for long term gains of the organisation as the business is there to be profitable*".

The line manager ensures that all enrolled employees were aware of the roles and responsibilities which were allocated to them, in the development and implementation of software within the organisation. Meetings were used to negotiate the allocation of tasks to individuals.

Employees who were involved in the development and implementation of software in the organisation relied on information that came from their superior in order to carry out their tasks diligently. Otherwise, the results of their participation might become costly for the organisation. Allocation of tasks to team members was as vital as the development and implementation of software. It was vital for

individuals to understand the tasks they had been allocated in the software development and implementation.

Even though they enrolled in the deployment process, there was at times where the software developers did not conduct unit testing which they were supposed to do before handing in the software to the quality assurance team. This resulted in many defects being logged against the software development team. According to one of the software developers, "*the software developers buy time by delivering non-working solution to the software testing team because they did not finish their work on time*".

Pressure to deliver software on time was even more challenging at the technical level. Prior to development, software developers were expected to compile the technical design specifications which involved the conversion of the functional design specifications into a technical language.

Also, training was required to upgrade the skills of some employees who enrol in software deployment in the organisation. This was in order for them to carry out their tasks competently in the development and implementation of software. One of the business analysts explained as follows: "*Doing proper analysis is a challenge on its own, it is a skill that we try to develop as we go along to do a thorough analysis of requirements*".

During the development and implementation of software within the organisation, employees learned different things. Therefore it did not necessarily mean that whatever they learnt in the previous projects would occur in the next project because software deployment projects differ. There was a lack of consistency in the distribution of resources in the development and implementation of software in the organisation. This had an impact on how employees participated in the activities of software deployment.

Lack of consistency in the distribution of resources affected the rate at which employees could develop, particularly those employees with limited skills. For example, some of the employees involved in software development often struggled to gather business requirements. As a result, it affected the quality of the software. One of the interviewees shared his experiences such as: *"During development we pick up some requirements which were over looked during the analysis or something that was not really noticed then it becomes a problem because sometimes you might find out that whatever you are picking up might be a show stopper (prevent us from continuing working)".*

The project managers and line managers acted as representatives for the focal actor by enrolling employees in the software development programme and its implementation. As a result, the enrolment process was a success because the employees agreed partake even though some of them had different interests.

Moments of Translation: Mobilisation

In the development and implementation of software in the organisation, employees acted on behalf of the organisation. Others were delegated to act as spokespersons to convince and persuade employees to participate in the delivery of systems (software) through its development and implementation. According to the organisational structure the senior managers were responsible for heading their domain (ISP domain). The line managers reported to the senior managers and the rest of the team members reported to their respective line managers. Mobilisations (pursuance) of employees were carried out along this structure. One of the line managers briefly explained the structure as follows: *"We have the following teams within the ISP domain: Project management, business analysis, soft-ware development, software testing, operations and support with their respective line managers".*

Once the project team was formed, the software development programme could start. However, business analysts needed to ensure that the analysis work was conducted for the system (software) to be developed. The business analysts became the spokespersons because they had to interview various project stakeholders to conduct the investigations concerning the development and implementation of software and the gathering of business requirements. Interviews, meetings and joint application design sessions (JAD) were used as a vehicle to mobilise the project stakeholders.

The software development and its implementation were headed by the project managers in the organisation. The project manager is supposed to coordinate all the project activities. As a result, the manager mobilises employees to engage in the activities involved in the software deployment. According to one of the project manager, *"my responsibilities is to ensure that timelines are met by resources, user requirements are documented, the system (software) is developed according to user requirements, the system (software) is tested and implemented successfully".*

The roles and responsibilities as mentioned above entitle the project manager to become a spokesperson because he or she is required to take care of the project team. The project manager mobilises the team by setting up team meetings to discuss project issues and progress.

As technology keeps on changing, it was vital for the project team members to constantly upgrade their technological skills. When the need arose those who required training were scheduled for training by their respective line managers. As a result, the line managers became spokespersons for their subordinates because they had to approve their training.

The line managers also mobilised their teams through team meetings to get feedback from their subordinates. The billing and operations manager explained that: *"Weekly meetings are used as checkpoints to verify progress on tasks assigned to individuals"*.

Once the mobilisation was successful, the actor-network would begin to function with the objective of implementing the proposed system (software). The marketing team was the focal actor mainly because the development and implementation of systems (software) was initially initiated by them. The management willing accepted the task of mobilisation which was linked to their performance appraisals. Mobilisation was a success and employees were enthusiastic about what they contribute in the development and implementation of software.

After the analysis through the lens of ANT, the disjoint that occurs between the development and implementation stages in software deployment became empirically clear. The factors of disjoint were identified and are presented in the next section.

THE CASE STUDY: FINDINGS

From the analysis of the data, six factors were found to have a critical influence on the software development and its implementation in the organisation. These factors include a lack of adequate skills, staff retention, a lack of continuity, competitiveness at individual levels, a negotiated scheme and performance appraiser. Figure 1 depicts the factors and how they relate to each other. These factors are discussed below:

Lack of Adequate Skill

There was lack of adequate skill for the development and implementation of software within

Figure 1. Human interaction in software deployment

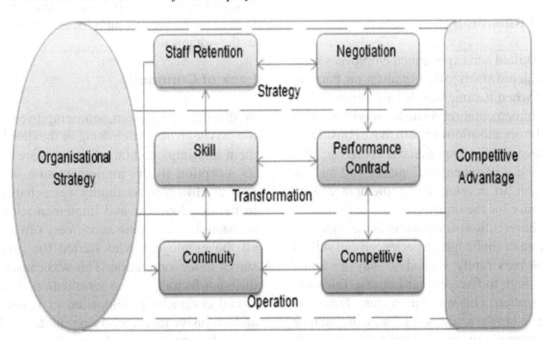

the organisation. As revealed in the analysis of the data, the training of both technical and business personnel was carried out on an on-going basis parallel to the development and implementation of the software. This was done because many of the personnel were not adequately skilled to fulfil their roles, and hence there was a need for their skills to be upgraded.

Due to a lack of adequate skills, some of the crucial business requirements were either missed or not properly articulated during the planning phase. This had an impact on the realisation of the objectives laid out for the system development phase. As a result of this shortage of skills, which were also often inadequate, the organisation was challenged when it came to system analysis. In an attempt to close the skills gap, the organisation invested more on training and development of identified employees. The exercise (training and development) was prohibitively expensive, but the organisation needed it for competitive advantage.

Staff Retention

The skilled and experienced employees were considered assets to the organisation, particularly when it came to software development and implementation. As such it was imperative for the organisation to retain those considered to have acquired specialised skills. The specialised skill-set ensured quality and made it possible to develop and implement quality software for the organisation.

The retention of employees in the organisation was a challenge. Experienced and skilled employees rarely wanted to stay for more than three to five years at Letsatsi Telecommunication. This was due to many factors as revealed in the analysis, such as an imbalance in remuneration and lack of recognition for efforts and excellence. The highly skilled

employees felt cheated and unfairly treated as the lesser skilled were paid equally or even higher than some of them. As a result, the highly skilled were not willing to assist the less skilled colleagues whenever they needed assistance. The lack of cooperation had an impact on the timeline, as well as the quality of software development and implementation.

Due to unfair remuneration some employees resorted to searching for new job opportunities because they were not fairly treated. Those considered to have fewer skills had no choice but to work harder in order to deliver the tasks which were assigned to them. This helped them to upgrade themselves. Once they were skilled enough, they also moved to other organisations for "greener pastures". It was a vicious circle. Even though the organisation realised that it was vital to retain experienced and highly skilled employees, they did not have a retention strategy. The competition for experienced and highly skilled people intensified, and the brain drain increased. This had an impact on the support systems (software), particularly because there was a lack of continuity in supporting the systems in the organisation.

Lack of Continuity

Within the organisation, some employees had the privilege of participating in the development and implementation of software from its inception to the implementation of the project. However, continuity was a challenge in the development and implementation of software in the organisation. Very often, not all the employees who started the project stayed to its completion. This was caused by different factors such as termination of individual contracts, re-allocation of resources and exclusiveness in carrying out tasks. For example, there was a financial constraint, which caused the organisation to feel some resources were too expensive and as a result the contract was terminated. Unfortunately, there were no employees who were sufficiently qualified to take over the work of the previous software developer. Even though the organisation later managed to get a replacement, the confidence of the software team was dented. It was a challenge to get a replacement primarily because it was difficult to understand how the previous software developer had structured and developed (write) the programming code of the systems (software). This resulted in wasted time, and sometimes the quality of the systems was compromised.

As mentioned earlier, the brain drain in the organisation had an impact on many processes and activities, including lack of continuity. As people left, it became difficult for the organisation to introduce and practice certain methodologies as the newer employees were not familiar or conversant with them.

Another factor that caused lack of continuity was the exclusive nature in which tasks were structured, shared and carried out during the software development and implementation. Some of the employees who participated in the completion were not involved in the beginning of many of the projects. Some employees (such as project managers and business analysts) had the opportunity to be part of the team that started the development and implementation of software in terms of planning in the organisation. Other employees such as the software developers and testers were not involved in the beginning, but were active participants in the middle of the projects. This resulted in the project stakeholders sometimes lacking the same understanding of what the systems development and implementation project required. It was the responsibility of the line managers to ensure that their subordinates got involved from the beginning of the project. A failure

to do so contributed to the lack of continuity in the development and implementation of software within the organisation.

Competitiveness at Individual Levels

In the development and implementation of software, there was competition amongst the employees. The competitiveness was both positive and negative to systems (software) deployment (development and implementation) in the organisation. The competition among the employees motivated individuals to deliver tasks allocated to them within a given timeframe. It also encouraged individuals to work on their own as well as in the teams. Working as a team allowed the managers to have a contingency plan in cases where employees took leave, were involved in accidents or resigned within 24 hours.

On the other hand, competition among employees affected productivity in the development and implementation of software in the organisation. Some employees were not willing to share information because they thought other employees were gaining recognition through their efforts. The employees who were able to express themselves better in the English language, in meetings, often high jacked the brilliant ideas of those who had difficulty in expressing themselves in English. This led to interpersonal team challenges and in-fighting among employees.

Negotiative Scheme

The project managers were responsible for compiling the project plans. The project managers' in conjunction with other stakeholders, agreed on timelines and individuals and teams deliverables in the development and implementation of software in the organisation. However, negotiations did not happen at all times, instead, some managers applied obligatory passage points (OPP).In actor-network theory, OPP is a state of little or no negotiation, where the focal actor is indispensable.Some of the managers used the mandate accorded to them by the organisation to get the employees to carry out certain tasks. This was attributed to the fact that some project managers suspected that employees (particularly, those considered to be highly skilled) were using their stock of knowledge to manipulate processes and events for their personal interest.

However, if the project managers were unable to detect lies from the truth as presented and reflected by the stakeholders who were involved, their project plans would reflect incorrect information to the management.This had an impact on decision-making, which the entire organisation relied upon for competitive advantage. Hence the performance contract and appraiser of individuals and teams was critical, and needed uniformity.

Performance Appraiser

Performance appraisal is often instituted to achieve organisational objectives.In the organisation, some challenges to the systems development and implementation were associated with how individuals and teams' performances were appraised.There was no system to appraise the performances of the employees during the software development and implementation in the organisation. Employees were appraised by the number of change requests they had worked on irrespective of whether they were implemented or not. This affected the moral of other employees because they needed to be appraised on the result and success of tasks during the project.

As revealed in the analysis, there was no performance evaluation matrix to follow

in appraising employees.The performance evaluation matrix enables the managers to assist employees in areas of weakness. The identification of employees' weaknesses often led to training. The performance evaluation also counts towards individuals' incentives such as annual financial increases. Employees expect to be remunerated based on their performances on various software development projects.For the employees to receive a financial increase they should fulfil a certain score in accordance with the matrix. Those employees whose scores reflect that they have underperformed should not get any financial increase.

The findings from the analysis highlighted certain acts and factors. These acts and factors were analysed to gain better understanding as to why those factors existed during the development and implementation of software in the organisation.

INTERPRETATION OF THE FINDINGS

The factors which influence and impact the development and implementation of computer software in the organisation as revealed in the analysis are now interpreted in this section, using the moments of translation. The factors include over dependency, complementary skills, skills alignment, reuse of resources, and human resources:

Over Dependency

The development and implementation of software within the organisation is a team effort.It begins as an initiative from the marketing department and grows into a software development project.These projects are logged against the Integrated Change Control Forum (ICCF), which are then assigned to various business units or departments.The department

Figure 2. Non-technical factors influencing software deployment

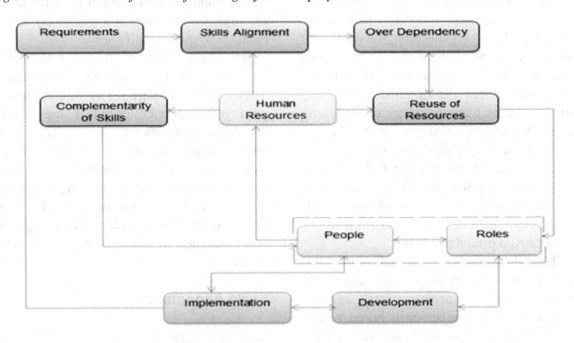

which is assigned a project is expected to put together a team which will be responsible for the development and implementation of software. The team consists of a variety of skilled personnel such as business analysts, software developers, software testers and system administrators.

The project plan is seen as a blue print of how the organisation's project is to progress from its initiation to its conclusion. A well-constructed plan influences the project outcome, whether it is ultimately a success or failure. The project plan entails the scope, goals, schedules, risks, quality control and milestones of the project. It also has a section which describes, in detail, the breakdown of each task and how its goals will be achieved. The plan when properly laid out, documents all the effort that is needed for the project to reach a successful conclusion. The main purpose of planning is to create specific deliverables for

the organisation. The reason the organisation is in business is to create a revenue stream and the successful completion of projects for competitive advantage.

The other team members are dependent on the business requirements to complete their tasks. The business analyst is responsible for putting the requirements together. Failure to do so makes it difficult for a software developer to write the program and the software tester to test it. The business requirement documentation dictates the functionality of the software. Without it, it is impossible to develop the software.

Complementarity of Skills

Complementarity of skills plays a vital role in the creation of a powerful and productive team. It is difficult for an organisation to perform well when it is dominated by a single member of the team. However, the team members need to be managed and controlled by the managers. In the organisation, the managers are able to administer the teams and ensure that the tasks that were assigned to them are completed within the timelines. However, domination by a few members of the teams creates behaviour that suppresses the overall skills of the bigger team. It is imperative to recognise each team member's ability as it encourages productivity. When individual skills of the team members are acknowledged, they complement each other and this result in an improved productive team.

The complementarity of skill among the team members has the ability to fill gaps within a team. Where one person is weak, another is strong. Where one person is bored with a particular task, that same task may excite someone else within the team. The overall goal is to discover the complementarity of skills for everyone in the team and acknowledge their abilities as equal to everyone else in the team. No task or skill should ever be discounted or seen as worthless. The complementarity of skills needs to be encouraged in order to have a productive team. Some team members are threatened by other skilled team members. However, a talent needs to be viewed positively and not as a threat to someone else's position.

Various skills were required to develop and implement software within the organisation. Such skills included project management, business management, software development, software testing, system administration and systems support. In the development and implementation of software within the organisation such skills complemented each other. The contribution of each team member was vital towards the development and implementation of software. Lack of some of the above mentioned skills in the devel-

opment and implementation of software in the organisation impacted negatively in the creation of quality software.

The main reason for having a team was to work together and help each other where assistance was required. Two or more minds put together were better than a single mind. The team members were expected to work independently as well as in a team. Team work enables all team members to expand their thinking. Creative thinking motivates productivity among team members. However, it was vital for the line managers to align the tasks assigned to the relevant skilled team members.

Skills Alignment

Each individual's skills and abilities play a crucial role in the effectiveness of the organisation's efforts to provide the best possible products and services. The organisations need to align each individual's unique training program with its future goals and performance requirements. Training assists team members in upgrading themselves and it eradicates a lack of adequate skills. Technology is continuously changing and it compels the team members to regularly go on training so they do not become obsolete. There have been so many improvements in the development and implementation of software since information technology gained popularity. In the early stages of IT, software developers were required to write many lines of code to write sophisticated software but lately there are ready made objects which are used to do the same thing. It is no longer necessary to write many lines of code.

Therefore it is the responsibility of the various team managers to utilise their subordinates efficiently and effectively in the development and implementation of software in order achieve positive results. The managers know their subordinates well enough to assign them tasks aligned to their skills. However, a high performance team is motivated by team spirit which inspires extraordinary performance in the development and implementation of software in the organisation. A high functioning team also needs to align the energies of its members in a way that moves them powerfully in a chosen direction.

A visionary leader needs to select staff wisely and deploy them where their skills can be utilised advantageously in the development and implementation of software in the organisation. As a result, it pulls different team members into a powerful alignment that gains momentum and motivates individual contribution in the development and implementation of software. The skills alignment allows the team members to be ready and able to put varying opinions aside and fully commit to the same direction chosen by the team. The fully aligned team takes the time up front to consider decisions that are healthy with the intention of producing effective results in the development and implementation of software.

However, lack of skill alignment, causes some individuals to struggle in terms of finishing their tasks on time. This is as a result of the managers not knowing their subordinates properly. Therefore, the development and implementation of software is negatively affected because certain skills are not utilised where they are supposed to be. For example, it is like assigning a support and maintenance individual a software development role. The support individual only knows how to monitor and support the software. He does not necessarily know how to perform software development duties. If any problems are encountered during maintenance, their duty is to report these incidents to the software developers.

Contingency plans play an important role in the organisation in case the highly skilled employees leave the organisation.

Reuse of Resources

The visionary manager needs to have a contingency plan in the development and implementation of software within the organisation. For example, suppose an employee dies, becomes sick or resigns with immediate effect, the organisation is faced with a huge challenge. The reuse of resources enables the development and implementation of software to continue as planned even though the team has to work extra hours to finish the work within the agreed duration.

Lack of reuse of resources brings the development and implementation of software to a halt when too many resources are affected by unforeseen circumstances. As a result, it affects the employees badly as some employees prefer to work in silos. The taking over of duties from absent employees is a challenge. Those who were supposed to take over had no idea about what the absent employees were doing. This results in some of the software development tasks needing to be reworked due to misunderstandings.

Human Resources

It is the responsibility of the human resource personnel, in conjunction with the managers of various teams, to employ suitable candidates to fulfil the duties of the development and implementation of software within the organisation. Among other things, they ensure that the performance appraisal contracts for employees are carried out.

All the employees involved in the development and implementation of software within the organisation have to be competitive enough to fulfil the tasks assigned to them. The tasks assigned to the employees correspond to the performance contract agreed upon between the employer and employees. The employees are expected to be competitive enough in order to make a positive contribution towards the development and implementation of software within the organisation. As a result recognition and salary increments are dictated by the individuals' performance.

However, some employees were not happy with the way performance appraisals were conducted. Instead of appraising the employees through their positive contribution towards the development and implementation of software, things were done the other way round. The employees are appraised according to the number of change requests they were involved in, irrespective of whether they failed or succeeded.

Staff retention plays a vital role within the organisation because it enables the employees to focus on their work rather than worrying about their remuneration. Performance evaluation matrix helps to detect gap for retention of employees. It serves as a contract between the employer and employee. It also encourages the employees to be productive, and makes the team members feel appreciated about the contribution they are making towards the software development. Failure to remunerate the employees properly affects productivity as some employees focus more on searching for better opportunities.

CONCLUSION

This paper has presented the conclusions made from the findings of this study. It has been established that the disjoint between software development and its implementation has an effect on IT projects. The organisations were somehow aware of the disjoint which occurs in their computing environment. But they continued to focus more on software deployment rather than lessening those disjoints in the early stages of the software project. The software development and its implementation within the organisation requires a diversity of skilled personnel which results in too much task dependency. Those people are more experienced able to offer assistance to the less skilled in order deliver the project within the estimated duration.

The difference mainly lies in assessing the disjoint which occurs in the software deployment in the organisation. The difference includes how actors connect, associate, relate and manifest themselves during the software development and its implementation. The theorised factors can now be put into practice by organisations which intend to improve their performance in this field.

REFERENCES

Avison, D., & Fitzgerald, G. (2006). *Information Systems Development Methodologies, Techniques & Tools* (4th ed.). London: McGraw-Hill.

Bergeron, F., & Raymond, L. (1992). Planning of information systems to gain a competitive edge. *Journal of Small Business Management, 30*(1), 21–26.

Bielski, L. (2008). Eight tech innovations that took banking into the 21st century. *ABA Banking Journal, 100*(11), 74–86.

Bistričić, A. (2006). Project information system. *Tourism and Hospitality Management*, *12*(2), 213–224.

Chen, D. Q., Mocker, M., Preston, D. S., & Teubner, A. (2010). Information systems strategy: Reconceptualization, measurement, and implications. *Management Information Systems Quarterly*, *34*(2), 233–259.

Goncalves, F. A., & Figueiredo, J. (2010). How to recognize an Immutable mobile when you find one: Translations on Innovation and design. *International Journal of Actor-Network Theory and Technological Innovation*, *2*(2), 39–53. doi:10.4018/jantti.2010040103

Goswami, D., & Ragahavendran, S. (2009). Mobile-banking: can elephants and hippos tango? *The Journal of Business Strategy*, *30*(1), 14–20. doi:10.1108/02756660910926920

Greenhalgh, T., & Stones, R. (2010). Theorising big IT programmes in healthcare: Strong structuration theory meets actor-network theory. *Social Science & Medicine*, *70*(9), 1285–1294. doi:10.1016/j.socscimed.2009.12.034 PMID:20185218

Iyamu, T., & Tatnall, A. (2009). An Actor-Network analysis of a case of development and implementation of IT strategy. *International Journal of Actor-Network Theory and Technological Innovation*, *1*(4), 35–52. doi:10.4018/jantti.2009062303

Luoma-Aho, V., & Paloviita, A. (2010). Actor-networking stakeholder theory for today's corporate communications. *Corporate Communications: An International Journal*, *15*(1), 49–67. doi:10.1108/13563281011016831

Macome, E. (2008). On Implementation of an information system in the Mozambican context: The EDM case viewed through ANT lenses. *Information Technology for Development*, *14*(2), 154–170. doi:10.1002/itdj.20063

Pappa, D., & Stergiouglas, L. K. (2008). The emerging role of corporate systems: An example from the era of business process-oriented learning. *International Journal of Business Science and Applied Management*, *3*(2), 38–48.

Paré, G. (2004). Investigating information systems with positivist case study research. *Communications of the Association for Information Systems*, *13*(1), 233–264.

Pawłowska, A. (2004). Failures in large systems projects in Poland: Mission impossible? *Information Polity*, *9*(4), 167–180.

Požgaj, Ž., & Sertić, H., & MarijaBoban, M. (2007). Effective information systems development as a key to successful enterprise. *Management*, *12*(1), 65–86.

Tetteh, G. K., & Snaith, J. (2006). Information system strategy: Applying galliers and Sutherland's stages of growth model in a developing country. *The Consortium Journal*, *11*(1), 5–16.

Wickramasinghe, N., Bali, R. K., & Goldberg, S. (2009). The S'ANT approach to facilitate a superior chronic disease self-management model. *International Journal of Actor-Network Theory and Technological Innovation*, *1*(4), 21–34. doi:10.4018/jantti.2009062302

Yin, R. K. (2009). *Case study research: Design and methods* (4th ed.). Thousand Oaks, CA: Sage.

KEY TERMS AND DEFINITIONS

Actor-Network Theory: Actor network theory is a sociotechnical theory that integrates both human and non-human actors to form or create networks.

Deployment: This includes the development and implementation of software.

Human Interaction: This refers to the way in which human beings relate with one another within context, in carrying out their activities.

Information Technology: Encompasses all forms of technology that are used to create, store, exchange, and use information in its various forms, which include text data, voice, image, and multimedia for different purposes, such as business, and social communication.

Non-Technology: This covers entities, such as human, process, rules and regulations that are involved in activities in the organisational context.

Software: Software consists of computer programs designed to execute specific objective.

Chapter 10
School Management Software in Australia and the Issue of Technological Adoption

Bill Davey
RMIT University, Australia

Arthur Tatnall
Victoria University, Australia

ABSTRACT

Approaches to innovation adoption often fail to explain why similar technologies in a single environment can have very different adoption outcomes. In this chapter, the single environment of education management systems in one country (Australia) are used to show how outcomes of similar technologies can be very different. An Actor-Network approach is used to explain how some technologies succeeded and others failed. Understandings reached in this case illuminate the power of the approach that includes listening to the technological actors in addition to the human. The chapter identifies actors and interactions and shows the connection between those interactions and the final outcomes of the innovations.

EDUCATIONAL MANAGEMENT IN AUSTRALIA

Australia is a federation of six states and two territories. This chapter uses two systems created for the Victorian State Education system and one Federal Government system to create an understanding of the networks in which those intended innovations were proposed for adoption. The Commonwealth (Federal) Government is based in Canberra but school

education is the province of the State Governments. The Federal Government from time to time provides money for nationally significant education projects.

The effective administration of both schools and school systems requires information systems capable of reporting on a large amount of data. The data for these systems is collected from formal and informal sources including, amongst other things: student enrolments, observational surveys, early year interviews,

DOI: 10.4018/978-1-4666-6126-4.ch010

running records, other formal testing and anecdotal notes (Tatnall and Tatnall 2007).

In this chapter we will examine three information systems for administrative use in schools in Victoria to investigate how they contribute to school community involvement (Davey and Tatnall 2012; Tatnall and Davey 2013). This analysis is facilitated by the fact that each of these systems is working in the same environment.

RESEARCH METHOD: THE CASES

Three information systems were analysed from an ANT perspective; CASES21, a Victorian Government financial system for schools; The Ultranet, a communications system intended to connect parents, schools, teachers and students; and MySchool, a Federal Government system intended to provide the general community with school performance data.

CASES21 (Victoria)

CASES21 is a successful information system at the Victorian Government level. To manage an education system at State level requires management of financial information. This has been difficult in systems such as the Victorian State system as schools were traditionally independent in the management of their administrative data. In particular, getting consistent financial data from schools when they each had their own accounting system was almost impossible (Birse 1994). This is why CASES (Computerised Administrative Systems Environment in Schools) was introduced in the mid-1980s and mandated for use in all Victorian government schools. Whereas, previously each school was able to keep their own financial and student records

in their own way, now they would have to use a common system provided by the Education Ministry. All Victorian Government schools now make use of the latest version, CASES21, for school management purposes including school administration and finance (Department of Education and Early Childhood Development 2011a; Tatnall, Michael and Dakich 2011).

The Administration Module offers management facilities such as: storing student and family data, maintaining attendance records, recording student achievement, detailing school activities, basic timetabling and daily organisation. The Finance Module creates and receipts family and student invoices, manages debtors and creditors, manages the school's asset register, processes and manages the school's payroll and manages school finances and budgets. It also generates appropriate financial reports. CASES was developed as a tool for overall school administration and as a means whereby schools reported back to the Department of Education (Tatnall 1995). Its prime purpose was to enable reporting from schools back to the Department of Education, rather than to directly assist schools with their own administration (Birse 1994; Tatnall 1995).

The Ultranet (Victoria)

'The Ultranet' is a local name given to an unsuccessful system at the Victorian Government level. Traditionally in Australia, parents have been kept informed about their child's progress at school through the use of school reports, parent-teacher evenings and an annual 'Open Day'. All parents want to know as much about the school where their children spend so much time as possible, and also how well they are progressing. Recently the Victorian

Education Department thought that there must be a better way, and began development of the Ultranet.

The Ultranet, launched in September 2010, was designed to support knowledge sharing across Victorian government schools with their half million students, along with their parents and teachers. It was intended to provide facilities for keeping parents informed about their children as well as for curriculum delivery and online teaching and learning (Department of Education and Early Childhood Development 2010). It is: "*a student centred electronic learning environment that supports high quality learning and teaching, connects students, teachers and parents and enables efficient knowledge transfer.*" (Department of

Education and Early Childhood Development 2011b). Although many education systems make use of tools for online teaching and learning, the Ultranet is quite different in that it also provides information to parents.

The Ultranet has many of the features of a business extranet in that it is closed to people outside the Victorian government school community and requires a username and password to gain access (Tatnall and Davey 2013). One major difference, however, is that with over half a million users, the Ultranet is larger than most business extranets (Tatnall and Dakich 2011).

The Ultranet was designed with three principle uses in mind: Firstly, to facilitate teachers' ability to create curriculum plans,

Figure 1. What is the Ultranet? (Department of Education and Early Childhood Development 2011b). (http://www.education.vic.gov.au/about/directions/Ultranet/default.htm).

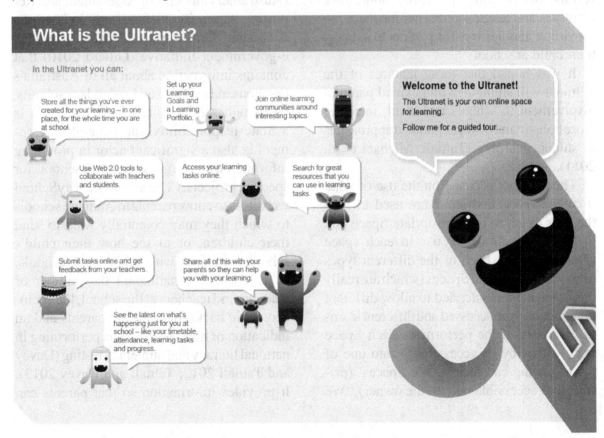

collaborate with other teachers, monitor student progress and provide student assessment. Secondly to allow students to access personalised learning activities and to keep an ongoing record of these activities. Students will be able to collaborate with students from their own school and with students from other Victorian government schools. They will be able to create learning portfolios and use online communication tools such as wikis, blogs and discussion boards. Thirdly, the Ultranet was also designed to assist parents in gaining the benefits of flexible access to student information and school resources. The idea was that they would be able to access the Ultranet to see information that would help them keep up-to-date with their child's learning (Tatnall, Dakich and Davey 2011; Tatnall, Michael et al. 2011). This dynamic student profile will include attendance records, timetables, test results and learning progress, homework activities and tasks, and teacher feedback, so providing another way for parents to support their child at school.

It was hoped that these features of the Ultranet will strengthen and extend parental involvement in schools and result in richer more holistic and better negotiated approaches to student learning (Tatnall, Michael et al. 2011).

The Ultranet is based on the use of what it calls Spaces, and Icons are used to help the user to get to the appropriate Space and to the Applications for use in each space (Tatnall 2011a). Each of the different types of specially designed Spaces (which are really mini-websites) is intended to allow different information to be accessed and different learning activities to be performed. Each Space is classified by its accessibility into one of the following categories: 'Me Spaces' (private, and accessible only to the owner), 'We

Spaces' (shared with permission) and 'See Space's (open, public access). The currently available spaces are: Home, eXpress Space, Design, Community, Collaborative Learning, Learner Profile, Learning Tasks, My Content and Connect (Tatnall, Michael et al. 2011; Tatnall and Davey 2013).

Unfortunately there is now little doubt that the Ultranet has been a significant failure, and the Victorian Government has recently sold the system to a private company (Department of Education and Early Childhood Development – Victoria 2013; Preiss 2013; Tatnall and Davey 2013), meaning that schools will have to pay a fee to use it.

MySchool (Australia)

MySchool is a system at the Federal government level that has been successfully adopted. (Australian Curriculum Assessment and Reporting Authority 2010a). Early in 2010 the Australian Government launched this new e-government initiative (Gillard 2010) that contains information about all of Australia's ten thousand primary and secondary schools. (Although school education in Australia is a State responsibility, the Federal Government is also a significant actor in providing information and targeted grants to schools for specific projects.) The aim of the MySchool website is to allow parents to compare schools to which they may potentially want to send their children, or to see how their child's school compares with other local schools. It includes information on the number of students and teachers at the school, the socio-economic background of the parents and an indication of how the school is performing in national literacy and numeracy testing (Davey and Tatnall 2012; Tatnall and Davey 2013). It provides information so that parents can

Figure 2. MySchool website (http://www.myschool.edu.au/) with information for Montmorency South Primary School

compare their school (or potential school) with neighbouring schools and with sixty statistically similar schools (Gillard 2010).

There has been a good deal of controversy about MySchool as some schools consider that they have been unfairly treated and that as a consequence they have lost potential students. Others have argued that the methodology for obtaining and showing the data is flawed.

INNOVATION AND ADOPTION

In this chapter we see innovation as including getting new ideas accepted or new technologies adopted and used (Tatnall 2005). After the invention and development of a new technology it does not automatically follow that this will be adopted by its potential users and this chapter traces three very similar systems in one environment where only two can be said to have been adopted. Whether or not a particular school management system is adopted, and how it is used is best considered in terms of innovation theory, but first it is worthwhile to distinguish between invention and innovation. While invention involves the discovery or creation of new ideas or technologies, innovation is the process of putting these ideas or technologies into commercial or organisational practice (Maguire, Kazlauskas and Weir 1994; Tatnall 2007b; Tatnall 2011b). The Oxford dictionary defines *innovation* as *"the alteration of what is established; something newly introduced"* (Oxford 1973) and is concerned with individual and business decisions to *adopt* new inventions (Tatnall 2007a; Tatnall 2011b). The study of innovation does not concern itself with inventors and the details of their inventions, but about

individual and organisational decisions to adopt these new inventions. Invention does not necessarily invoke innovation, nor is invention necessary or sufficient for innovation to occur (Tatnall 2005).

One critical question that must be asked before attempting to investigate the adoption process in each specific case is whether or not the potential adopters *had some choice* in making the adoption. If this is not so then the discussion would need to be quite different. It would then centre not so much on adoption, but on the degree of adoption and whether all aspects of the technology were adopted, as well as on the degree of use. One of the difficulties faced in investigating the adoption of technological innovations is that not all of these innovations are adopted in the form in which they were proposed – not all are adopted without change. This raises the question of just what was adopted in each case if it was in some way different from what was proposed by its instigator. While many of the models or approaches to the theorising of socio-technical innovation (including Innovation Diffusion (Rogers 2003) and the Technology Acceptance Model (Davis 1989)) find partial adoption difficult to explain, in this article we will make use of Innovation Translation which has no trouble in doing so.

Perhaps it could be said that we are being a little unfair, but personally we cannot see much value in the explanatory power of the Technology Acceptance Model (TAM), as reliance on perceived usefulness and perceived ease of use seems rather obvious and of little value in understanding what is really going on in a very complex situation and why some innovations are adopted in one way in one place but another way in a different situation. In this article we will thus make use of Innovation Translation as a means of investigating adoption and use of school management systems.

The view of innovation proposed by actor-network theory, sometimes also known as a Sociology or Translations, considers the world to be full of hybrid entities (Latour 1993) containing both human and non-human entities and elements to help in the explanation of technology adoption (Tatnall and Davey 2007). Innovation Translation (Latour 1986; Law and Callon 1988; Latour 1996) makes use of a model of technological innovation that gives due credit to the contributions of both human and non-human actors. It also notes that innovations are often not adopted in their entirety but only after 'translation' into a form that is more appropriate for use by the potential adopter. Callon et al. (1983) propose that translation involves all the strategies through which an actor identifies other actors and arranges them in relation to each other, while Latour (1996) speaks of how 'chains of translation' can transform a global problem, such as the transportation needs of a city like Paris, or the use of a new school management system, into a local problem like continuous transportation, or using the Ultranet to inform a parent about their child's progress. Latour (1986) maintains that the adoption of an innovation comes as a consequence of the actions of everyone in the chain of actors who has anything to do with it. He suggests that each of these actors shapes the innovation to their own ends, but if no one takes up the innovation then its movement simply stops. Here, instead of a process of transmission, we have a process of continuous transformation (Latour 1996) where faithful acceptance involving no changes is a rarity requiring explanation. "*Instead of the trans-mission of the same token – simply deflected or slowed down by friction – you get ... the continuous transformation of the token.*" (Latour 1986:286).

It is not so much the actors themselves that are important though, but rather their interactions with each other. In ANT terms a network is simply an interconnection of actors, both human and non-human, and what is particularly important is how these actors associate with each other. Actor-network theory sees the process of innovation in terms of *translation* from one state to another and Callon (1986) suggests that the process of translation has four aspects or 'moments': Problematisation, Interessement, Enrolment and Mobilisation. We will make use of these concepts of innovation translation in the article.

ADOPTION AND USE: AN ANT ANALYSIS

The first step in any ANT analysis is identification of the actors and we will look at the human and non-human actors involved in the creation and use of each of these administrative systems. We will then look at the interactions between these actors.

The Actors

This study is especially interesting as most of the human actors (schools, teachers, students and parents) and many of their interactions, are the same for each system. Many of the non-human actors (broadband connections, web 2.0 technologies, schools, school computers, home computers, policies, privacy laws, School Councils) are also common. It is thus primarily the interactions between these human actors and with the non-human actors that may lead to any new insights as the main difference in each case is the software applications used.

CASES21

Here the human actors include: teachers, school principals, school administrators, teacher educators and the CASES21 Developers. The non-human actors include: broadband connections, Web 2.0 technologies, schools, school computers, policies, privacy laws, School Councils, the Victorian Department of Education and Early Childhood Development, the Victorian Government and the technology of CASES21.

The Ultranet

Human actors: students, teachers, school principals, parents, teacher educators, pre-service teachers and the Ultranet developers. Non-human actors: broadband connections, Web 2.0 technologies, schools, school computers, home computers, the emerging National Curriculum, policies, privacy laws, School Councils, the Victorian Department of Education and Early Childhood Development, the Victorian Government and the technology of the Ultranet.

MySchool

Although at a different governmental level MySchool has similar human actors as there are no Federal government schools. For MySchool the human actors include: students, teachers, school principals, parents, teacher educators, pre-service teachers and the MySchool Developers. The non-human actors include: broadband connections, Web 2.0 technologies, schools, school computers, home computers, the emerging National

Curriculum, policies, privacy laws, School Councils, the Commonwealth Department of Education Employment and Workplace Relations, the Australian Government and the technology of the MySchool website.

Significant Interactions

Governments with School Communities

In Victoria, State Schools are largely self-governing (- Independent Schools, by their very nature are also self-governing, but in a rather different way). Part of the decentralisation effort of government has been to grant more power to schools to employ their own individual teachers, although this is constrained by budgets. This decentralisation has been accompanied by attempts to make schools accountable for performance and a number of techniques such as a National Curriculum initiative and the National Testing Plan are being employed (Australian Curriculum Assessment and Reporting Authority 2010b) for this purpose (Tatnall and Davey 2013). For a long period, interpretation and delivery of the curriculum has been largely a school matter with government maintaining only a general oversight. The political dichotomy between making schools self-governing while allowing governments to still have sufficient control so that they can fulfil their accountability requirements to the electorate for education expenditure takes some juggling. In recent years however, governments (both federal and state) have made an attempt to standardise aspects of the curriculum. In 2008 the Australian Government instituted the National Assessment Program - Literacy and Numeracy (NAPLAN) in Australian schools. NAPLAN consists of national tests in reading, writing, language conventions (spelling, grammar and punctuation) and numeracy and all students in Years 3, 5, 7 and 9 are assessed on the same days every year (Australian Curriculum Assessment and Reporting Authority 2010b). All this has, so far, had little effect on the way in which curriculum is delivered, but will MySchool and the Ultranet change this?

All this is relatively new in Australia and the interactions between governments and schools are changing. At this time, the relationship between governments and schools could not be said to constitute one of complete trust.

Teachers and Teacher Unions with Governments

In Victoria, school teachers are represented by the Australian Education Union (AEU). Teacher unions are concerned not just with employment conditions, but also with matters relating to the interrelationships between schools. The MySchool website has been seen to threaten this relationship by providing easy access to information about all schools and the possibility of a 'leagues table' of the 'best' schools in the country. The AEU has been concerned that although the MySchool website does not do this, perhaps newspapers and other media will make use of the data it contains to do so. One of the recent campaigns by the AEU has been the 'Stop league tables campaign' (Australian Education Union 2011), resulting in the AEU submission to the 'Senate Education Employment and Workplace Relations Committee' into the 'Administration and Reporting of NAPLAN Testing' (Australian Education Union 2010). The Union campaign is directed at preventing the Government having a more intrusive influence on teaching. One effect of the MySchool website seen as being a significant problem by the union was the possibility of a school

changing its delivery as a result of poor showing in the National Tests (Davey and Tatnall 2012; Tatnall and Davey 2013).

Teachers have been told by the Victorian Education Department to "explicitly teach for NAPLAN", focus on literacy and numeracy and give students a "daily NAPLAN item" in class ... The directive has led to accusations that Victorian education authorities are pressuring schools to "teach to the test" to lift their performance on the new website at the expense of a broader curriculum. (Perkins 2010)

Most teachers believe that the teacher is the best qualified person to know what happens in the school community and also the person with understanding of the local educational environment. This can be seen to mitigate against anyone else having the right to significant input into school level decisions, and the relationship between government and teachers can be seen as being very important to the fate of the three administrative systems.

CASES21 (and its predecessor) was built to provide a clear single means of gathering the data required by government from schools. The original version did not give any recognition to the needs of the schools, just the government. In the knowledge that schools saw themselves as having significant data needs the first version changes in CASES were to include modules that answered local school needs for the data being collected by government (Davey and Tatnall 2012; Tatnall and Davey 2013), and in this way the CASES software immediately began 'co-operating' with the schools and teachers. MySchool similarly recognised the power of teachers at the local level and so completely bypassed them, communicating through the public website rather than through normal school communication channels, all of which required teacher involvement. The Ultranet, on the other hand, has little government involvement or control and was designed to involve teachers as a part of the value of the system.

Teachers with Students

This very important interaction has a very long history. It has not been materially affected by CASES21 or MySchool. How it will be affected by the use of the Ultranet is not yet completely clear.

Teachers with Parents

The interviews we have conducted show that teachers see several different types of relationships with parents. Firstly with the enthusiastic parents who want to be part of the school and will do whatever they can to do so. At the primary school level these parents are involved with as many voluntary jobs such as canteen worker, working bee participant and provider of reading help as they can. Some of these parents seek election to the School Council which is an official, unpaid position with some governance duties for the school. School Council members, however, do not have a direct method for communicating with other parents, and teachers do not see them as having the influence of elected officials with the backing of an electorate. By the secondary school level parental involvement generally diminishes to just some working bees and other volunteer jobs, and has a much lower proportion of families involved.

Next there are those parents who see the teachers as domain experts who should (and do) take sole responsibility for the student learning experience. These parents will sign off homework and dutifully attend the occasional Parent Teacher meeting which (in the past) have general been held once per term (semester) on a single evening for all families. These parents are happy to be directed as to what role to play and will openly become involved if they feel their 'consumer rights' have been infringed when their children perform less well than expected (Davey and Tatnall 2012; Tatnall and Davey 2013).

There are also parents who seen to be disinterested in the school. Teachers interviewed indicated that some parents were never seen and could not be relied upon to support school or teachers' decisions.

These relationships have a significant impact on the chances of acceptance of MySchool and the Ultranet. CASES21 has an impact on parents to the extent that the system provides the school with fast access to important information (from the Education Department) through a single system, but they only have access to its output in the form of lists and reports. MySchool, on the other hand talks directly to parents without the involvement of teachers, informing them of the results of the NAPLAN tests in their child's school and by comparison to those of other schools. One possible consequence of this MySchool information mentioned in interviews with parents is that parents may consider moving their children to another school if their current school's performance is below par. The Ultranet seeks to make the task of communicating with parents very much easier for a teacher, and teachers could become more aware, with little effort, of the home circumstances and some of the other things that affect students if parents were enticed to use the system. In our interviews, however some teachers saw the Ultranet as presenting an extra workload burden in which they saw little value and some parents were not able to see anything on the system that would entice them to look further, let alone contribute. Some doubt must however be placed on drawing conclusions from this as the system was not then fully functional. Many of the functions that have been present for some time however, do seem to be being ignored by both parties. At this stage, the relationship between parents and teachers would lead us to conclude that unless new actors emerge or new associations develop the system will not progress far past the present stage. It would not be in the interests of a teacher who claims power over the education process to be completely candid with parents as this would presuppose that parents could do something with any information they receive. Similarly a system that does not provide useful information is unlikely to have parents contributing.

Teachers with MySchool, CASES21 and the Ultranet

Teachers are not involved in any way with the operation of the MySchool website other than to see its output. MySchool is aimed at parents and not at teachers. Most teachers, except those in administrative positions, have little direct interaction with CASES21 other than making use of its output. Our interviews suggest that teachers do not tend to make suggestions for improvement to CASES21 as they think that these will not be given serious consideration. The Ultranet has been designed to involve teachers in performing their daily work and to assist them in working with students. It does, however, have the mixed problems of providing little help to teachers in their daily work and exposing them to local accountability that they have not experienced before.

Students with the Ultranet

School students have no direct involvement with CASES21. The same is true of the MySchool website except perhaps to hear their parents discussing this. They will however, have considerable involvement with the Ultranet if their school is using it to its full potential. The Ultranet offers the possibility of a significantly changed learning experience for students by allowing them to access personalised learning activities such as the creation of learning portfolios and use of online communication tools such as wikis, blogs and discussion boards. It will also enable them to keep an ongoing record of these activities and to collaborate with students from their own school and with students from other Victorian government schools.

Parents with MySchool with the Ultranet

Using the MySchool website, parents are able to compare their child's school with other schools. How this interaction will play out is not yet clear but could be most significant for the future of schooling in Victoria. The Ultranet aims to assist parents in gaining access to student information and school resources by providing additional information about their child's learning. If parents adopt this to the full it could have a profound effect on education.

Administrators and School Principals with MySchool, CASES21 and the Ultranet

School bursars and other administrative staff make considerable use of CASES21. It has had a considerable effect on how they perform their work, but as it is a closed system they are not able to have any effect on it. School principals make good use of the output of the CASES21 system normally provided for them by their administrative staff. Like their teachers, school principals have a profound interest in the MySchool website as potentially it could affect the number of new students wanting to attend their school and to the retention of current students. At the time of writing it was not clear how school principals would interact with the Ultranet other than to encourage their teachers and the parents of their students to work with it.

Our ANT approach to this investigation considered all of the interactions between these actors as being very significant.

ADOPTION AND TRANSLATION

It would be good both to project the future adoption and adaptation of these technologies and also to improve their chances of successful adoption. Borrowing a little from Diffusion of Innovations by Rogers (2003), Perez (2002) views the adoption and life cycle of a new technology as consisting of a number of phases from introduction to full expansion. Innovation Translation, however, would consider this rather differently as consisting of a number of translations from the original technology to one in the form it is finally adopted. Many ANT articles look at how a technology has been translated before leading to adoption, but we will attempt to go a little further. In this section we will consider only the Ultranet and propose some possible future translations based on our interviews with the principal actors regarding the different ways they saw this as possibly developing.

Possible Translations and the Likelihood of Adoption

We have described three situations where a new Information Systems has been introduced to the educational environment of Australia. Our contention is that this puts us in the enviable situation of being able to see the actors

and interactions common to all three, to see how these interactions have led to particular translations and hence enable us to make sensible suggestions as to how the new (Ultranet) system might become translated in the future.

ANT is normally seen as a post-facto research approach that yields an explanatory story of a translation. In this case the attempt will be made to identify *possible* stories in advance of the completion of any translation. Of course the 'predictions' can only be of *possible* translations. These discussions will, hopefully be a useful guide to those responsible for the implementation and adoption of the Ultranet and the technique seems fruitful for any other studies where prior knowledge of actors and their interactions is extensive.

The first translation presented is founded upon the characteristics of the software: if the Ultranet makes it difficult for parents to interact with each other, then the translation would not be one in which the system promoted action by parents, for example, for them to band together to effect the National Curriculum.

An alternative translation would be one in which a network could be formed at the school level where teachers, parents and students are encouraged to make the curriculum local and living. This type of interaction has been seen in indigenous communities where traditional education informs schooling. An examination of the supporting literature for the Ultranet project suggests that this is an intended aim of the new system.

From the ANT analysis of the three systems we can see that the Ultranet could be adopted in any one of a number of effective translations. These examples show the way in which these translations would be consistent with the already identified actors and their interactions:

A System Promoting Involvement of Parents in the Life of Schools

This translation would see the Ultranet as producing a community wide educational environment. Student education would be seen as part of their whole life and family rather than a disjoint time during semester holidays. This translation could result in a reduced face-to-face interaction between parents and schools, but it offers new avenues for flexible partnerships between teachers and parents giving asynchronous communications, better alignment of goals and aims between families and teachers and a shared understanding of the desired outcomes of the school. Digital literacy could cause concerns as parents with lack of digital skills may find it difficult to engage with the innovation. This is likely to have a number of adverse effects on the adoption of the Ultranet and could result in widening the gap between those who have and those who have not, initiating further inequalities in the acquisition of social capital and access to digital citizenship. This translation is prompted by the interactions shown in previous cases between the information systems, parents, teachers, communities.

A Social Networking System and New Learning Platforms for Students

This translation sees the Ultranet as a system for creating effective local and global networks of learners who can communicate, exchange information and collaborate in augmented realities (real and virtual), allowing for new learning and teaching practices to emerge. The system would provide students with an innovative, multidimensional eLearning environment. It is unlikely that students will bring in the personal dimension of social networking because of the closely monitored nature of the Ultranet. This translation also carries the possibility that educational direction could be lost. It is prompted by the interactions shown in previous cases between information systems, students from schools in Victoria and all around the world.

A Community Networking System for Community Directed Education

The Ultranet has the capacity to create a social network for all those interested in their local schools. Parents, teachers and students could be enabled to take an active role. Educational decisions, resource allocation, individual student progress and teacher employment could be the subject of community discussion and decision making. The idea of communities deciding what happens in the detail of schooling forms the basis for several independent schools in Victoria. Some of these matters have theoretically been devolved to local communities in State Schools, but the practicalities of democratic decision making normally preclude all but a select group doing the decision making. The Ultranet could provide the platform for community directed schooling. This again raises the issue of digital divide. Communities poor in resources and those with low expectation of schooling could suffer disadvantage over a system with minimum standards.

A System for Monitoring Student Progress

Teachers will be more informed about individual students through systemic information collected by other teachers. While open communication between teachers has positive possibilities for individual treatment of students, such a system can confirm expectations of student ability which have been shown to be detrimental. This translation is predicted from studying the past interactions of information systems, students, parents and teachers.

A System for Teacher Collaboration and Professional Development

The Ultranet may provide an online platform for teachers to collaborate, share learning objects and access professional development programs. It could also provide teachers with opportunities for online professional learning, trouble-shooting and technical support. A missing actor here is that teacher education institutions are not considered to be part of the password-protected networks. This limits the agency of these actors to successfully prepare student teachers for their future workplace by not having good access to the Ultranet. This has been the case with other Victorian Government initiatives and this anomaly should be addressed. This translation is prompted by the interactions shown in previous cases between teachers, schools, school leadership(s), teacher education institutions, Victorian Institute of Teaching, pre-service teachers and teacher educators.

CONCLUSION

This chapter used three administrative systems in the same Australian context to illustrate the power of the ANT approach to understanding adoption. At the time of writing, two of those systems had been implemented and the third, the Ultranet, was in the process of implementation. We used the knowledge of actors and their interactions from the known translations of two systems to provide insight into the possible translations of the new system. This process yielded five alternative translations that could form part of the eventual translation of the Ultranet.

This chapter has attempted to provide two contributions. Firstly the analysis seeks to provide information to the providers of the Ultranet as to what might come of their innovation. This includes both positive and problematic outcomes arising from the translations that might occur under the influence of the actors. This knowledge should be useful in helping to provide positive direction from the providers and to help them anticipate possible outcomes and mitigate any adverse outcomes.

The second contribution is to the ANT community. For as long as qualitative research has been used it has been criticised as not being predictive by those of a positivist bent. We know that a simple quantitative prediction is seldom of importance in situations involving humans and complexity. Instead the qualitative researcher seeks wisdom and the explanatory stories of ANT studies certainly provide us with that wisdom. In this chapter we have suggested an approach, at least in special circumstances, where an ANT lens yields a wisdom that can then be applied to an unfinished project. We suggest here that knowledge of a set of actors and their interactions should allow us to become informed of a likely set of possibilities. We do not contend that these are inevitable or even point to their likelihood, but suggest that this process provides a means of applying the wisdom gained through an ANT examination of an environment and using it to guide the future of the new innovation.

REFERENCES

Australian Curriculum Assessment and Reporting Authority. (2010a). *My School*. Retrieved November 2010, from http://www. myschool.edu.au/SchoolSearch.aspx

Australian Curriculum Assessment and Reporting Authority. (2010b). *Naplan - National Assessment Program*. Retrieved Jan 2012, from http://www.naplan.edu.au/home_page.html

Australian Education Union. (2010). *Australian Education Union Submission to the Senate Education Employment and Workplace Relations Committee into the Administration and Reporting of NAPLAN Testing*. Melbourne, Australia: Australian Education Union.

Australian Education Union. (2011). *Stop League Tables*. Retrieved Jan 2012, from http://www.aeufederal.org.au/LT/index2.html

Callon, M. (1986). *Some Elements of a Sociology of Translation: Domestication of the Scallops and the Fishermen of St Brieuc Bay. Power, Action & Belief. A New Sociology of Knowledge?* London: Routledge & Kegan Paul.

Callon, M., Courtial, J. P., Turner, W. A., & Bauin, S. (1983). From translations to problematic networks: An introduction to co-word analysis. *Social Sciences Information. Information Sur les Sciences Sociales, 22*(2), 191–235. doi:10.1177/053901883022002003

Davey, W., & Tatnall, A. (2012). Using ANT to Guide Technological Adoption: the Case of School Management Software. *International Journal of Actor-Network Theory and Technological Innovation, 4*(4), 47–61. doi:10.4018/jantti.2012100103

Davis, F. (1989). Perceived Usefulness, Perceived Ease of Use, and User Acceptance of Information Technology. *Management Information Systems Quarterly, 13*(3), 319–340. doi:10.2307/249008

Department of Education and Early Childhood Development. (2010). *Ultranet - Students@ Centre Trial*. Retrieved November 2010, from http://www.education.vic.gov.au/about/directions/ultranet/trial.htm

Department of Education and Early Childhood Development. (2011a). *CASES21*. Retrieved January 2012, from http://www.education.vic.gov.au/management/ictsupportservices/cases21/functionality.htm

Department of Education and Early Childhood Development. (2011b). *Ultranet*. Retrieved January 2012, from http://www.education.vic.gov.au/about/directions/ultranet/default.htm

Department of Education and Early Childhood Development – Victoria. (2013). *S245-2013 DEECD and NEC sign agreement to facilitate continued provision of the Ultranet*. Author.

Gillard, J. (2010). *My School website launched*. Retrieved May 2010, from http://www.deewr.gov.au/ministers/gillard/media/releases/pages/article_100128_102905.aspx

Latour, B. (1986). *The Powers of Association. Power, Action and Belief. A New Sociology of Knowledge?* London: Routledge & Kegan Paul.

Latour, B. (1993). *We Have Never Been Modern*. Cambridge, MA: Harvester University Press.

Latour, B. (1996). *Aramis or the Love of Technology*. Cambridge, MA: Harvard University Press.

Law, J., & Callon, M. (1988). Engineering and Sociology in a Military Aircraft Project: A Network Analysis of Technological Change. *Social Problems, 35*(3), 284–297. doi:10.2307/800623

Maguire, C., Kazlauskas, E. J., & Weir, A. D. (1994). *Information Services for Innovative Organizations*. Sandiego, CA: Academic Press.

Oxford. (1973). *The Shorter Oxford English Dictionary*. Oxford, UK: Clarendon Press.

Perez, C. (2002). *Technological revolutions and financial capital: The dynamics of bubbles and golden ages*. Cheltenham, UK: Edward Elgar Publishing. doi:10.4337/9781781005323

Perkins, M. (2010). *Teach for tests, teachers told*. Retrieved January 2012, from http://www.theage.com.au/national/education/teach-for-tests-teachers-told-20100204-ng6o.html

Preiss, B. (2013). Plan to Recoup Funds from Troubled Ultranet System. *Age*, 5.

Rogers, E. M. (2003). *Diffusion of Innovations*. New York: The Free Press.

Tatnall, A. (1995). Information Technology and the Management of Victorian Schools - Providing Flexibility or Enabling Better Central Control? In *Information Technology in Educational Management*. London: Chapman & Hall. doi:10.1007/978-0-387-34839-1_13

Tatnall, A. (2005). Technological Change in Small Organisations: An Innovation Translation Perspective. *International Journal of Knowledge. Culture and Change Management*, *4*(1), 755–761.

Tatnall, A. (2007a). *Business Culture and the Death of a Portal*. Paper presented at the 20th Bled e-Conference - eMergence: Merging and Emerging Technologies, Processes and Institutions. Bled, Slovenia.

Tatnall, A. (2007b). Innovation, Lifelong Learning and the ICT Professional. In *Education, Training and Lifelong Learning*. Laxenburg, Austria: IFIP.

Tatnall, A. (2011a). *Information Systems Research, Technological Innovation and Actor-Network Theory*. Melbourne, Australia: Heidelberg Press.

Tatnall, A. (2011b). Innovation Translation, Innovation Diffusion and the Technology Acceptance Model: Comparing Three Different Approaches to Theorising Technological Innovation. In *Actor-Network Theory and Technology Innovation: Advancements and New Concepts*. Hershey, PA: IGI Global.

Tatnall, A., & Dakich, E. (2011). *Informing Parents with the Victorian Education Ultranet*. Novi Sad, Serbia. *Informing Science*.

Tatnall, A., Dakich, E., & Davey, W. (2011). The Ultranet as a Future Social Network: An Actor-Network Analysis. In *Proceedings of 24th Bled eConference*. University of Maribor.

Tatnall, A., & Davey, W. (2007). *Researching the Portal*. Vancouver, Canada: Information Management Resource Association.

Tatnall, A., & Davey, W. (2013). The Ultranet and School Management: Creating a new management paradigm for education. In *Next Generation of Information Technology and Educational Management*. Heidleberg, Germany: Springer. doi:10.1007/978-3-642-38411-0_15

Tatnall, A., Michael, I., & Dakich, E. (2011). *The Victorian Schools Ultranet - An Australian eGovernment Initiative*. Paper presented at International Information Systems Conference - iiSC 2011. Muscat, Oman.

Tatnall, C., & Tatnall, A. (2007). Using Educational Management Systems to Enhance Teaching and Learning in the Classroom: An Investigative Study. In *Knowledge Management for Educational Innovation*. New York: Springer. doi:10.1007/978-0-387-69312-5_10

KEY TERMS AND DEFINITIONS

Actors, Human and Non-Human: Actor-network theory attempts to treat the interactions and contributions of both human and non-human actors equally and in the same way.

Administrative Use of ICT in Schools: The use of computer software in the management and administration of schools tasks such as finance, record keeping and communication.

Approaches to the Theorising of Socio-Technical Innovation: These include – Innovation Diffusion, Technology Acceptance Model and Innovation Translation.

CASES21 (Computerised Administrative Systems Environment in Schools): A Victorian Government school management software package for purposes including school administration, finance and reporting.

Innovation and Adoption: Innovation involves the Adoption of new technologies or ideas and getting them accepted and used.

MySchool: An Australian Federal Government Internet-based system intended to provide the general community with school performance data. Its aim is to allow parents to compare schools to which they may potentially want to send their children, or to see how their child's school compares with other local schools.

NAPLAN: Australian Government National Assessment Program - Literacy and Numeracy.

The Ultranet: A communications system intended to connect parents, schools, teachers and students. It is "a student centred electronic learning environment that supports high quality learning and teaching, connects students, teachers and parents and enables efficient knowledge transfer".

Chapter 11
Studying DNA:
Envisioning New Intersections between Feminist Methodologies and Actor–Network Theory

Andrea Quinlan
Lakehead University, Canada

ABSTRACT

Feminist methodologies and Actor-Network Theory (ANT) have often been considered opposing theoretical and intellectual traditions. This chapter imagines a conversation between these seemingly divergent fields and considers the theoretical and methodological challenges that ANT and particular branches of feminist thought raise for the other. This chapter examines an empirical project that calls for an engagement with both ANT and feminist methodologies. Through the lens of this empirical project, four methodological questions are considered, which an alliance between ANT and feminist methodologies would raise for any research project: 1) Where do we start our analysis? 2) Which actors should we follow? 3) What can we see when we begin to follow the actors? 4) What about politics? The potential places where ANT and feminist methodologies can meet and mutually shape research on scientific practice and technological innovation are explored. In doing so, this chapter moves towards envisioning new intersections between feminist methodologies and ANT.

INTRODUCTION

According to Sandra Harding (2008), a well-known feminist scholar in science studies, feminist scholarship and Science and Technology Studies (STS) have a lot to learn from one another. Despite the past few decades of feminist scholars' efforts to integrate gender, race, and class into analyses of science and technology, these themes remain largely absent in mainstream STS. Feminist scholarship, Harding suggests, still has much to teach STS about how science and technology are shaped by gender, race, and class and the importance of politically informed methodology. Conversely, Harding

DOI: 10.4018/978-1-4666-6126-4.ch011

argues that feminist scholarship has much to learn from the questions that STS poses about how science and technology are organized, practiced, and constructed within the Western world. Through this, Harding offers a hopeful vision of further dialogue between STS and feminist scholarship.

Actor-Network Theory (ANT) is a theoretically informed approach in STS that has been used to study the rapidly changing worlds of science and technology. Despite its apparent success in STS, feminist scholars in the 1990s and early 2000s raised sharp criticisms of ANT for its disregard of gender (Wajcman, 2000) and power inequalities (Casper & Clarke, 1998; Star, 1991), and for its apolitical and "insufficiently radical" orientation (Wajcman, 2000, p. 452). From these critiques, it would seem that feminist traditions and ANT are incommensurable. But perhaps, as this paper will explore, this need not be the case.

This paper takes up Harding's (2008) hopeful projection and imagines a meeting between the seemingly divergent fields of feminist methodologies and Actor-Network Theory. I will examine an empirical project that calls for an engagement with both ANT and feminist scholarship. Through the lens of this empirical project, I will consider what these diverging theoretical and methodological traditions can learn from each other. I will consider four methodological questions that an alliance between ANT and feminist methodologies would raise for any research project: 1) Where do we start our analysis? 2) Which actors should we follow? 3) What can we see when we begin to follow the actors? 4) What about politics? Through these questions, I will explore how ANT and feminist methodologies challenge and potentially speak to one another.

FEMINIST METHODOLOGIES AND ANT

Actor-Network Theory and most feminist theoretical and methodological approaches stem from very different political, intellectual, and historical traditions. ANT grew predominately out of the intellectual movements of post-structuralism, constructivism, and ethnomethodology (Law, 1999). Feminist scholarship, on the other hand, grew from social and political movements that hinged on eradicating gendered, raced, and classed inequalities (MacKinnon, 2005). While some branches of feminist theory have, similar to ANT, drawn on post-structuralism and constructivism (e.g. Haraway, 1991; Butler, 2004; Mohanty, 2003), others have been built from theoretical traditions such as Marxism, existentialism, and psychoanalytic theories (de Beauvoir, 1957; Davis, 1983; Benjamin, 1988).

Feminist methodology is a diverse field made up of many distinct empirical, theoretical, and methodological approaches. To refer to feminist methodology as a unified tradition of thought and practice is a vast over-simplification of its diverse history. In a similar way, the diversity in ANT studies makes it difficult, if not impossible, to define in broad terms (Law, 1999). The diversity within these two fields presents a challenge for creating dialogue between them. This paper will therefore draw on very particular traditions within feminist methodologies and ANT, and will employ working definitions of each. These definitions are not intended to erase the diversity *within* these fields, but rather, allow for an exploration of what lies between them.

In this paper, I draw most significantly on Feminist Standpoint Theory, a theoretically informed methodology that is deeply rooted in Marxist epistemology and hinges on the epistemological assumption that oppressed and/or marginalized people see relations of power most clearly (Harding, 1991a). Harding defines Standpoint Theory as "thought that begins with the lives of the oppressed" (Harding, 1991a, p. 56). Women, people of colour, people living with disabilities and/or experiencing poverty, and other marginalized groups are assumed to have the capacity to see mechanisms of oppression in ways that dominant groups cannot. The oppressed have, what some theorists have called "epistemological privilege" on dominant, oppressive relations of power (Brooks, 2007, p. 69). As this paper will show, this branch of feminist methodology raises particular challenges for ANT.

I define ANT as a deeply theoretical methodology that has been commonly used to explore the "messiness" (Law, 2004, p. 4) of scientific practice and technological innovation. ANT, like feminist Standpoint Theory, does not provide an explanatory framework for empirical realties, as social theories often do (Law, 2007). Nor does it provide a rigid set of methodological rules for studying associations (Latour, 2005). Rather, as Law (2007) suggests, ANT is a "toolkit for telling interesting stories" (p.1). ANT provides tools for tracing "actors", both human and non-human, and the ways they work collectively in "networks" of action (Latour, 2005, p.5). Research in the field of ANT often tells stories of the "complexities", "translations", and "multiplicities" found in science and technology (Law & Mol, 2002, p.7).

While ANT and Feminist Standpoint Theory represent very distinct traditions of thought and practice, both provide useful tools for studying empirical realities. As this paper will show, creating a dialogue between these fields has the potential to expand the toolkits of each.

AN EMPIRICAL PROJECT

Since its development in the 1980s, forensic DNA analysis has grown to be one of the most predominant scientific technologies used in the Canadian legal system (Gerlach, 2004). The growing public excitement around forensic technologies has supported a rush of resources into this growing field of forensic work. In the context of sexual assault, various professional groups have formed to accommodate this new legal emphasis on forensic DNA evidence collection and analysis (Bumiller, 2008). Sexual Assault Nurse Examiners (SANE) are a relatively new professional group responsible for collecting DNA samples from sexual assault survivors'/victims'[1] bodies (Du Mont & Parnis, 2001; Parnis & Du Mont, 2006). The DNA samples that are collected through a forensic exam can be used to identify sexual offenders through DNA profiling methods (Quinlan, Fogel, & Quinlan, 2010).

The process of DNA extraction is incredibly invasive and can be traumatic and at times painful for the survivor/victim (Doe, 2003). A forensic examination can last up to 3 hours and often includes vaginal, anal, rectal, and/or oral swabbing (Parnis & Du Mont, 2006). When a survivor/victim consents to the forensic exam, they have little control over which body parts are examined, how long the exam lasts, and what kind of evidence is taken (Doe, 2003). Some, although certainly not all, survivors/victims experience a sense of vulnerability and powerlessness during the process of DNA collection and the legal investigation and prosecution that sometimes follows (Du Mont, White, & McGregor, 2009; Doe, 2012).

Scholars in Feminist Standpoint Theory have argued that social relations can be most clearly and objectively seen from the perspective of those that hold the least amount of power (Harding, 1991b, 2008; Hartsock, 2004; Smith, 2005). This methodological tradition in the context of the empirical project on DNA would remind us to be attentive to a survivor's/victim's experiences of having little power within medical and legal institutions (Smith, 2005). It would remind us to take seriously how these social relations look "from below" (Harding, 2008, p. 1). Using this approach, we could see the survivor's/victim's experiences of being disempowered in relation to DNA evidence extraction and construction.

While a feminist standpoint analysis may help to uncover an important part of the story of DNA analysis, there are complexities that a view from below cannot capture. The technologies and practices of DNA analysis are in a constant state of evolution (Quinlan, Fogel, & Quinlan, 2010). New approaches for DNA testing are increasingly being introduced to the field of forensics. For example, there is a growing interest in mitochondrial DNA analysis, which constructs DNA profiles from the mitochondria of the cell verses the nucleus, as older DNA testing techniques did (Quinlan, Fogel, & Quinlan, 2010). These new approaches to DNA testing are often incompatible with existing resources for perpetrator identification, such as the National DNA Databank, which stores only DNA profiles that have been constructed from cell nuclei. To use the words of Actor-Network Theorist, John Law (2004), this is a "messy" (p. 1) and complex ground to trace.

Actor Network Theory would remind us to be attentive to these brewing controversies and the complexity of negotiating scientific practices and technological development. More specifically, ANT in this study would call our attention to the network of humans and non-humans, texts, tools, and technologies that are working to assemble DNA evidence in the legal system. Using ANT, these aspects of DNA's story could become visible.

This empirical project begs an alliance between ANT and feminist methodology. Both draw attention to different, yet equally important dimensions of DNA evidence construction in cases of sexual assault. But can these divergent approaches be brought together? It is to this query that we now turn.

WHERE DO WE START?

Bruno Latour (2005), a well-known Actor Network Theorist, suggests that a "slogan for ANT" (p. 12) is to follow the actors and "grant them back the ability to make up their own theories of what the social is made of" (p. 11). The sociologist, he claims, must begin the analysis by listening to actors' and following them through the networks they inhabit. He criticizes mainstream sociology, what he calls "sociology of the social" (p. 25), for not listening closely enough to actors when they describe their networks. While Latour raises an important point about increasing our listening skills as sociologists, he says very little about *which* actors we should be listening to. This begs some important methodological questions: Which actors should be telling the story of the network? And what informs this choice? Do we simply listen to those actors who have the loudest voice? Or those that are the most visible? And if so, what are the consequences of doing so?

Feminist scholar, Leigh Star (1991), criticizes ANT for focusing its analysis on the most powerful actors in the network, or as she calls them, the "executives" (p. 29) or the "victors" (p. 33). Judy Wajcman (2000) makes a similar observation when she says, the "agents in ANT are most commonly male heroes, big projects and important organizations" (p. 453). In Latour's (2010) most recent work on the Conseil de-Etat in France, he describes his adventures in the worlds of law and science, two fields that he describes as being made up of male judges and "elite men in white coats" (p. 221). In Latour's narrative of law and science, there are only powerful men and their tools and texts. Women, racialized people, people living with disabilities and experiencing poverty are nowhere to be found.

The study of forensic DNA analysis could begin in the same place that Latour (2010) began his most recent study: we could allow the loudest and most powerful actors to describe the network. We could speak to forensic scientists, lawyers, and police officers, gathering their tales of constructing DNA evidence. However, in taking this approach, is there a piece of the network that would be ignored?

Star (1991) and Wajcman (2000) both argue that ANT's focus on powerful actors often leads to the erasure of actors that sit on the margins of networks. Star (1991) asserts that if ANT were to pay attention to the marginal actor, an entirely different actor-network could be drawn.[2]

Deciding on a methodological entry point for empirical analyses is for many feminist scholars, an important political decision. Embedded in Feminist Standpoint Theory's move to take up the 'view from below' is a political and methodological critique of social science traditions that do not take seriously the perspectives of marginalized and/or oppressed groups (Harding, 1991b, 2008; Hartsock, 2004; Smith, 2005). These theorists' contention that the 'view from below' provides a more accurate vantage point on social reality is an inherently political one. Standpoint Theory could be employed in the

study on DNA analysis. To do this, we might begin with the assumption that the survivor/victim of violence is a marginalized actor who has a privileged vantage point on the institutional practices of DNA evidence collection. However, if the aim is to study the material messiness of forensic science and technology, does the survivor/victim provide the best vantage point?

Apart from the forensic exam, survivors/victims are completely excluded from the production of forensic DNA evidence. What goes on in the lab, and any other site beyond the exam room, such as the police processing and evidence holding centre, the DNA Databank, and lawyers' offices, remains completely invisible to most survivors/victims. The only human actors in these sites are those who hold varying degrees of institutional credibility and power to participate in institutional practice. This raises the question: Is it possible to maintain a focus on the materiality of DNA evidence construction, without resorting to a tale told by the powerful?

Star (1991) suggests this is indeed possible. Loosely drawing on Standpoint Theory, Star proposes that the marginalized actor's perspective can be "a point of departure" (p. 38) for the study of an actor-network. She argues, the "voices of those suffering from the abuses of technological power are among the most powerful analytically" (p. 30). As an example, Star uses her own experience of contending with onion allergies at McDonalds and describes the difficulties she has had ordering a hamburger without onions. She describes herself as an actor that is on the margins of the McDonalds network. From her experience, she argues that it is possible to see how the McDonalds network is built in a way that does not easily accommodate onion allergies. Taking the marginalized actor as an entry point into the McDonalds network, she argues, would produce a radically different network than one that could be drawn from the experiences of powerful actors, such as the owners and originators of McDonalds. It would be possible to see a network 'from below' and examine how it is built to include and exclude particular actors.

Star (1991) distances her approach from Feminist Standpoint Theory when she argues that it is not necessary to be limited to a 'view from below'. Instead, this view can be used as merely an entry point into the complexity and multiplicity of an actor network. Being a marginal actor, for Star, involves holding multiple locations: the "marginal person [is] the one who both belongs and does not belong" (p. 50). Beginning with the marginal actor, who is simultaneously inside and outside of the network, we come to understand "multiple membership in many worlds" (p. 30). Through this entry point, it becomes possible to see how networks are "stabilized and standardized" (p. 44) for some actors but not for others. It becomes possible to see how the network itself can be multiple.

In proposing this approach, Star illustrates how feminist methodological traditions might inform a new starting place for ANT studies. With this new entry point, Star argues that it will be possible to draw new actor networks, ones that are attentive to the complexity that power inequality introduces to any network of associations.

Using Star's (1991) understanding of marginality, it might be said that a survivor/victim of sexual violence sits simultaneously inside and outside of the legal system's institutional network. The survivor/victim is 'inside' in the sense that their body serves as the "crime scene" from which DNA evidence is gathered (Mulla, 2008, p. 301). However, the survivor/victim is simultaneously 'outside', as they are excluded from many of the practices within the legal system. The survivor's/victim's 'multiple membership' in and outside of the legal system

could illuminate a new level of complexity in the legal network. We could see how the network is organized in ways that disempower many survivors/victims and privilege scientific practice and the actors engaged in it. Or, to use Star's language, it would be possible to see how the legal system is 'stabilized' for some actors and not for others.

While Star (1991) provides some useful insight into how feminist methodology might inform a new starting place for an ANT study, she leaves many remaining questions. How do we identify or find marginal actors? Once we have found them, how do we decide which marginal actor to begin our analysis with? What are the empirical and political consequences of choosing one marginal actor over another? In the context of the DNA study, two important questions would be: what other actors in the network, besides the survivor/victim, might provide a useful perspective and what would be the consequences of choosing those actors over the survivor/victim? All these questions collectively invite another: which actors should we follow when tracing actor-networks?

WHICH ACTORS SHOULD WE FOLLOW?

For Latour (2005), sociologists must follow the actors who *act*. Active actors, according to Latour, leave visible "traces" (p. 29) of their action, which an interested sociologist can follow. Star's (1991) description of the marginalized actor reveals, however, that not all actors have equal capacities, resources, and the power necessary to leave visible traces – or at least traces that are visible to the sociologist. In the context of forensic DNA analysis, some visible traces that a sociologist might follow are institutional protocols, reports, case files, forensic samples and forensic technologies. All of these traces are assembled and/or authored by actors who have varying degrees of authority, credibility, and power in the medical and legal practice. Survivors/victims have little power to influence the content and shape of these traces. In this context, Latour's dictum of only following actors who leave visible traces risks erasing survivors/victims from medical and legal actor-networks, which survivors/victims are implicated in when they report a sexual assault to physicians, nurses, and/or police. In other words, Latour's rule of thumb in this context carries risks of producing a one-dimensional network of actors who all have the power to create observable traces.

If we take Star's (1991) suggestion to follow marginal actors seriously, then we must consider not only the actors who have the power to create visible traces, but also those who do not. But how do we identify all of these actors? If an actors' capacity to create a trace is no longer the distinguishing feature of an actor, then how do we see all the actors who are relevant to an actor network? Most importantly, to use Star's (1991) terms, how do we identify marginal actors?

Some feminist scholars may propose that marginal actors are self-evident and can be easily identified through their lack of access, resources, and power. However, by labelling actors as marginal before the analysis begins, what is lost? I suggest that there may be a distinct analytical disadvantage of identifying marginal actors a priori and using them as an analytical entry point, as some Standpoint Theorists suggest. By assuming marginality we forgo an empirical opportunity to describe *how marginality is made*. We lose the chance to describe how a marginal actor *comes to be* marginal through practice in actor-networks. In the context of forensic DNA analysis, *how* survivors/victims come to be marginal actors who are placed both inside and outside

of medical and legal practice is theoretically interesting and empirically significant. This kind of analysis would reveal the institutional practices through which survivors/victims come to be disempowered. If our aim is to investigate the 'messiness' of science and technology, these aspects of practice in which actors are made cannot be overlooked. Designating survivors/victims as inherently marginal before the investigation begins risks ignoring these intricacies of practice.

Perhaps this is where ANT can provide some guidance. In his well-known article on scallops, Actor-Network theorist Michael Callon (1986) argues that "science and technology are dramatic 'stories' in which the identity of actors is one of the issues at hand" (p. 198). This contention can be usefully put in alternate terms: the identity of the marginal actor is something that is not self-evident, but instead something that needs to be described. While some feminist scholars may view this as a threat to the recognition of systemic power, it need not be the case. If a marginal identity becomes an object of investigation, we can see how marginality is made through practice. I propose that illustrating how actors come to be marginal through practice is a more optimistic stance. If marginality is a product of practice, it could be otherwise.

Perhaps our analytical starting point need not be a particular set of actors, but instead, scientific practice. Our questions become who is doing what, when, where, how, and with what consequences to whom? If we trace scientific practice, while remaining attentive to questions of access and power, marginal and non-marginal actors will become visible alongside the practices through which their identities are made. With this as a point of departure, a subsequent question is raised: what can we see once we have begun to follow the actors?

WHAT CAN WE SEE?

Latour (1987, 2005), along with other Actor-Network Theorists, have argued that concepts like gender, power, and inequality have become reified within Sociology. Sociologists, he asserts routinely misuse these concepts to explain away the realities that they observe. Instead of relying on preconceived concepts, Latour (2005) argues that sociologists should simply describe what they see.

ANT's criticism of concepts like gender, power, and inequality, have been suggested by some feminist scholars, like Star (1991) and Wajcman (2000), to be one of the greatest clashes between feminist scholarship and ANT. While many feminist traditions take up gendered, classed, and raced social structures and power inequalities within them, Latour's ANT labels them as preconceived concepts and attempts to replace them with description. These opposing positions appear to leave little room for dialogue.

In the study of DNA analysis, gender, power, and inequality cannot be ignored. Women are the most common survivors/victims of sexual assault (Brennan & Taylor-Butts, 2008), a gendered reality that undoubtedly shapes medical and legal practice around sexual violence. Compounding this reality, women of colour, women living with disabilities, and women living in poverty are more likely to be survivors/victims of sexual violence and more likely to face discrimination and disbelief by the nurses, lawyers, and police that they go to for help (Doe, 2012). Alongside these realities, feminist scholars have repeatedly revealed the many ways in which survivors/victims of violence are systematically and structurally disempowered in the legal system (Smart, 1989; Doe, 2003; Bumiller, 2008; Sheehy, 2012). To ignore these experiences and the systemic dimensions of power that they represent would be to miss a crucial part

of the story of DNA analysis in sexual assault cases. Is there a way to see gender, power, and inequality with ANT?

Although Latour's (2005) emphasis on description initially appears to preclude the analysis of gender, race, power, and inequality, this need not be the case. Instead, perhaps we could consider Latour's cautionary note as a reminder of the dangers of jumping to explanatory concepts without spending enough time describing their origins. And perhaps this could be made useful for a feminist reading of DNA analysis.

Taking Latour's (2005) criticism of preconceived concepts seriously in the study of forensic DNA analysis may inspire a detailed description of how gender and raced inequalities are systematically made and/or organized in the medical and legal actor-network. Instead of labelling legal practices as automatically producing inequalities, the emphasis might instead be put on describing the practices that produce inequality. In describing the work of powerful actors in the network and the tools, technologies, and texts they use to do their work, it would become possible to see how gendered and raced survivors/victims *become* silenced (or even better, are *made* silent) and stripped of the power to define their own experiences of sexual violence. By taking description seriously, gender, race, power, and inequality are not necessarily erased, but are instead seen as forces that are produced through the actors' work. Describing medical and legal actors' practices that produce inequality between themselves and survivors/victims would be a crucial part of the story of DNA analysis in cases of sexual assault.

ANT's emphasis on description over analytical concepts does not necessarily need to halt a feminist analysis. Instead, it can remind us to question what it is that we can see: instead of preconceived concepts, we can attempt to see relations between actors that are organized in particular ways. This may be another mutual ground for ANT and feminist methodologies. However, some new challenges arise when introducing the question of politics.

WHAT ABOUT POLITICS?

Latour (2005) argues that the problem with Critical Sociology, an umbrella term he uses to describe all sociology that is informed by a critical politic, is that it mistakenly places politics before empirical description. He suggests that description becomes impossible if it is preceded with a political agenda. In arguing this, Latour seemingly sweeps critical scholarship of all stripes, including the many branches of feminist theory that have taken seriously gendered, raced, and classed inequalities, into the same corner of irrelevance. Feminist Standpoint Theory, with its deep roots in Marxist epistemology, is seemingly no exception to Latour's generalizations. While Latour (2005) criticizes Critical Sociologies for being *too* political, feminist scholars have criticized ANT for being *a*political (Wajcman, 2000). From this, it might seem that the differing historical roots and political orientations of ANT and feminist methodologies put them at odds. However, this is not necessarily the case.

Law (2004), another well-known Actor Network Theorist, offers a slightly more generous approach to politically informed analysis than Latour. Law begins his discussion of politics by making a claim that seemingly attacks many feminist projects. He states that he wishes to "divest [methods] of a commitment to a particular version of politics: the idea that unless you attend to certain more of less determinate phenomena (class, gender or ethnicity…), then your work has no political relevance" (2004, p. 9). Despite arguing this, Law does not erase the potential of a guid-

ing political view, as Latour does. Instead he suggests that our research must "imagine and participate in politics and other forms of the good in novel and creative ways" (p. 9). Law suggests that "political goods" (p. 9) cannot be decided from the outset of a research study, but are instead arrived at through the particular contexts in which they are situated. What may be a political good in one context may not be in another.

While Law's (2004) acknowledgement of 'political goods' and their importance in research moves a tentative step closer to an analysis informed by some feminist traditions, it still leaves many unanswered questions. In divorcing himself from the "particular version of politics" (p. 9) that takes inequalities of power seriously, Law leaves us with little direction on *whose* political goods we should be tethering our analysis too.

In the study on DNA analysis, a research finding that scientific facts in the forensic DNA lab are messy and sometimes multiple might be a 'political good' for defence lawyers who are seeking to delegitimize the DNA evidence against their clients. Whereas for survivors/victims of violence, this same finding may be far from what could be considered to be a 'political good'. A feminist politic would remind us of the importance of questioning *whose* political goods should motivate our research. Instead of political goods being abstract and determined contextually, as Law suggests, they could more usefully be defined as goods that serve the interests of those who are made marginal in particular actor-networks. Orienting an analysis to this end would ensure that our investigation is geared towards making a valuable contribution to groups that have been disempowered in scientific practice. The study of DNA could be done with the aim of forwarding the interests of those who become marginalized, traumatized, and made invisible within the legal system. In doing so, we would not be sacrificing description, as Latour might suggest, but would instead be orienting our analysis to what would be useful for those who are disempowered.

A feminist politic does not necessarily impede description of an actor-network. Rather, it can drive description in useful ways. In doing so, feminist methodologies can inspire a more honest and responsible handling of politics in ANT.

INTERSECTIONS BETWEEN ANT AND FEMINIST METHODOLOGIES

Despite how they are often cast, feminist methodologies and ANT need not be enemies. They can instead challenge one another and inform each other's methodological practices. As this paper has shown, there are potential places where ANT and feminist methodologies can meet and mutually shape different aspects of research on scientific practice and technological innovation. In being attentive to the possible dialogue between these seemingly divergent approaches, new ways of studying and understanding science and technology come to the fore.

Feminist methodological traditions like feminist Standpoint Theory remind us of the political importance of the experiences of those who are on the margins of scientific and institutional practice. They open the possibility of understanding how marginality is made in a network of action and encourages us to examine networks from different vantage points. Analyses informed by feminist methodological traditions remind us of *whose* political goods our research should be oriented to. In the study on DNA analysis in cases of sexual assault, feminist methodologies remind us to listen to survivors/victims of sexual violence to understand how networks of practices are organized to (re)produce their experiences of disempowerment.

ANT heightens our attention to the work that human and non-human actors do to maintain networks of action. It draws our attention to the material 'messiness' of scientific practice and technological artifacts. It can open the possibility of examining the actors that collectively reproduce inequalities. In the study on DNA analysis, ANT reminds us to pay attention to the messy, complex work that people and technologies do to maintain the legal network of action.

Both ANT and feminist methodologies have something to offer to this project. Without feminist methodologies, we may lose the voice of the survivor/victim and without ANT, we may lose the multitude of actors and messy practices within the network of the legal system. By drawing these approaches together, it becomes possible to see forensic DNA in sexual assault cases in more complex ways.

As Harding's (2008) assertion allowed us to hope, ANT and feminist traditions can indeed learn from one another. With further dialogue between these approaches, it may be possible to move from a place where feminist methodologies and ANT are treated as opposing forms of analysis, to a place where intersections become visible and they can be interwoven. Through this, a feminist-informed ANT may become possible.

REFERENCES

Benjamin, J. (1988). *The bonds of love: Psychoanalysis, feminism, and the problem of domination*. New York: Random House.

Brennan, S., & Taylor-Butts, A. (2008). *Sexual assault in Canada (Catalogue 85F0033M - No. 19)*. Ottawa, Canada: Canadian Centre for Justice Statistics Profile Series, Statistics Canada.

Brooks, A. (2007). Feminist standpoint epistemology: Building knowledge and empowerment through women's lived experience. In S. Nagy Hesse-Biber, & P. Leavy (Eds.), *Feminist research practice: A primer* (pp. 53–82). Thousand Oaks, CA: Sage. doi:10.4135/9781412984270.n3

Bumiller, K. (2008). *In an abusive state: How neoliberalism appropriated the feminist movement against sexual violence*. Durham, NC: Duke University Press.

Butler, J. (2004). *Undoing Gender*. New York: Routledge.

Callon, M. (1986). Some elements of a sociology of translation: Domestication of the scallops and the fisherman of St. Brieue Bay. In J. Law (Ed.), *Power, action, and belief: A new sociology of knowledge?* (pp. 196–223). London: Routledge.

Casper, M., & Clarke, A. (1998). Making the pap smear into the 'right tool' for the job: Cervical cancer screening in the USA, circa 1940-95. *Social Studies of Science, 28*(2), 255–290. doi:10.1177/030631298028002003 PMID:11620085

Clarke, A., & Montini, T. (1993). The many faces of RU486: Tales of situated knowledges and technological contestations. *Science, Technology & Human Values, 18*(1), 42–78. doi:10.1177/016224399301800104 PMID:11652075

Davis, A. (1983). *Women, race & class*. New York: Vintage Books.

de Beauvoir, S. (1957). *The second sex*. New York: Vintage Books.

Doe, J. (2003). *The story of Jane Doe*. Toronto, Canada: Vintage Canada.

Doe, J. (2012). Who benefits from the sexual assault evidence kit? In E. Sheehy (Ed.), *Sexual assault law in Canada: Law, legal practice and women's activism* (pp. 355–388). Ottawa, Canada: University of Ottawa Press.

Du Mont, J., & Parnis, D. (2001). Constructing bodily evidence through sexual assault evidence kits. *Griffith Law Review, 10*, 63–76.

Du Mont, J., White, D., & McGregor, M. (2009). Investigating the medical forensic examination from the perspectives of sexually assaulted women. *Social Science & Medicine, 68*, 774–780. doi:10.1016/j.socscimed.2008.11.010 PMID:19095341

Gerlach, N. (2004). *The genetic imaginary: DNA in the Canadian criminal justice system*. Toronto, Canada: University of Toronto Press.

Haraway, D. (1991). *Simians, cyborgs, and women: The reinvention of nature*. New York: Routledge.

Harding, S. (1991a). *Whose science? Whose knowledge: Thinking from women's lives*. New York: Cornell University.

Harding, S. (1991b). Rethinking standpoint epistemologies: What is strong objectivity? In L. Alcoff, & E. Potter (Eds.), *Feminist Epistemologies* (pp. 49–82). New York: Routledge.

Harding, S. (2008). *Sciences from below: Feminism, postcolonialities, and modernities*. Durham, NC: Duke University Press. doi:10.1215/9780822381181

Hartsock, N. (2004). The feminist standpoint: Developing a ground for a specifically feminist historical materialism. In S. Harding (Ed.), *The feminist standpoint reader: Intellectual and political controversies* (pp. 283–310). New York: Routledge. doi:10.1007/0-306-48017-4_15

Latour, B. (1987). *Science in action: How to follow scientists and engineers through society*. Cambridge, MA: Harvard University Press.

Latour, B. (2005). *Reassembling the social: An introduction to actor-network theory*. Oxford, UK: Oxford University Press.

Latour, B. (2010). *The making of law: An ethnography of the council d'etat*. Cambridge, MA: Polity Press.

Law, J. (1999). After ANT: Complexity, naming and typology. In *Actor network theory and after* (pp. 1–14). Oxford, UK: Blackwell Publishing.

Law, J. (2004). *After method: Mess in social science research*. New York: Routledge.

Law, J. (2007). *Actor Network Theory and Material Semiotics*. Retrieved from http://www.heterogeneities.net/publications/Law-ANTandMaterialsSemiotics.pdf

Law, J., & Mol, A. (Eds.). (2002). *Complexities: Social studies of knowledge practices*. Durham, NC: Duke University Press. doi:10.1215/9780822383550

MacKinnon, C. (2005). *Women's lives, men's laws*. Cambridge, MA: Belknap Press of Harvard University Press.

Mohanty, C. T. (2003). *Feminism without borders: Decolonizing theory, practicing solidarity*. Durham, NC: Duke University Press. doi:10.1215/9780822384649

Mulla, S. (2008). There is no place like home: The body as the scene of the crime in sexual assault intervention. *Home Cultures*, *5*(3), 301–325. doi:10.2752/174063108X368337

Parnis, D., & Du Mont, J. (2006). Symbolic power and the institutional response to rape: Uncovering the cultural dynamics of a forensic technology. *The Canadian Review of Sociology and Anthropology. La Revue Canadienne de Sociologie et d'Anthropologie*, *43*(1), 73–93. doi:10.1111/j.1755-618X.2006.tb00855.x

Quinlan, A., Fogel, C., & Quinlan, E. (2010). Unmasking Scientific Controversies: Forensic DNA Analysis in Canadian Legal Cases of Sexual Assault. *Canadian Women's Studies*, *21*(1), 98–107.

Sheehy, E. (Ed.). (2012). *Sexual assault in Canada: Law, legal practice, and women's activism*. Ottawa, Canada: University of Ottawa Press.

Sismondo, S. (2010). *Introduction to science & technology studies*. Oxford, UK: Wiley-Blackwell.

Smart, C. (1989). *Feminism and the power of law*. New York: Routledge. doi:10.4324/9780203206164

Smith, D. (2005). *Institutional ethnography: A sociology for people*. New York: Altamira Press.

Star, S. L. (1991). Power, technologies and the phenomenology of conventions: On being allergic to onions. In J. Law (Ed.), *A sociology of monsters? Essays on power, technology and domination*. London: Routledge.

Star, S. L., & Griesemer, J. (1989). Institutional ecology, 'translations' and boundary objects: Amateurs and professionals in Berkeley's museum of vertebrate zoology, 1907-1937. *Social Studies of Science, 19*(3), 387–420. doi:10.1177/030631289019003001

Wajcman, J. (2000). Reflections on gender and technology studies: In what state is the art? *Social Studies of Science, 30*(3), 447–464. doi:10.1177/030631200030003005

KEY TERMS AND DEFINITIONS

Actor-Network Theory: A theoretically informed methodology that has been commonly used to explore scientific practice and technological innovation.

Feminist Methodologies: A diverse field of methodological approaches that have been influenced by feminist theoretical perspectives.

Feminist Standpoint Theory: A theoretically informed feminist methodology that hinges on the epistemological assumption that marginalized and/or oppressed groups have a privileged perspective on social relations of power. With feminist standpoint theory, women, people of colour, and other marginalized groups are granted epistemological privilege and social relations are studied from their perspective.

Forensic DNA Analysis: The analysis of DNA for the purposes identifying suspects of criminal acts.

Sexual Assault Forensic Examination: A medical examination that involves collecting forensic evidence from the body of a survivor/victim of sexual assault.

ENDNOTES

[1]   Here, I purposely adopt the imperfect term *survivor/victim*. The forward slash serves a distinct symbolic purpose: it acknowledges that identities in relation to violent experiences can be complex, multiple, and changing. The term survivor/victim proposes the possibility that victim and survivor identities can be simultaneously claimed, while the forward slash conveys the possibility of identities in between or beyond survivor and victim.

[2]   Star is not alone in arguing this. Other feminist STS scholars have similarly argued that if ANT were to grant epistemic privilege to marginalized actors, the shape of actor-networks would inevitably change (Star & Griesemer, 1989; Star, 1991; Clarke & Montini, 1993; Casper & Clarke, 1998).

Chapter 12
Achieving E–Health Success:
The Key Role for ANT

Nilmini Wickramasinghe
Epworth Healthcare, Australia & RMIT University, Australia

Arthur Tatnall
Victoria University, Australia

ABSTRACT

Healthcare delivery continues to be challenged in all OECD countries. To address these challenges, most are turning their attention to e-health as the panacea. Indeed, it is true that in today's global and networked world, e-health should be the answer for ensuring pertinent information, relevant data, and germane knowledge anywhere anytime so that clinicians can deliver superior healthcare. Sadly, healthcare has yet to realize the full potential of e-health, which is in stark contrast to other e-business initiatives such as e-government and e-education, e-finance, or e-commerce. This chapter asserts that it is only by embracing a rich theoretical lens of analysis that the full potential of e-health can be harnessed, and thus, it proffers Actor-Network Theory (ANT) as such a lens.

INTRODUCTION

Superior access, quality and value of health-care services have become a global priority for healthcare to combat the exponentially increasing costs of healthcare expenditure. E-Health in its many forms and possibilities appears to offer a panacea for facilitating the necessary transformation for healthcare. While a plethora of e-health initiatives keep mushrooming both nationally and globally, there exists to date no unified system to evaluate these respective initiatives and assess their relative strengths and deficiencies in realizing superior access, quality and value of health-care services. Our research serves to address this void. Specifically, we focus on three key components namely: 1) understanding the web of players (regulators, payers, providers, healthcare organizations, suppliers and last but not least patients) and how e-health can modify the interactions between these players as well as create added value healthcare services. 2) the development of an e-health preparedness grid that provides a universal assessment tool for all e-health initiatives and

DOI: 10.4018/978-1-4666-6126-4.ch012

3) the development of an e-health manifesto, a declaration of policy, intent and the necessary components of successful e-health initiative. Taken together and applied systematically this will then enable a critical assessment of the areas that e-health initiatives should best target as well as the necessary steps and key success factors that must be addressed such as technological, infrastructure, education or policy elements. However, the paper goes further and notes that simply being e-health prepared is a necessary but not sufficient condition. It identifies the need to incorporate a rich theoretical lens of actor-network theory (ANT) in order to truly uncover all key issues and thereby ensure successful realization of the full potential of any e-health solution.

HEALTHCARE

Healthcare is a growing industry. Between 1960 and 1997 the percentage of Gross Domestic Product (GDP) spent on healthcare by 29 members of the Organization for Economic Cooperation and Development (OECD) nearly doubled from 3.9 to 7.6% while the growth between 1995-2005 was on average 4% with the US spending the most (nearly 2.5 times more than any other country) and this is expected to reach 19.5% GDP by 2017[1]. Since 2000, total spending on healthcare in these countries has been rising faster than economic growth, which has resulted in an average ratio of health spending to GDP of 9.0% in 2008 (OECD, 2010). Hence, reducing this expenditure as well as offering effective and efficient quality healthcare treatment is becoming a priority globally as is reflected in the fact that all OEDC countries are looking seriously into healthcare reform and especially the role for e-health solutions (OECD, 2010). Technology and automation have the potential to reduce these costs (Ghani et al., 2010; America In-

stitute of Medicine, 2001; Wickramasinghe, 2000); thus, e-health, specifically the adoption and adaptation of web based technologies and advancements through Web 2.0, appears to be a powerful force of change for the healthcare industry worldwide.

Such external environmental forces are translating into numerous changes with regard to the role of technology for healthcare delivery at the organizational level. So much so that we are witnessing, healthcare providers grasping at many opportunities, especially in response to legislative mandates, to incorporate IT (information technology) and telecommunications with web based strategies to improve service and cost effectiveness to their key stakeholders; most notably patients. Many such e-initiatives including the e-medical record which in some form or other is currently being implemented in various countries. However these do not seem to represent a coherent and universal adoption of e-health.

To date, healthcare has been shaped by each nation's own set of cultures, traditions, payment mechanisms and patient expectations. Therefore, when looking at health systems throughout the world, it is useful to position them on a continuum (Figure 1) ranging from high (essentially 100%) government involvement (i. e. a public healthcare system as can be seen in the UK or Canada) at one extreme to little (essentially 0%) government involvement (i. e., private healthcare system as can be seen in the US) at the other extreme with many variations of a two tier system (i. e. mix of private and public as can be seen in countries like Australia and Germany) in between. However, given the common problem of exponentially increasing costs facing healthcare globally, irrespective of the particular health system one examines, the future of the healthcare industry will be partially shaped by commonalties such as this key unifying problem

Figure 1. Continuum of healthcare systems

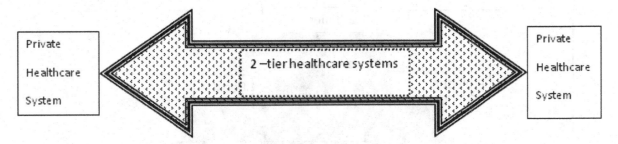

and the common forces of change including: i) empowered consumers, ii) e-health adoption and adaptability and iii) shift to focus on the practice of preventative versus cure driven medicine, as well as four key implications, including: i) health insurance changes, ii) workforce changes and changes in the roles of stakeholders within the health system, iii) organizational changes and standardization and iv) the need for healthcare providers and administrators to make difficult, yet necessary choices regarding practice management.

E-HEALTH

E-health is a very broad term that encompasses various activities related to the use of many e-commerce technologies and infrastructures most notably the Internet for facilitating healthcare practice. The World Health Organization (WHO, 2005) defines e-health as "a new term used to describe the combined use of electronic communication and information technology in the health sector" or "as the use, in the health sector, of digital data-transmitted, stored and retrieved electronically-for clinical, educational and administrative purposes, both at the local site and at a distance". What is significant to note from this definition is that e-health entails the delivery of health services and health information enhanced through the Internet and other related e-commerce tech-

nologies. Moreover, the term characterizes not only a technical development, but also a paradigm shift to focus on networked, global thinking, to improve healthcare locally, regionally, and globally by using information and communication technologies which impact a web of players as depicted in Figure 2. E-health then is an emerging field at confluence of medical informatics, technology, public health and business.

In addition to the definition of e-health, we believe it is important when examining e-health initiatives to focus beyond the "electronic: component of the "e" in e-health and in fact think about this in a broader manner; i.e. to think of the 8 "e's" in e-health (i.e. efficiency, enhancing quality, evidence-based, empowerment, education, extending the scope, ethics and equity)(Eysenbach, 2001; Wickramasinghe et al, 2005; Wickramasinghe & Schaffer, 2010). Table 1 serves to summarize these important "e's" of e-health.

BEING E-HEALTH PREPARED

In order for successful e-health solutions to ensure a necessary first step is for a state of e-health preparedness to be achieved (Wickramasinghe et al., 2012; Wickramasinghe and Schaffer, 2010). An appropriate level of preparedness can only be achieved through the assessment and analysis of broadly-based

Figure 2. Web of healthcare players (adapted from Wickramasinghe and Schaffer, 2010)

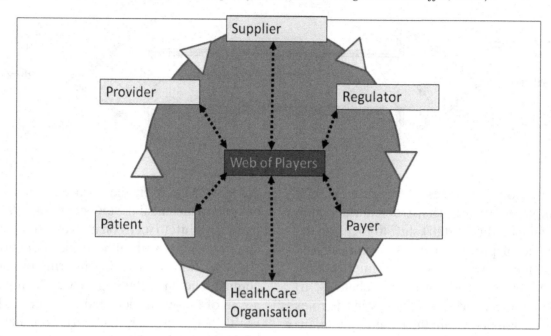

knowledge and key facts (Wickramasinghe and Schaffer, 2010; Wickramasinghe et al., 2012). Thus, development of appropriate preparedness is predominantly a strategic task that requires intimate knowledge of several aspects of the environment.

By taking into consideration all the e's of e-health and after a thorough analysis of various e-health initiatives as well as an in depth assessment of critical success factors necessary to effect successful e-health projects work by Wickramasinghe et al. (2005; Wickramasinghe and Schaffer, 2010) has resulted in the development of a framework to assess the e-health potential and preparedness of a country with regard to its e-health initiatives (Figure 3). In particular, the framework highlights the key elements that are required for

Table 1. Summary of the "e's" in e-health (Eysenbach, 2001; Wickramasinghe et al, 2005; Wickramasinghe & Schaffer, 2010)

E's in E-Health	Description
Efficiency	Support cost effective healthcare delivery
Enhancing quality	Reduce medical errors
Evidence based	Support evidence based medicine
Empowerment	Help patients to be more active and informed in their healthcare decisions and treatments
Education	Help physicians and patients understand the latest techniques and healthcare issues
Extending the scope	Do not limit healthcare treatment to conventional boundaries
Ethics	Including but not limited to privacy and security concerns
Equity	Decrease rather than increase the gap between "haves" and "have nots"

Figure 3. A framework for assessing a country's/region's e-health potential (adapted from Wickramasinghe et al. 2005; Wickramasinghe and Schaffer, 2010)

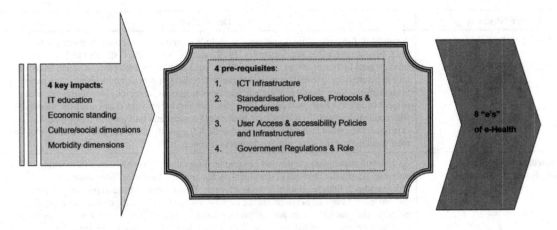

successful e-health initiatives and therefore provides an elegant tool that allows analysis beyond the quantifiable data into a systematic synthesis of the major impacts and pre-requisites. Moreover, the framework contains four main pre-requisites, four main impacts, and the implications of these parameters to the eight e's of e-health. Tables 2 and 3 serve to summarize the four pre-requisites and impacts respectively. By examining both the pre-requisites and the impacts, not only is it possible to assess the potential of a country and its preparedness for e-health but it is also possible to assess its ability to maximize the eight "e's" of e-health.

In assessing ones level of preparedness for any e-health initiative, a good first is to examine ones standing with respect to the four pre-requisites and four impacts discussed above and from this then evaluate if it will be possible to evaluate to move forward successfully (ref). In the following section, we

will attempt to provide a guideline that will facilitate such an evaluation. A systematic way to do this is via the e-health preparedness grid.

THE E-HEALTH PREPAREDNESS GRID

By taking the four main pre-requisites as well as the four major impacts identified in the e-health framework in Figure 3; namely, the information communication technology infrastructure, the standardization policies, protocols and procedures, the user access and accessibility policies and infrastructures, governmental regulations and role as well as the impact of IT education, the impact of morbidity rate, the impact of world economic standing and the impact of cultural/social dimensions, it is possible to develop a grid for assessing e-health preparedness (Figure 4; Wickramasinghe et al., 2005; Wickrama-

Table 2. Pre-requisites for e-health (adapted from Wickramasinghe et al., 2005; Wickramasinghe and Schaffer, 2010)

Pre-Requisite	Description
ICT Infrastructure	A typical ICT infrastructure is envisaged to include a combination of the following: phone lines, fibre trunks and submarine cables, T1, T3 and OC-xx, ISDN, DSL other high-speed services used by businesses as well as satellites, earth stations and teleports. A sound technical infrastructure is an essential ingredient to the undertaking of e-health initiatives by any nation. Such infrastructures should also include telecommunications, electricity, access to computers, number of Internet hosts, number of ISP's (Internet Service Providers) and available bandwidth and broadband access. To offer a good multimedia content and thus provide a rich e-health experience, one would require a high bandwidth. ICT considerations are undoubtedly one of the most fundamental infrastructure requirements (Samiee, 1988).
Standardization Policies, Protocols and Procedures	As e-Health spans many parties and geographic dimensions, to enable such a far reaching coverage, significant amounts of document exchange and information flows must be accommodated. Standardization is the key for this. Hence, standardization polices, protocols and procedures must be developed at the outset to ensure the full realization of the eight e's of e-health. The standardization polices, protocols and procedures play a significant role in the adoption of e-health and the reduction of many structural impediments (Samiee, 1998).
User Access and Accessibility Policies and Infrastructure	Access to e-commerce is defined by the WTO (World Trade Organization) as consisting of two critical components 1) access to Internet services and 2) access to e-services (Panagariya, 2000). The user infrastructure includes a number of Internet hosts and number of web sites, web users as a percent of the population as well as ISP(internet service providers) availability and costs for consumers, PC penetration level etc.
Governmental Regulation and Control	The key challenges regarding e-health use include; 1) cost effectiveness; i. e. less costly than traditional healthcare delivery, 2) functionality and ease of use, i.e., they should enable and facilitate many uses for physicians and other healthcare users by combining various types and forms of data as well as be easy to use; and 3) they must be secure. One of the most significant legislative regulations in the US is the Health Insurance Portability and Accountability Act (HIPPA, 2001). Given the nature of healthcare and the sensitivity of healthcare data and information, it is incumbent on governments not only to mandate regulations that will facilitate the exchange of healthcare documents between the various healthcare stakeholders but also to provide protection of privacy and the rights of patients (Samiee, 1998; Goff, 1992; Gupta, 1992). Irrespective of the type of healthcare system; i.e., whether 100% government driven, 100% private or a combination thereof, it is clear that some governmental role is required to facilitate successful e-health initiatives.

singhe and Schaffer, 2010). The grid consists of four quadrants that represent the possible states of preparedness with respect to the key parameters for e-health success. The low preparedness quadrant identifies situations that are low with respect to all four pre-requisites for e-health potential. The medium preparedness quadrant identifies two symmetric situations; namely, a combination of high and low positioning with respect to the four pre-requisites for e-health potential. Finally, the high preparedness quadrant identifies situations that are high with respect to all four pre-requisites for e-health potential. This grid not only shows the possible positioning of a given e-health initiative with respect to its e-health preparedness (i. e. low, medium or high) but also the path that should be taken, and more specifically the pre-requisite factors must be focused on, to migrate to the ideal state of preparedness.

However, being prepared is only the first necessary step. It is also essential to be ready for an upcoming-health initiative. Readiness must be based on germane knowledge. Furthermore it requires the background of germane knowledge that will dictate the nature of the subsequent response. Readiness is

Table 3. Key impacts for e-health (adapted from Wickramasinghe et al., 2005; Wickramasinghe and Schaffer, 2010)

Impacts	Description
IT Education	A sophisticated, well educated population boosts competition and hastens innovation. According to Michael Porter, one of the key factors to a country's strength in an industry is strong customer support (Porter, 1990). Thus, a strong domestic market leads to the growth of competition which leads to innovation and the adoption of technology enabled solutions to provide more effective and efficient services such as e-health and telemedicine. As the health consumer is the key driving force in pushing e-health initiatives, a more IT educated healthcare consumer would then provide stronger impetus for e-health adoption.
Morbidity Rate	There is a direct relationship between health education and awareness and the overall health standing of a country. Therefore, a more health conscious society, which tends to coincide with a society that has a lower morbidity rate, is more likely to embrace any e-health initiatives. Furthermore, higher morbidity rates tend to indicate the existence of more basic health needs (WHO, 2003) and hence treatment is more urgent than the practice of preventative medicine and thus e-health could be considered an unrealistic luxury and in some instances such as when a significant percentage of a population is suffering from malnutrition related diseases is even likely to be irrelevant at least in the short term. Thus, the modifying impact of morbidity rate is to prioritize the level of spending on e-health versus other basic healthcare needs.
Cultural/Social Dimensions	Healthcare has been shaped by each nation's own set of cultures, traditions, payment mechanisms and patient expectations. While the adoption of e-health, to a great extent, dilutes this cultural impact, social and cultural dimensions will still be a moderating influence on any countries e-health initiatives. Another aspect of the cultural/social dimension relates to the presentation language of the content of the e-health repositories. The entire world does not speak English so the e-health solutions have to be offered in many other languages. The e-health supporting content in web servers/sites must be offered in local languages, supported by pictures and universal icons. This becomes a particularly important consideration when we look at the adoption and diffusion of evidence-based medicine as it will mean that much of the available evidence and case study data will not be easily accessible globally due to language barriers.
World Economic Standing	Economies of the future will be built around the Internet. All governments are very aware of the importance and critical role that the Internet will play on a country's economy. This makes it critical that appropriate funding levels and budgetary allocations become a key component of governmental fiscal policies so that such initiatives will form the bridge between a traditional healthcare present and a promising e-health future. Thus, the result of which would determine success of effective e-health implementations and consequently have the potential to enhance a country's economy and future growth.

therefore context dependent. Readiness is thus a most essential tool in response to the embracement of a new e-health initiative. While intuitively obvious, the practical development of readiness is not an easy task. Possession of knowledge is not equivalent to the ability to employ it under the stress of less-than-routine circumstances. Thus to establish an appropriate level of readiness the first step is to assess the situation and key aspects through an appropriately robust and rich theoretical lens. We believe actor-network theory because of its ability to blend and combine social and technical perspectives is a most suitable lens.

ACTOR-NETWORK THEORY

Actor-Network Theory (ANT) is based on a recursive philosophy (Latour 1992). Its fundamental stand is that technologies and people are linked in a network. ANT tries to bridge the gap between a socio-technical divide by denying the existence of purely social or technical relations. In doing so it takes a very radical stand and assumes that each entity (such as technologies, organization) are actors therefore have the potential to transform and mediate social relationships (Cresswell et al. 2010). It also emphasizes

Figure 4. E-Health preparedness grid (adapted from Wickramasinghe et al., 2005)

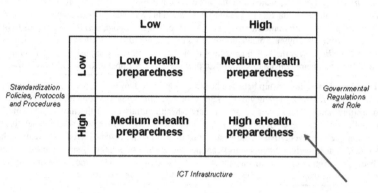

the concept of heterogeneous networks because of the non-similar nature of elements and their relationship in network makes these networks open and evolving systems (Hanseth 2007). Therefore Actor- networks are highly dynamic and inherently unstable in their nature; and a better understanding of how alignment between people, technology, their roles, routines, values, training and incentives as well as understanding of the role of technology that how it can facilitate or negatively impact the work processes and tasks in an organization can stabilize these network to some extent (Greenhalgh & Stones 2010; Wickramasinghe et al. 2011). For this reason, ANT can be a material-semiotic approach and can provide an appropriate lens to study the ordering of scientific, technological, social, and organizational processes and events (Wickramasinghe et al. 2011). To realize the importance of the application of ANT it is important to understand the key concepts of

ANT and then map them to the critical issues concerning e-health initiatives.

Key Stages of ANT

In addition to the key concepts of ANT described in Table 4, there are also three critical stages of ANT which need to be considered as discussed in turn below.

Stage 1: Inscription

A process of creating technical text and communication artifacts to protect actor's interests in a network is described as an inscription (Leila 2009; Wickramasinghe & Bali 2009; Latour 2005). This is a term used for all texts and communications in different mediums including but not limited to journal articles, conference papers and presentations, grants proposals and patents. The idea of Inscription also relates to the notion of durability;

Table 4. Key ANT concepts and their mapping to e-health initiatives

Key Constructs of ANT	Mapping to E-Health Initiatives
Actor/Actant: Actors are the web of participants in the network including all human and non-human entities. Because of the strong biased interpretation of the word actor towards human; a word actant is commonly used to refer both human and non-human actors.(Wickramasinghe et al. 2011).	In any e-health context Actors/Actants include different stakeholders in healthcare delivery settings such as Technology (Web 2.0, Databases, Graphical User Interfaces and different Computer hardware and Software) and People (service providers, healthcare funders, healthcare service recipients, healthcare organizations, suppliers and private health insurers) as well as clinical administrative technologies, work process and health records in the form of paper or electronic.
Heterogeneous Network: is a network of aligned interests formed by the actors. This is a network of materially heterogeneous actors that is achieved by a great deal of work that both shapes those various social and non-social elements, and "disciplines" them so that they work together, instead of "making off on their own" (Latour, 1996, Latour, 2005; Wickramasinghe et al. 2011).	The specific technology clearly forms the main network of different applications in this context. But it is important to understand that the heterogeneous network in ANT requires conceptualizing the network as aligned interest including people, organizations, standards and protocols and their interaction with technology.
Tokens/Quasi Objects: are created through the successful interaction of actors/actants in a network and are passed between actors within the network. As the token is increasingly transmitted or passed through the network, it becomes increasingly punctualized and also increasingly reified. When the token is decreasingly transmitted, or when an actor fails to transmit the token, punctualization and reification are decreased as well (Wickramasinghe et al. 2011).	In e-health contexts this translates to successful cost effective and efficient healthcare delivery. It is important to understand that to maintain the integrity of the network at all times is very important because if wrong information is passed through the network, the errors would be devastating and can propagate quickly and will multiply.
Punctualization: Within the domain of ANT every actor in the web of relations is connected to others and as a whole it will be considered as a single object or concept same as the concept of abstraction is treated in Object Oriented Programming. These sub-actors are sometime hidden from the normal view and only can be viewed in case of the network break-down; this concept is often referred as a depunctulization. Because ANT require all actors or sections of network to perform required tasks and therefore maintain the web of relations. In case of any actor cease to operate or maintain link the entire Actor-Network would break down resulting in ending the punctualization. Punctualization is a process and cannot be achieved indefinitely rather is a relational effect and is recursive that can reproduce itself (Law 1997)	For example, a computer on which one is working would be treated as a single block or unit. Only when it breaks down and one needs help with spare parts can reveal the hidden chain of network consist of different actors made up of (People, Computer parts and organizations). Similarly in an e-health context, uploading the health record of a patient is in reality a consequence of the interaction and coordination of many sub-tasks. This will only reveal itself if some kind of breakdown at this point occurs and depunctualization of the network happens and all sub-tasks then would need to be carefully examined.
Obligatory Passage Point (OOP): broadly refers to a situation that has to occur in order for all the actors to satisfy the interests that have been attributed to them by the focal actor. The focal actor defines the OPP through which the other actors must pass through and by which the focal actor becomes indispensable (Callon, 1986).	In e-health contexts, we can illustrate this by taking the example of access and user rights. The interface of the system is developed in a way that no service can access any record without using secure logins, which in this case constitute an obligatory passage point through which they have to pass for their everyday activities.
Irreversibility: Callon (1986, p. 159) states that the degree of irreversibility depends on (i) the extent to which it is subsequently impossible to go back to a point where that translation was only one amongst others and (ii) the extent to which it shapes and determines subsequent translations.	In the context of the very complex nature of healthcare operations, irreversibility is less likely to occur and would be more dependent on social networks and the nature of the interaction between human and non-human actors in the network. Here it is important to remember though the chain of events needs to be monitored carefully so the future events can be addressed in the best possible manners.

for instance a general discussion would be less durable as compared to a recorded meeting. Therefore, the idea behind Inscription is to enhance the durability of the network by associating them with durable material. Actors can use Inscription as a path to gain credibility in enrolment and the co-optation process during translation.

Stage 2: Translation

Translation is a very important and vital concept in ANT. This term is used to explain the process of creation of Actor-Networks and the formation of ordering effects (Callon 1986; Law 1992). This stage can help researchers in providing the insight into how the software system can be integrated into the very complex environment of healthcare. The process of Translation can also be called the process of negotiation because after the creation of the network in the presence of many actors a strong or primary actor would translate interests of other actors into his/her own by negotiating with them. At this stage all actors decide to be part of network if it is worthwhile to build it (Wickramasinghe & Bali 2009).

The process of Translation of Actors/Actants is achieved through a series of four moments of translations (Callon, 1986). Figure 3 depicts the key ANT concepts along with these four moments of Translation.

Stage 3: Framing

Framing is an operation that can help to define actors and distinguish different actors and goods from each other (Callon, 1986). This last and final stage in the ANT process can help network to stabilize. At this stage key issues occurred throughout the e-health adoption /implementation should already

have been negotiated within the network and technologies can become more stable over time (Wickramasinghe & Bali 2009).

DISCUSSION

ANT is considered an appropriate choice to facilitate a state of readiness in the context of e-health initiatives because it can identify and acknowledge any impact of human and non-human social or policy issues within the healthcare setting (Latour et al. 1996). Moreover, it is robust enough to accurately capture all the complexities, nuances and richness of healthcare operations. In so doing, it can also help to investigate and theorize the question of why and how networks come into existence, what sort of associations and impact they can have on each other, how they move and change their position in a network, how they enrol and leave the network and most importantly how these networks can achieve stability (Doolin & Lowe 2002; Callon 1986; McLean & Hassard 2004). ANT's assumption is that if any new actor is enrolled in the network or an old actor leaves the network it would affect whole network (Cresswell et al. 2010; Doolin & Lowe 2002). These considerations are naturally most relevant in the context of any e-health scenario.

In addition, ANT also can help to understand the active role of objects in shaping social realities by challenging assumptions of the separation between non-human and human worlds (Walsham 1997; Greenhalgh & Stones 2010; Tobler 2008; Law & Hassard 1999; Rydin 2010). This helps researchers to study the complexities of the relationships between human and non-human actors, the sustainability of power relationships between human actors and what kind of influence artifacts can have on human actors relationships in transforming healthcare (Cresswell et al. 2010).

The rationale to choose ANT to assess the state of readiness for any e-health initiative is thus connected with the strength of ANT to identify and explore the real and perceived complexities involved in the healthcare service delivery sector. Indeed several scholars have noted the benefits of ANT and it has been applied in the implementation and adoption of different healthcare innovation studies (Berg, 2001; Cresswell et al. 2010; Cresswell et al. 2011; Bossen, 2007; Hall, 2005). To date it has yet to be embraced as a tool to facilitate a macro level analysis of e-health readiness as we have suggested in the proceeding discussion.

CONCLUSION

E-commerce, as noted by the UN Secretary General's address, is an important aspect of business in today's 21st century. No longer then is it a luxury for nations, rather it is a strategic necessity in order for countries to achieve economic and business prosperity as well as social viability. One of the major areas within e-commerce that has yet to reach its full potential is e-health. This is due to the fact that healthcare generally has been slow in adopting information technologies. In addition most e-health initiatives to date have been less than successful. Not only is there a shortage of robust normative frameworks that may be used as guidelines for assessing e-health preparedness and identifying the key areas and deficiencies that need to be addressed in order for successful e-heath initiatives to ensue, but also there is a lack of appropriate analytic frameworks to address the level of readiness. Moreover, e-health is more than a technological initiative; rather it also requires a major paradigm shift in healthcare delivery, practice and thinking. We have attempted to address this gap by discussing the need to be both prepared and ready. To address the needs of preparedness we have discussed the importance of – and presented a useful framework to- assessing e-health preparedness and thereby, facilitating the focus of efforts and resources on the relevant issues that must be addressed in order that successful e-health initiatives follow (i. e., the eight e's of e-health are in fact realized). For readiness, we have presented ANT as an appropriate, robust and rich, theoretical framework. The first step in the development of any viable e-health strategy is to make an assessment of the current state of e-health preparedness and then how to either move to a state of higher preparedness (i. e., the high quadrant) or focus on maintaining a current high quadrant status – both of these will be possible through the use of our framework and thus its value. It is advised that next an assessment of readiness is made which by definition must be context dependent. We believe that taken together the assessments of preparedness and readiness will enable a comprehensive assessment for e-health initiatives, that to date has been lacking. More importantly such a comprehensive assessment will ensure that successful outcomes and the full potential of the e-health solution is indeed realized.

REFERENCES

America Institute of Medicine. (2001). *Crossing the Quality Chasm - A New Health System for the 21st Century Committee on Quality of Health Care*. National Academy Press.

Eysenbach, G. (2001). What is e-health? *Journal of Medical Internet Research*, *3*(2). doi:10.2196/jmir.3.2.e20

Ghani, M., Bali, R., & Naguib, I., Marshall, & Wickramasinghe, N. (2010). Critical issues for implementing a Lifetime Health Record in the Malaysian Public Health System. [IJHTM]. *International Journal of Healthcare Technology and Management, 11*(1/2), 113–130. doi:10.1504/IJHTM.2010.033279

Goff, L. (1992). Patchwork of Laws Slows EC data Flow. *Computerworld, 26*(15), 80.

Gupta, U. (1992). Global Networks: Promises and Challenges. *Information Systems Management, 9*(4), 28–32. doi:10.1080/10580539208906896

Health Insurance Portability and Accountability Act (HIPPA). (2001, May). *Privacy Compliance Executive Summary.* Protegrity Inc.

OECD. (2010). *OECD health data 2010.* Retrieved from http://stats.oecd.org/Index.aspx?DatasetCode=HEALTH World Health Organization (WHO). (2005). *What is e-health?* Retrieved from http://www.emro.who.int/his/ehealth/AboutEhealth.htm

Panagariya, A. (2000). E-commerce, WTO and developing countries. *World Economy, 23*(8), 959–978. doi:10.1111/1467-9701.00313

Porter, M. (1990). *The Competitive advantage of Nations.* New York: Free Press.

PricewaterhouseCoopers Healthcare Practice. (n.d.). Retrieved from www.pwchealth.com

Samiee, S. (1998). The Internet and International Marketing: Is There a Fit? *Journal of Interactive Marketing, 12*(4), 5–21. doi:10.1002/(SICI)1520-6653(199823)12:4<5::AID-DIR2>3.0.CO;2-5

United Nations Conference on trade and Development. (2002). Retrieved from http://r0.unctad.org/ecommerce/ecommerce_en

Wickramasinghe, N. (2000). IS/IT as a Tool to Achieve Goal Alignment: A theoretical framework. *International Journal of Healthcare Technology and Management, 2*(1-4), 163–180. doi:10.1504/IJHTM.2000.001089

Wickramasinghe, N., Bali, R., Kirn, S., & Sumoi, R. (2012). *Critical Issues for the Development of Sustainable E-health Solutions.* New York: Springer. doi:10.1007/978-1-4614-1536-7

Wickramasinghe, N., Fadllala, A. M. A., Geisler, E., & Schaffer, J. L. (2005). A framework for assessing e-health preparedness. *International Journal e-Health, 1*(3), 316-334.

Wickramasinghe, N., & Schaffer, J. (2010). *Realzing Value Driven e-health Solutions.* Washington, DC: IBM Center for the Business of Government.

World Health Organization (WHO). (2003). Retrieved February 23, 2010 from http://www.emro.who.int/ehealth/

KEY TERMS AND DEFINITIONS

E-Health Potential: The initial ability and structures that an entity possess before implementing an e-health solution. The better suited the underlying structure is and the better skilled the key stakeholders are regarding e-health the more prepared the entity is said to be.

E-Health: The use of computer and web based technologies to provide healthcare delivery.

Healthcare Challenges: Key areas that are impacting the ease of healthcare delivery e.g. escalating financial pressures, rapid increase of technology solutions, an aging population and the growth of chronic diseases.

Preparedness: Is predominantly a strategic task that requires intimate knowledge of several aspects of the environment.

Readiness: Is based on germane knowledge and it requires the background of germane knowledge that will dictate the nature of the subsequent response. Readiness is therefore context dependent.

ENDNOTES

[1]   OECD Health Data 2009

Chapter 13

Evolving Digital Communication:
An Actor–Network Analysis of Social Networking Sites

Mohini Singh
RMIT University, Australia

Jayan Kurian
RMIT University, Vietnam

ABSTRACT

This chapter analyses elements of social networking sites to establish how a combination of heterogeneous elements of technology, media, language, users, data, and information are networked together to provide this new communication media. Social networking sites are also referred to as social media sites, which can be explained using the Actor-Network Theory. Social networking sites have clearly achieved widespread adoption as a new means of communication in a very short time around the globe. An analysis of literature on social networking sites is included in this chapter to reflect the new social networking language and style, the content shared via this media, the mode of use, and the language used for communication, which is a combination of a number of technological and social entities. This chapter explains how the Actor-Network Theory (ANT) can be used to explain social networking and includes some issues for research on this topic.

INTRODUCTION

Social networking sites are fast becoming the principal communication and information sharing tool used by people of all ages, and backgrounds in all regions of the world. Social networking sites also referred to as

social media sites are developed on the Web 2.0 platform, which offers an architecture for participation and allows users to control their own data and information (Kim et al., 2010). Web 2.0 is an extension of Web 1.0 on which individuals deliver content and services in the public domain creating a network effect

DOI: 10.4018/978-1-4666-6126-4.ch013

through which others can access, update and combine content (Cummings et al., 2009). Characteristics of Web 2.0 enable formation of communities via collaboration and information sharing; novel methods of data presentation with 'mashed up' (combined) information from different sources; and with Ajax supported creative and responsive interfaces (Ankolekar, et al., 2007). Social networking applications therefore developed on the Web 2.0 platform are designed around architecture of participation and communal collaboration (Sena, 2009).

Individuals are using social networking sites for communication, collaboration, information sharing, networking, finding 'lost' friends and forming communities. Although business organisations are also resorting to social networking sites for advertising, marketing and engaging employees (Singh et al., 2010), the focus of this chapter is on 'social' user issues and characteristics of social networking for users. Web 2.0 based social networking sites are widely used by all age groups and their adoption is increasing by the day. In this chapter the aim is to establish the characteristics of social networking from the most popular sites (each with over 30 million users) and determine their 'dimensions' of networking and communication.

These are noted in Table 1.

Adoption of social networking sites by individuals is so significant that users of these sites range from 34 million on ask.fm to more than a billion on Facebook (Table 1). The growing number of users indicates the importance of social networking sites and their relevance to society. Due to social networking sites being so widely adopted in a very short period of time, and accessible on a variety of platforms such as personal computers, mobile phones, laptops and other ubiquitous technology Turban et al., (2011), it is considered essential to explore heterogeneous elements of social networking that are making them so prevalent. Although there are numerous publications on one or more aspect of social networking sites, significant earlier studies on this topic are focussed on its taxonomy (Kim et al, 2010), definition, history and scholarship (Boyd, 2007) risk, trust and privacy concerns (Fogel and Nehmad, 2009), changes in user behaviour (Patchin and Hinduja, 2010)., self-disclosure model (Posey et al., 2010); (Krasnova et al., 2010), and impact on business environment (Sena, 2009); (Singh et al., 2010). More recently, Kietzmann, et al., (2010) addressed social media applications in organisations from the perspective of how individuals in the organization use these tools.

Table 1. Social networking sites with over 30 million users - February, 2014

Social Networking Site	Date of Origin	Registered Users in 2013
Facebook	Feb 2004	1.15 Billion
Twitter	July 2006	500 million
Google+	June 2011	500 million
Linkedin	May 2003	238 million
Instagram	October 2010	130 million
Pinterest	March 2010	70 million
Meetup	June 2002	35 million
ask.fm	June 2010	34 million

Source: http://www.ebizmba.com/articles/social-networking-websites

Turban, et. al (2011) established a framework for recognising social technology opportunities, adoption and risk mitigation, Wu (2013) explored social network effect on productivity and job security, Goh et al (2013) established the impact of market generated content on consumers, Haefliger et al (2011) explored social software interactions strategies, and Aral et al (2013) focused on a framework for research on social media related business transformation. A generic consideration and understanding of why they are so widely used, what is the content shared on it, who are the users, where is it used, in which time zones they can be used and how do they support communication are not known. This chapter attempts to highlight the elements of social networking sites, the content shared on these sites, potential participants, temporal aspects of communication, as well as the linguistic elements of social networking by critically reviewing literature on these topics.

We structure the literature in the following section under purpose, content, participants, technology, time and linguistic elements to establish the components of social networking sites and their increased adoption around the world. This enables an understanding of social networking sites as a network of a number of heterogeneous elements forming the social networking sites used as a communication and collaboration tool, the content shared on this media, the users of this media, the mode of use, and linguistics elements to establish how these are networked together to support the social networking via this media.

LITERATURE REVIEW

A review of literature on purpose, content, potential participants, technology platform, time and the language used for communication using social networking sites are presented in the next section.

Purpose

Users collaborate and communicate via social networking sites by taking advantage of Web 2.0 capabilities to author and edit content (Nardi et al., 2004), to communicate in real time (Madhavan and Goasgen, 2007), whilst being mobile (Bolter and Mcintyre, 2007). The growth in social networking is so significant that not only are there web sites such as Ning and KickApps supporting their creation, but a search engine is now available for finding them (Kim et al., 2010). Although there are a very large number of social networking sites available, the most popular ones are listed in Table 1 with the year of their origin to demonstrate the wide adoption rates of these social networking sites (SNS) as communication tools in such a short span of time. Most of these are now available in different languages enabling people in different parts of the world, as well as people of all ages to be users of this media. Social networking sites in this paper are based on Kim et al. (2010) description comprising of both social web sites and social media sites. Social network sites connect people to other people in online communities such as MySpace, Facebook, Windows Live Spaces, and other similar sites. The communities can range from a network of offline friends connected via social network sites to online acquaintances which may include one or more interest groups such as school attended, hobby, support, profession, ethnicity, gender, etc. These are further described as general or vertical, both of which are either open or closed (Conry-Muray, 2009). Open sites are open to general users, vertical sites connect people with a common trait such as schools attended, ethnicity or a common interest such as arts, whereas closed sites are for members of particular groups only. Social media sites such as YouTube, Flickr, Digg, Instagram, and others support user created content made up of photos, videos, book marks of web pages, user profiles, user activities and text. They are used as a new asynchronous means of communication between a user and the user's online community to broadcast messages or any other user created content (UCC) to a large number of people, and to view and respond at any time to messages posted by other people (Kim et al., 2010). These sites support data from different sources (mashed up data) from instant messaging, chat, blogs, multimedia, video clips, tagging, tasks and calendars, scrapbooking, hobbies, interests and photographs (Madhavan and Goasgen, 2007) to be presented in meaningful ways for better communication. The main reason why people use social networking sites are for communication, to connect with existing members, stay in contact with others and sometimes to kill time (Fitzimmons, 2008). Other researchers identified the reasons for social website use to be for enjoyment (Rosen and Shermon, 2006), self-presentation (Boyd, 2007), to maintain social ties (Ellison et al., 2007) crowd sourcing and political campaigns.

Content

An important characteristic of social networking site is the ability of users to generate their own content (user created content - UCC) without any programming or computing knowledge (Kim et al., 2010). Although the content on each social web site can vary, Kim, et al (2010) explain that common types of user created content shared on these sites include basic information such as name, photo, birthday, gender, relationship status, type of

relationship desired, email address, phone numbers, address, school attended, education and work, current employer, and friends. In addition, it may include member's interests, hobbies, favourite movies and shows, favourite music and books, quotations, travel experience, photos of destinations visited, opinions, comments and a diary. Content shared via social networking sites according to Turban, et al (2011) may include a photo album, film repository, information collection, news collection, or downloaded TV and film. Through *file sharing* applications data such as personal photographs (Van House, 2009), music video (Cohen, 2009) and film and photographs (Naim, 2007) are transferred, downloaded or posted online for access by a wider audience. Users can share personal information, gossip as well as promote a self-inflated profile on social networking sites (Solove, 2007). Self-disclosure of personal and private information by users on social networking sites has been explained by Krasnova (2010) to be based on social exchange theory (Homans, 1958) which is founded on a subjective evaluation of benefits and costs. Benefits of personal information posted on social networking sites include trust building, empathy and reciprocation which generally outweigh costs associated with vulnerability (Johnson and Paine, 2007).

All users can post new UCCs on social networking sites and view at any time UCCs created by users' friends and their network. Although many of the UCCs are irrelevant to a user, some can be helpful and a new source of knowledge (Kim et al., 2010). Some social web site users are celebrities who use this medium for self-promotion or for communicating with their fan bases (Metcalfe, 2009) and (Wortham, 2009). Although a few of the celebrities might communicate directly with their fans via their social website, many have other people to *ghost write* their UCCs (Cohen, 2009).

Patchin and Hinduja (2010) suggest that in addition to the types of information above, some irresponsible information such as photos in different costumes, swear words, evidence of alcohol, tobacco and drug use and other personal information is also shared on social networking sites. Problematic material including pornography, politically sensitive material that is prohibited in some countries, and comments that are defamatory or violation of privacy have also been identified on social networking sites. Posey et al. (2010) describe the content shared on social networking sites to be social in nature, open and self-disclosing. Disclosing and sharing information on social networking sites foster relationships (Chiu, et al. (2006) and Web 2.0 authoring tools enable users to collaboratively create, share and recreate knowledge from multiple sources for collective intelligence.

Types of information shared on social networking sites include professional (DiMicco et al., 2009; Skeels and Grudin, 2009); recommendations (Subramani and Rajagopalan, 2003; Morris et al., 2010); opinions (Harper et al., 2009; Naaman et al., 2010); Morris et al., 2010); information (Valenzuela et al., 2009; Harper et al., 2009); conversation (Boyd and Heer., 2006; Harper et al., 2009); invitations (Papacharissi, 2009; Morris et al., 2010); parenting (MacPhee et al., 1996; Walker and Riley, 2001); anecdotes (Binder et al., 2009; Naaman et al., 2010); random thoughts or statements (Hinduja and Patchin, 2008; Naaman et al., 2010); diaries (Kaplan and Haenlein, 2010; Naaman et al., 2010); membership (Spertus et al., 2005; Papacharissi, 2009); advice (Constant et al.,1996; Harper et al., 2009); announcements (Brandtzæg and Heim, 2011; Honeycutt and Herring, 2009); comments (Subrahmanyam et al., 2008; Honeycutt and Herring, 2009); reflection (Steinfield et al., 2009; Honeycutt and Herring, 2009); greetings (Chun et al.,

2008; Honeycutt and Herring, 2009); referrals (Baum et al., 2013; Neumann et al., 2005) and numerous other categories. Information presented on SNS are viewable by the public or restricted to approved friends.

Content presented on social networking sites can also prove to be harmful if targeted by spammers, phishing and malware attacks (Kim et al., 2010). Irresponsible data on social networking sites has led to some legal implications resulting in court actions, and the issue of truthfulness on social networking sites such as fake names, ages, schools attended, qualifications achieved, etc which is becoming an issue for debate and the need for regulations and standards in the use of social networking sites. Misuse of social networking sites to present false and other undesirable information can also result in damage to self or damage to others. Examples of self-damage caused by information presented on SNS include denied admission to universities, lost employment opportunities, arrest of criminals and lost court cases (Ruber and Kogler, 2001). Some users engage in spreading false rumours, participate in cyber bullying and cyber stalking (Krasnova et al., 2010) and some have become less productive due to the excessive time they spend on social networking sites (Kim et al., 2010).

User Profiles

Although social networking sites are also gaining importance in government and business organisations, the discussion in this chapter is on individual users. The millions of SNS users range from teenagers (Patchin and Hinduja, 2010) to older people (Fraser and Dutta, 2008), who are either individual users, members of networks or community groups, professionals, special interest groups, classmates, ethnic diasporas, and other groups with similar attributes. Fogel and Nehmad (2009) are of the opinion that there are approximately equal percentages of men and women users and most users spend an average of one hour a day of their time on social networking sites. The online community for a user of social networking sites may include offline friends who are members of the same site, as well as new online friends and acquaintances (Ruber and Kogler, 2001). Fogel and Nehmad (2009) also suggest that users of social networking sites have a greater risk taking attitude since in spite of knowing the possible risks of their information being seen by others they are comfortable placing personal information on these sites. This is supported by Acquisti and Gross (2006) who reported that users of social network sites were freely providing their personal contacts and details on these sites. The types of information presented on social network sites generally portray user behaviour which is largely being scrutinised by prospective employers, teachers, counsellors, and law enforcement agencies to determine the character of these users (Fitzimmons, 2008; Schweitzer, 2005).

Although most users take advantage of social network sites privacy settings to control who may access parts of their profiles, many users are not concerned about the '*digital print*' they leave on social networking sites (Stross, 2009). Dwyer et al. (2006) refer to trust and trust in other social website members to be factors that supersede privacy concerns of social website users.

SNS Technology

SNS is supported by Web 2.0 technology platform that brings about collaboration and communication (Li and Bernoff, 2008) and enables users to create their own content (Korica et al., 2006). Most of the social networking sites are IT entrepreneurial outcomes started by one or small groups of engineers similar to the devel-

opment of eBay, Fastflowers.com and Hotmail (Singh and Waddell, 2006). Web 2.0 being an extension of Web 1.0 includes all the features of Internet access, web servers, firewalls, http protocols and traffic manager, accessible on PCs, lap tops and other portable technologies such as iPads and mobile phones (Turban et al., 2010). *Filesharing* (Oberholzer-Gee and Strumpf, 2007) distributes and provides access to digitally stored information, such as computer programs, multi-media (audio, video), documents, or electronic books (Wikipedia, 2010) by enabling content sharing.

Most social networking sites have similar features although new features are created or added continuously to these sites to make them more dynamic, attractive to users, combat security issues, and to increase participation. These features within a site facilitate interaction amongst a population that is constantly engaged in multiple tasks (Patchin and Hinduja, 2010). Some social networking sites were developed with the main purpose of supporting information sharing (Crowston and Williams, 2000) such as Wikipedia, and *Mashups* make available combined information such as travel information together with Google maps (Singh et al., 2010).

Social networking sites are accessible by members on different platforms, regions of the world and in all time zones. These channels are internet and mobile based technologies (Furuta and Marshall, 1996) that allow sharing of user created content across time and space. Communication on social networking sites can be synchronous allowing users to respond and comment concurrently, simultaneously or all together at once. However, the broad application of social networking sites is based on asynchronous responses, that is, not requiring real time responses giving users the freedom to respond at their convenience (Kim et al., 2010). Originally, the World Wide Web (WWW) was intended to be used to share ideas and promote discussion within a scientific community. Web 2.0 heralds a return to these original uses, and prompts significant changes in the ways the World Wide Web is now being used especially for SNS extending collaboration and communication to users from any part of this world, on any platform and in any language. How widely a social web site is adopted is easily established by the traffic on the site.

Language

Communication via social media has created a new type of language (English slang) due to the amount of space the website allows (Korhan, 2010), young users concealing content form adults (Whittaker, 2008), the language is considered 'cool' and innovative (Zimmer, 2009), or adding an element of style to the communication (Korhan, 2010). Besides the icons such as a thumb to indicate like on facebook, Whittaker's research on Bebo included the use of phrases such as '*Getting MWI*' – to be *Getting Mad With It* meaning getting drunk; '*Legal*' – suggests person posting is 16 and legally allowed to have sex; and '*taken*' or '*Ownageeee*' conveying *being in a relationship*. A Twitter allows 140 character space in which the message is referred to as '*tweets*'; people signed up to get them are '*followers or tweeple*'; the sphere in which it operates is '*Twittersphere or Twitterverse*'. A '*twoosh*' is a message that fits the maximum of 140 characters exactly, without editing (Zachary, 1999). Korhan (2010) is concerned about language on blogs using words such as '*folks*' or '*dude*' instead of people. The good part of this trend is that it is collective language, inclusive of all e.g, our friends and colleagues, however, the adverse effect of it is that the audience for whom the message is intended, may not always know it is for them. This can result in a hanging message

with no one reading it. Zimmer explains a '*LiveJournal*' to be a virtual community of bloggers and diary-keepers, '*friending*' is marking something as a favourite, getting '*tagged*' in a photo on Facebook is getting your name associated with that photo, and '*untagging*' is to avoid anyone coming across the evidence of a particularly embarrassing moment. Another common thread in social media communication is the language that originated from chat rooms and text messaging. Although there is a long list on the URL http:www.netlingo.com/acronyms.php some common ones include '*411*' meaning information; '*@TEOTD*' meaning at the end of the day; '*459*' meaning 'I love you'; '*B4YKI*' meaning 'before you know it'; and '*LLT*' meaning 'looks like trouble'.

Finally, Whittaker (2008) suggests that recognition has a very important role within young people's identity construction and social-networking pages provide the opportunity to gain both positive and negative recognition. She argues that young people are responding to the technological age they are living in thus becoming very creative in their use of language. However, Whittaker (2008) emphasises that social media slang is unprofessional and detrimental to one's professional etiquette.

The above literature discussion on social networking sites clearly indicates that social networking sites are an innovation in communication made possible by a combination of heterogeneous elements. Elements of social networking sites identified from literature are discussed in the next section. These include:

Technology

Web 2.0, an extension of Web 1.0 led to the development of social networking sites offering a new medium of communication, networking and collaboration to users of all ages, in different languages and in all regions of the world. It enables delivery of content that is either user created or presentation of existing information in new ways by combining different forms of data and allowing others to access, respond and comment on the subject matter. Technology capabilities have been ingeniously combined on this platform to deliver new experiences to users leading to a very wide participation and collaboration among myriads of user groups. Web 2.0 capabilities of file sharing, *mashed up* data from different sources, multimedia enriched information, multi-platform access and easy to use, promote interaction, mobility and a very wide user participation.

Communication Media

The wide adoption of social networking sites reveals its exponential growth as an important communication media with millions of users around the world. This new media also comes with easy to create content and a search engine for finding a relevant application for users. Different social networking sites are popular in different parts of the world connecting communities that are either open to all or closed to particular groups. Social networking sites are general as well as vertical based on a common trait of users. Access to social networking sites are achieved from a variety of platforms with richness added to information including photos, video clips, calendar, scrapbook and book marks leading to creative, useful and interesting content. Social networking sites are fast replacing other communication tools such as emails and phone calls for a large number of people.

Standards for User Created Content

Social networking sites are user friendly with high level authoring tools allowing users to create content without the need for technical knowledge or programming. Social networking sites carry user information that ranges from personal data, experiences, professional achievements, hobbies, interests, quotations, to political opinions and some undesirable information that can have problematic implications. Although some users are careful with the type of information they present on social networking sites, many are not concerned about the evidence they leave on this new media. Problematic content can have long term implications, therefore standards are required for UCC to provide guidance to users on types of information presented on their SNS.

Digital Identity

The millions of social website users are people from different age groups, ethnic backgrounds, speakers of different languages, both male and female or community groups. Self-disclosure of information by individuals on social networking sites is influenced by trust building, empathy and reciprocation although sometimes this leaves the users vulnerable to those seeking information on them. All information on social networking sites is cached or copied somewhere else. Although all user created content is not of interest to all users, some are knowledge enhancing and generate collective intelligence. By placing information on social networking sites one leaves a digital identity that can be used to determine the character of users especially sought by employers, law enforcement units, counsellors, teachers and others.

Linguistic Elements

Communication via social networking sites has resulted in a new *networking/web language* that is either an acronym for an English phrase, words to reflect a position or state, a symbol for a word or extending a human situation to technology. Social website terminology is made up of a combination of characters that are numeric, alphanumeric and symbols. Some believe this to be a response to this new technological era, while others consider it detrimental to professional etiquette. How will this shape future communication trends is yet to be determined.

The above entities are multidimensional involving a heterogeneous amalgamation of textual, conceptual, social and technical actors. Therefore social networking sites is analysed in the following section using the Actor Network Theory.

ACTOR NETWORK THEORY

According to the Actor Network Theory (ANT) entities are hybrid, made up of dissimilar elements including people, objects and organisations (Latour, 1986); (Callon, 1986); (Tatnall and Guilding, 1999). ANT entails dissimilar elements that are both social and technical that are inseparable. Social networking sites discussed above are a heterogonous combination of textual, conceptual, social, human and technical actors. Therefore Actor Network Theory is an apposite analytical framework to establish how social networking sites are socially constructed and to identify global research issues for this innovative and advanced communication and knowledge sharing tool.

ANT is based on *agnosticism* promoting analytical impartiality towards all actors (technical and non-technical); *general symmetry* explaining the conflicting viewpoints of different actors in the same terms with neutral vocabulary for human and non-human actors; and *free association* meaning elimination of priori distinctions between technical and social (Callon, 1986). ANT is an approach to structuring and explaining the links between society and technology. It has the ability to illustrate how technology becomes acceptable and is taken up by groups in society. ANT has been used to explain casemix systems in hospitals (Bloomfield, et al., 1992), EDI networks and standards (Monteiro and Hanseth, 1996), B2B SME portal (Tatnall and Guilding, 1999), communication network (McBride, 2011). The primary focus of ANT is on stakeholders (actors) and how they are involved in shaping the social networking sites. With social networking sites the actors are both human and non-human who pursue interests which may encourage or constrain technology (Monteiro and Hanseth, 1996). These actors include Web 2.0 based technology and social networking sites, user created content, communication, SNS language with symbols, media and a set of users. These actors make up a network of interests that become stable as they are aligned to the technology (McBride, 2011). Alignment is achieved via translation of interests and enrolment of actors into the network. Translation indicates how an actor's interests become aligned, and alignment is established in inscriptions such as communication due to technological developments, peer pressure to use technology, partnership demands, community networks, innovation, to pass time and knowledge sharing. These social website entities support communication traits that are irreversible. Therefore evolving digital communication, Web 2.0 technology, the users and their networks, communication symbols and the social networking sites form an actor-network.

IMPLICATIONS FOR RESEARCH

As discussed above, Web 2.0 based social networking sites are technical as well as human combination of entities. Without UCC (created by people/humans) it is non-functional, at the same time without the support of technology (Internet) it is non-functional. Social networking sites entail technological support required for such a development, user profiles, user generated content, linguistic elements of communication on this new media, and user networking requirements. Although research issues for this topic are wide and numerous, an important research issue derived from the literature analysis above and the Actor Network Theory is: 'determining the role of each element in social networking sites. Will social networking sites be functional if one of the elements in this network is removed, changed, increased or decreased? To establish this, the following is suggested.

Identify the Actors in different social networking sites. A list of social networking actors that enable communication via social networking sites is to be identified for a sample of widely used SNS.

Establish the role of each actor to note the changes (evolvement) by investigating content, technologies used for access, reasons for use, linguistic elements of communication, types of data in user created content, use of multimedia, frequency of use and profiles of users.

Build the Actor Network Model. The identification of links between the actors will develop the actor network model for communication via social networking sites. This will help establish the alignment of actors by establishing the strength of the connections.

Establish Irreversibility. This will help understand the importance and role of each of the actors, how fixed are these and what can or cannot be changed. This will require an understanding of the types of communication that takes place and types of information shared via this media.

Identify factors that support communication via social networking sites and those that deter the adoption of this media for communication.

Identify communication innovation. The ANT analysis will establish a list of social networking features, communication etiquette, technological requirements, training and orientation needs for users, global and cultural dimensions and further acceptance of social networking sites as a communication tool.

Other issues that will be established from the above will include technological developments, security of data, universal/international standards for UCC, digital divide, linguistic issues and implications for future communication and community development. The need for universal standards for UCC to avoid personal harm, bullying, plagiarism, professionalism and impact on personal and professional opportunities is anticipated to be an important outcome. Although there are so many aspects to consider on the language used for SNS communication, the important considerations are the need for a computer application to convert this new language into proper English and other languages. This chapter highlights the importance of social networking as the new communication and collaboration tool and the application of Actor Network Theory to analyse the acceptance of social networking sites as a digital communication media.

REFERENCES

Acquisti, A., & Gross, R. (2006). *Imagined Communities: Awareness, information sharing, and privacy on the Facebook*. Paper presented at the 2006 privacy enhancing technologies (PET) Workshop. Cambridge, UK. Retrieved from http://privacy.cs.Cmu.edu/dataprivacy/projects/facebook/facebook2.pdf retrieved 2/2/2011

Ankolekar, A., Krotzsch, M., & Tran Tand Vrandecic, D. (2007). The Two Cultures, Mashing up Web 2.0 and Semantic Web. In Proceedings of IW3C2, (pp. 825 – 834). IW3C2.

Aral, S., & Weill, P. (2007). IT Assets, Organizational Capabilities, and Firm Performance: How Resource Allocations and Organizational Differences Explain Performance Variation. *Organization Science*, *18*(5), 763–780. doi:10.1287/orsc.1070.0306

Baum, D., Spann, M., Füller, J., & Pedit, T. (2013). *Social Media Campaigns for New Product Introductions*. ECIS 2013 Completed Research.

Binder, J., Howes, A., & Sutcliffe, A. (2009). *The Problem of Conflicting Social Spheres: Effects of Network Structure on Experienced Tension in Social Network Sites*. Paper presented at CHI 2009. Boston, MA.

Bloomfield, B. P., Coombs, R., Cooper, D. J., & Rea, D. (1992). Machines and manoeuvres: Responsibility accounting and construction of hospital information systems. *Accounting Management and Information Technologies*, *2*, 197–219. doi:10.1016/0959-8022(92)90009-H

Bolter, J. D., & Macintyre, B. (2007). Is It Live or Is It AR? *IEEE Internet Computing*. Retrieved from http://www.spectrum.ieee.org/aug07/5377

Boulos, M. G., & Wheeler, S. (2007). The emerging Web 2.0 social software: An enabling suite of sociable technologies in health and health care education. *Health Information and Libraries Journal*, *24*, 2–23. doi:10.1111/j.1471-1842.2007.00701.x PMID:17331140

Boyd, D. (2007). Why Youth (Heart) Social Network Sites: The Role of Networked Publics in Teenage Social Life. In D. Buckingham (Ed.), *MacArthur Foundation Series on Digital Learning – Youth, Identity, and Digital Media Volume*. Cambridge, MA: MIT Press.

Boyd, D., & Ellison, N. B. (2007). Social Network Sites: Definition, History, and Scholarship. *Journal of Computer-Mediated Communication*, *13*(1), 210–230. doi:10.1111/j.1083-6101.2007.00393.x

Boyd, D., & Heer, J. (2006). Profiles as Conversation: Networked Identity Performance on Friendster. In *Proceedings of Thirty-Ninth Hawai'i International Conference on System Sciences*, (pp. 59–69). Los Alamitos, CA: IEEE Press.

Brandtzæg, P.B., & Heim, J. (2011). A typology of social networking sites users. *Int. J. Web Based Communities, 7*(1).

Callon, M. (1986). Some Elements of a Sociological Translation: Domestication of the Scallops and Fishermen of St Brieu Bay. In *Power Action and Belief: A New Sociology of Knowledge*. London: Routledge and Kegan Paul.

Chiu, C.-M., Hsu, M.-H., & Wang, E. T. D. (2006). Understanding knowledge sharing in virtual communities: An integration of social capital and social cognitive theories. *Decision Support Systems*, *42*(3), 1872–1888. doi:10.1016/j.dss.2006.04.001

Chun, H., Kwak, H., Eom, Y. H., Ahn, Y. Y., Moon, S., & Jeong, H. (2008). *Comparison of Online Social Relations in Terms of Volume vs. Interaction: A Case Study of Cyworld*. Paper presented at IMC'08. Vouliagmeni, Greece.

Cohen, N. (2009, March 27). When stars Twitter, a ghost may be lurking. *The New York Times*.

Conry-Murray. (2009, March 21). *Can enterprise social networking pay off?* Retrieved from http://internetrevolution.com/document.asp?Docid=173854

Constant, D., Sproull, L., & Kiesler, S. (1996). The kindness of strangers: The usefulness of electronic weak ties for technical advice. *Organization Science*, *7*, 119–135. doi:10.1287/orsc.7.2.119

Crowston, K., & Williams, M. (2000). Reproduced and emergent genres of communication on the World-Wide Web. *The Information Society*, *16*(3), 201–215. doi:10.1080/01972240050133652

Cummings, J., Massey, A. P., & Ramesh, V. (2009). Web 2.0 Proclivity: Understanding How Personal Use Influences Organisational Adoption. [GIGDOC.]. *Proceedings of GIGDOC*, *09*, 257–263.

DiMicco, J. M., Geyer, W., Millen, D. R., & Dugan, C. (2009). People Sensemaking and Relationship Building on an Enterprise Social Network Site. In *Proceedings of Hawaii International Conference on System Sciences*, (pp. 1-10). IEEE.

Dwyer, C., Hiltz, S. R., & Passerini, K. (2006). Trust and Privacy Concern within Social Networking Sites: A comparison of Facebook and MySpace. In *Proceedings of the AMCIS*. AMCIS.

Dwyer, C., Hiltz, S. R., & Passerini, K. (2007). Trust and privacy concern within social networking sites: A comparison of Facebook and MySpace. In *Proceedings of the Thirteenth Americas Conference on Information Systems*. Keystone, CO: Academic Press.

Ellison, N. B., & Steinfield, C. (2007). The Benefits of Facebook Friends: Social Capital and College Students' Use of Online Social Network Sites. *Journal of Computer-Mediated Communication*, *12*(4), 1143–1168. doi:10.1111/j.1083-6101.2007.00367.x

Enders, A., Hungenberg, H., Denker, H. P., & Mauch, S. (2008). The long tail of social networking: Revenue models of social networking sites. *European Management Journal*, *26*, 199–211. doi:10.1016/j.emj.2008.02.002

Fang, L., & LeFevre, K. (2010). *Privacy Wizards for Social Networking Sites*. Paper presented at WWW 2010. Raleigh, NC.

Fitzsimmons, E. G., & Rubin, B. M. (2008). *Social networking sites viewed by admission officers: Survey shows some use of Facebook, MySpace as another aspect to college application*. Retrieved from http://www.chicagotribune.com/news/local/chi-facebook

Fogel, J., & Nehmad, E. (2009). Internet social network communities: Risk taking, trust, and privacy concerns. *Computers in Human Behavior*, *25*, 153–160. doi:10.1016/j.chb.2008.08.006

Fraser, M., & Dutta, S. (2008). *Throwing Sheep in the Boardroom*. London: John Wiley & Sons.

Furuta, R., & Marshall, C. C. (1996). Genre as Reflection of Technology in the World Wide Web. In *Hypermedia Design*. London: Springer-Verlag. doi:10.1007/978-1-4471-3082-6_19

Goh, K., Heng, C., & Lin, Z. (2013). Social Media Brand Community and Consumer Behavior: Quantifying the Relative Impact of User- and Marketer-Generated Content. *Information Systems Research*, *24*(1), 88–107. doi:10.1287/isre.1120.0469

Haefliger, S., Monteiro, E., Foray, D., & Von Krogh, G. (2011). Social Software and Strategy. *Long Range Planning*, *44*(5–6), 297–316. doi:10.1016/j.lrp.2011.08.001

Harper, F. M., Moy, D., & Konstan, J. A. (2009). *Facts or Friends? Distinguishing Informational and Conversational Questions in Social Q&A Sites*. Paper presented at CHI 2009. Boston, MA.

Hinduja, S., & Patchin, J. W. (2008). Personal information of adolescents on the Internet: A quantitative content analysis of MySpace. *Journal of Adolescence*, *31*(1), 125–146. doi:10.1016/j.adolescence.2007.05.004 PMID:17604833

Honeycutt, C., & Herring, S. C. (2009). Beyond Microblogging: Conversation and Collaboration via Twitter. In *Proceedings of the 42nd Hawaii International Conference on System Sciences*. IEEE.

Hoover, J. N. (2007, September 22). Social Networking: a time waste or the next big thing in collaboration?. *InformationWeek*.

Johnson, A. N., & Paine, C. B. (2007). Self-Disclosure, privacy and the Internet. In A. N. Johnson, K. McKenna, & U. Reips (Eds.), *Oxford Handbook of Internet Psychology*. Oxford University Press.

Kaplan, A. M., & Haenlein, M. (2010). Users of the world, unite! The challenges and opportunities of Social Media. *Business Horizons*, *53*, 59–68. doi:10.1016/j.bushor.2009.09.003

Kietzmann, J. H., Hermkens, K., McCarthy, I. P., & Silvestre, B. S. (2011). Social Media? Get serious! Understanding functional building blocks of social media. *Business Horizons*, *54*, 241–251. doi:10.1016/j.bushor.2011.01.005

Kim, W., Jeong, O., & Lee, S. (2010). On Social networking sites. *Information Systems*, *35*, 215–236. doi:10.1016/j.is.2009.08.003

Korhan. (2010). Retrieved from http://ezinearticles.com/?Social-Media-Vocabulary---Language-and-Slang- Corruption&id=1807049

Krasnova, H., Spiekermann, S., Korolevu, K., & Hidebrand, T. (2010). Online social networks: Why we disclose. *Journal of Information Technology*, *25*, 109–125. doi:10.1057/jit.2010.6

Latour, B. (1986). *The Powers of Association: Power, Action and Belief, A new sociology of knowledge?* Routledge and Kegan Paul.

Li, C., & Bernoff, J. (2008). Harnessing the Power of the Oh-So-Social Web. *MIT Sloan Management Review*, *49*(3), 36–42.

MacPhee, D., Fritz, J., & Heyl, J. M. (1996). Ethnic Variations in Personal Social Networks and Parenting. *Child Development*, *67*(6), 3278–3295. doi:10.2307/1131779

Madhaven, K. P. C., & Goasguen, S. (2007). Integrating Cutting-edge Research into Learning through Web 2.0 and Virtual Environments. *International Workshop on Grid Computing Environments*. Retrieved from http://www.collabogce.org/gce07/images/1/15/Madhavan.pdf

McBride, N. (2011). Using Actor Network Theory to Predict the Organisational Success of a Communications Network. In *Proceedings of the World Telecommunications Congress*, (pp. 1 – 6). Academic Press.

Metcalfe, J. (2009, March 29). The celebrity Twitter ecosystems. *The New York Times*

Monteiro, E., & Hanseth, O. (1996). Social shaping of information infrastructure: On being specific about technology. In *Information Technology and Changes in Organisational Work*. Chapman and Hall.

Morris, M. R., Teevan, J., & Panovich, K. (2010). *What Do People Ask Their Social Networks, and Why? A Survey Study of Status Message Q&A Behavior*. Paper presented at CHI 2010. Atlanta, GA.

Naaman, M., Boase, J., & Lai, C. H. (2010). *Is it Really About Me? Message Content in Social Awareness Streams*. Paper presented at CSCW 2010. Savannah, GA.

Naím, M. (2007). The YouTube Effect. *Foreign Policy*, *158*, 103–104.

Nardi, B. A., Schiano, D. J., & Gumbrecht, M. (2004). Blogging as social activity, or, would you let 900 million people read your diary. In *Proceedings of the 2004 ACM Conference on Computer Supported Cooperative Work*. Chicago: ACM.

Neumann, M., O'Murchu, I., Breslin, J., Decker, S., Hogan, D., & MacDonaill, C. (2005). Semantic social network portal for collaborative online communities. *Journal of European Industrial Training*, *29*(6), 472–487. doi:10.1108/03090590510610263

Oberholzer-Gee, F., & Strumpf, K. (2007). The Effect of File Sharing on Record Sales: An Empirical Analysis. *The Journal of Political Economy*, *115*(1), 1–42. doi:10.1086/511995

Papacharissi, Z. (2009). The virtual geographies of social networks: A comparative analysis of Facebook, LinkedIn and ASmallWorld. *New Media & Society*, *11*, 199. doi:10.1177/1461444808099577

Patchin, J. W., & Hinduja, S. (2010). Changes in adolescent online social networking behaviours from 2006 to 2009. *Computers in Human Behavior*, *26*, 1818–1821. doi:10.1016/j.chb.2010.07.009

Pempek, T. A., Yermolayeva, Y. A., & Calvert, S. L. (2009). College students' social networking experiences on Facebook. *Journal of Applied Developmental Psychology*, *30*, 227–238. doi:10.1016/j.appdev.2008.12.010

Posey, C., Lowry, P. B., Roberts, T. L., & Ellis, T. S. (2010). Proposing the online community self-disclosure model: The case of working professionals in France and the U. K. who use online communities. *European Journal of Information Systems*, *19*, 181–195. doi:10.1057/ejis.2010.15

Randerson, J. (n.d.). *Social networking sites don't deepen friendships*. Retrieved from http://www.guardian.co.uk/science/2007/sep/10/socialwork

Rauber, A., & Kogler, A. (2001). Integrating automatic genre analysis into digital libraries. In *Proceedings of the First ACM/IEEE-CS Joint Conference on Digital Libraries*. Roanoke, VA: ACM Press.

Rosen, P., & Shermon, P. (2006). Hedonic Information Systems: Acceptance of social networking websites. In *Proceedings of AMCIS*. Acapulco, Mexico: AMCIS.

Schweitzer, S. (2005). When students open up – a little too much. *The Boston Globe*.

Sena, J. A. (2009). The Impact of Web 2.0 on Technology. *IJCSNS*, *9*(2), 378–385.

Singh, M., Davison, C., & Wickramasinghe, N. (2010). Organisational Use of Web 2.0 Technologies: An Australian Perspective. In *Proceedings of AMCIS*. AMCIS. Retrieved from http://aisel.aisnet.org/amcis2010/

Singh, M., & Waddell, D. (2006). Innovation and Entrepreneurial Opportunities with Internet Ventures. In *Proceedings of the European Conference on Entrepreneurship and Innovation*, (pp. 247–254). Academic Press.

Skeels, M. M., & Grudin, J. (2009). *When Social Networks Cross Boundaries: A Case Study of Workplace Use of Facebook and LinkedIn*. Paper presented at GROUP'09. Sanibel Island, FL.

Solove, D. J. (2007). *The future of reputation: Gossip, Rumor, and Privacy on the Internet*. New Haven, CT: Yale University Press.

Spertus, E., Sahami, M., & Buyukkokten, O. (2005). Evaluating similarity measures: A large-scale study in the Orkut social network. In *Proc. SIGKDD'05* (pp. 678-684). ACM.

Steinfield, C., DiMicco, J. M., Ellison, N. B., & Lampe, C. (2009). *Bowling Online: Social Networking and Social Capital within the Organization*. Paper presented at C&T'09. University Park, PA.

Stone, B. (2008, July 2). Our practical attitudes toward privacy. *The New York Times*.

Stross, R. (2009, March 8). When everyone's a friend, is anything private? *The New York Times*.

Subrahmanyam, K., Reich, S. M., Waechter, N., & Espinoza, G. (2008). Online and offline social networks: Use of social networking sites by emerging adults. *Journal of Applied Developmental Psychology*, *29*, 420–433. doi:10.1016/j.appdev.2008.07.003

Subramani, M. R., & Rajagopalan, B. (2003). Knowledge-Sharing and Influence in Online Social Networks via Viral Marketing. *Communications of the ACM*, *46*(12). doi:10.1145/953460.953514

Tatnall, A., & Guilding, T. (1999). Actor-Network Theory and Information Systems Research. In *Proceedings of the 10th Australasian Conference on Information Systems*. Academic Press.

Turban, E., Bolloju, N., & Liang, T. (2011). Enterprise Social Networking: Opportunities, Adoption, and Risk Mitigation. *Journal of Organizational Computing and Electronic Commerce*, *21*(3), 202–220. doi:10.1080/10919392.2011.590109

Turban, E., Lee, J. K., King, D., Liang, T. P., & Turban, D. (2010). *Electronic Commerce: A managerial perspective* (6th ed.). Prentice Hall.

Valenzuela, S., Park, N., & Kee, K. F. (2009). Is There Social Capital in a Social Network Site? Facebook Use and College Students' Life Satisfaction, Trust, and Participation. *Journal of Computer-Mediated Communication*, *14*(4), 875–901. doi:10.1111/j.1083-6101.2009.01474.x

Van House, N. A. (2009). Collocated photo sharing, story-telling, and the performance of self. *International Journal of Human-Computer Studies archive*, *67*(12), 1073-1086.

Walker, S. K., & Riley, D. A. (2001). Involvement of the Personal Social Network as a Factor in Parent Education Effectiveness. *Family Relations*, *50*(2), 186–193. doi:10.1111/j.1741-3729.2001.00186.x

Ward, J. C., & Ostrom, A. L. (2006). Complaining to the masses: The role of protest framing in customer-created complaint web sites. *The Journal of Consumer Research, 33*(2), 220–230. doi:10.1086/506303

Whittaker. (2008). Retrieved from http://www.webuser.co.uk/news/top-stories/452168/teens-create-secret-social-network-language

Whitworth, B., & de Moor, A. (2004). Legitimate by Design: Towards Trusted Virtual Community Environments. In *Proceedings of 35th Hawaii International Conference on System Sciences*. IEEE.

Wikipedia. (2010). *Wikipedia About*. Retrieved from http://en.wikipedia.org/wiki/Help:About

Wortham, J. (2009, April 17). With Oprah onboard, Twitter grows. *The New York Times.*

Wu, L. (2013). Social Network Effects on Productivity and Job Security: Evidence from the Adoption of a Social Networking Tool. *Information Systems Research, 24*(1), 30–51. doi:10.1287/isre.1120.0465

Zachry, M. (1999). Constructing usable documentation: A study of communicative practices and the early uses of mainframe computing in industry. In *Proceedings of the 17th Annual International Conference on Computer Documentation* (pp. 22-25). New Orleans, LA: ACM Press.

Zhang, W., Johnson, T. J., Seltzer, T., & Bichard, S. L. (2010). The Revolution Will be Networked: The Influence of Social Networking Sites on Political Attitudes and Behavior. *Social Science Computer Review, 28*, 75. doi:10.1177/0894439309335162

Zimmer, B. (2009). *The Language of Social Media: Unlike Any Other*. Retrieved from http://www.visualthesaurus.com/cm/wordroutes/1845/

KEY TERMS AND DEFINITIONS

Actor Network Theory (ANT): People, technology and the social media platform networked together for social networking.

Elements of Social Networking Sites: Technology, communication media, standards for user created content, digital identity and linguistic elements.

Social Networking Sites: A web-based platform built on the Internet to facilitate information sharing, profile based communication and collaboration among like-minded people.

User Created Content: Types of information such as opinion, recommendation, and comments shared by users on social networking sites.

User Generated Content: Types of information such as photos, images, and videos shared by users on social networking sites.

Web 2.0: An architecture which incorporates active and dynamic collaboration, sharing, and communication among internet users.

Chapter 14
Disturbing Practices in Engineering Design Projects

Fernando Abreu Gonçalves
CEG-IST, Portugal

José Figueiredo
CEG-IST/DEG, Portugal & Technical University of Lisbon, Portugal

ABSTRACT

Most references to innovation relate to the development of new products. In this chapter, the authors do not address innovation in these terms; instead, they address them as changes in practices an engineer creatively adopts during engineering design projects. They adopt Actor-Network Theory as a way to understand these change processes (translations). The authors design a perturbation index inspired in Earned Value management to measure translation effort, having in mind the management of scope. Then they assess changes of regime in resource allocation of tasks and conclude some changes can lead to innovative results. That means we gain a wider view about scope, and scope management, being able to observe and change good practices, something crucial in engineering design projects where requirements and goals drift.

INTRODUCTION

In engineering processes adherence to good practices is positive for replication and efficiency, but limits innovation. In his day-to-day work an engineer has always opportunities to innovate, but this means challenging good practices. In this paper we address a way to challenge innovation in engineering design. Good practices sometimes crystalize losing their edge and missing innovation opportunities. Innovation in engineering design is not only about the design of new products. How thus the engineers commit themselves with innovative practices? Can we manage the adoption of innovative practices? Is there a possible compromise between good practice and project (design-projects) goals? Can we manage the degree of perturbation in project translations in an advantageous way? These questions define the focus of our quest.

Project Management standards cover different issues on project management within a roll of defined contexts. Maybe it is difficult

DOI: 10.4018/978-1-4666-6126-4.ch014

to generalize but engineering design and technological innovation (R&D) are topics not specifically addressed in the project management core standards (body of knowledge). In fact the uncertainty about the outcomes implies the scope to be dynamic (Pons, 2008) and scope management to be influenced and fertilized by different techniques.

Engineering design is characterized by complex interrelated activities and large uncertainties about precisely which solution path will be taken, such that the full scope of the project can often not be anticipated beforehand (Pons and Raine, 2005). When dealing with engineering design projects a main problem is related with evolving uncertainty (Sonnemans et al, 2003). Uncertainty at the formulation level, *what* the problem is, with all the negotiations with the key stakeholders in order to formulate the problem well. Uncertainty at the resource level, how can we grab and convince some crucial resources to be part of our project. And uncertainty at the execution level, how can we estimate things that were never done before, and how can we cope with such an amount of different solicitations and demands (strategic nature, operational type, product portfolio)? (Pich et al, 2002).

The concept of design drift can enhance the chance of achieving effective design goals and in most situations efficiency can be asserted in other phases of the project lifecycle. Design drift could traditionally be intended as a negative prospect in project management, a way some projects slip away from what the customer actually wanted or simply away from requirements. With a different approach, much inspired by Ciborra (2004), Monteiro & Hepso (2000), Orlikowski (1996) and Hanseth and Braa (2000), we understand design drift as a strategy to take advantage from the fuzzy front end of requirements in engineering design projects. Design drift is a

process in which what we obtain at the end of the implementation process is not what was intended originally. In fact, for us, design drift is a way of dealing with positive design risk.

The view of scope as the result of a Work Breakdown Structure (WBS) with its hierarchical formalism may convene impressions of a neat way to go about the scope management of a project. But as all project managers are well aware these impressions are illusory except for trivial cases. Ford and Chris Coulston (2008), recognizing a socio-technical co-evolution of products and markets, deal with the need to manage the "fuzzy front end of product development".

'So the phenomenon we are tackling is not inscription per se, but the cascade of ever simplified inscriptions that allow harder facts to be produced at greater cost', Latour (1990), see section 2. A fact is harder if the number of aligned allies is bigger, as is the case of a project where the cost of change builds up with time.

This is a very general description of innovative practice in engineering design but we argue that scope management can be entailed as an actor-network building, where things are negotiated with stakeholders, environment and restrictions. These negotiations are mainly translations.

After this introduction we begin by a brief explanation of a few relevant Actor-Network Theory (ANT) fundamentals. Then we address performance in projects using concepts of Earned Value Management (EVM). Acknowledging a social-technical view centered in ANT we then develop an approach that looks at scope management in engineering design projects as the management of a perturbation regime. This perturbation regime is used as a metaphor to understand allocation of resource efforts. This is described with a practical illustration and we conclude with a framework that explores the possibility of managing a fruitful balance between good practices and innovative practices.

REVISITING ANT

This section only intends to state a common ground to some fundamentals of ANT strictly related to what we are dealing with and that can be expressed as circulations, 'the social is a certain type of circulation that can travel endlessly' (Latour, 1997). For a comprehensive understanding of ANT terms we refer to Lancaster University, Centre for Science Studies webpage referenced in reference list.

The basic process underlying ANT is 'translation'. An actor translates another and defines-it but in doing so it also redefines itself. The process is reciprocal, with the actors defining one another in interaction. These definitions are inscriptions (Akrich, 1992), the activity of actors defining each other. This process of inscription gives rise to actor-networks. Inscriptions can be materialized in technological artifacts. In this process we choose focal actors, for example a project manager. Focal actors initiate the process of translation with different actors such as line managers, team members, clients).

Translation can also be understood as displacement, "to translate is to displace" (Callon, 1986); "Like Michel Serres, I use translation to mean displacement, drift, invention, mediation, the creation of a link that did not exist before and that to some degree modifies two elements or agents."

Posing the question 'how to act at distance over things that are not familiar with us?', Latour (1987), answers by defining Immutable Mobiles (IMs): 'By conceiving (constructing) means that are able to transform those things as mobile things while maintaining them immutable and making them combinable in order to be cumulated.' IMs are a kind of go-betweens in the space of Obligatory Points of Passage OPPs, good enough to find ways to capitalize to a centre. IMs are said to be immutable but in translations (displacements) there are always transformations and we are particularly interested in the possibility to manage those changes in pursuing innovative practices, (Gonçalves and Figueiredo, 2009, 2010a).

When one actor speaks on behalf of others, an Actor-Network can be seen as "geography of obligatory points of passage" (Callon, 1986). OPPs become what can be named as "centers of calculation", translating actor's interests and behaviors (matters of concern) into alignment. The process of alignment can proceed further into "blackboxing, a process that makes the joint production of actors and artifacts entirely opaque", (Latour, 1994). Centers of calculation are never alone. They compete and transform themselves by way of circulating inscriptions in actor-networks. As IMs accelerate capitalization to some center (some OPP), we want to investigate how IMs play in the success of OPPs.

For practical reasons we must find economic ways to follow actors. Time, Cost, and Scope are the main actors to follow in projects. 'Time' (duration of a task) should be particular useful to follow the work of centers of calculation.

To further explore the idea of materiality above mentioned, translation is "inscribed into durable material" (Law, 1992), we now turn our attention to Annemarie Mol and John Law (1994) that contributed with an interpretation of "space and spacialities" addressing displacements in different types of spaces: fluids - a kind of "Mutable Mobile" object;

regions - "Immutable Immobile" object; network – the place for IMs; and fires - "Mutable Immobiles".

Returning to the idea of circulation we surely retain the fluid space as a good metaphor for innovative practices in design projects (Law and Mol, 2000) and (Laet and Mol, 2000) but, more generally, we may consider that the four basic addressed "spatialities" are not frozen as if they were some fixed categorizations or 'ideal types' of some universal patterns (Boudourides, 2001). Our aim is to look at translations under the view of fluid circulation: 'instead of a fine laminar flow, we will most often get a turbulent flow of whirlpools and rapids. Time becomes reversible instead of irreversible', Latour, 1993).

A way (Latour *et al.*, 1992) to quantify the evolution of networks is the Program of Action (PA). A PA is a semiotic device to describe translations, a two-dimensional chart depicting a succession of proposals (paradigms) along a vertical axis (logical OR axis) where different proposals substitute one after another and a succession of syntagms along a horizontal axis (logical AND axis) where associations among actors align, Figure 1.

We may represent a PA as a flat Table, see Table 1, where in the first column we inscribe a focal proposal and in the next columns we inscribe the answers (agreement/disagreement) of the actors involved.

One can describe an actor-network in many different ways, and even numerically. Bruno Latour (1992) proposes some numerical indices, one of them being the Reality of the Project as the ratio of the number of associated elements that remain stable in a paradigmatic version by size of previous version. The greater this index the more real is the project, that is, the more extended and stronger is the network. To express this numerically Latour used the formula shown in Box 1.

We are interested in engineering design projects. To find a way of assessing performance in translations in the network we view innovative practices as a particular regime of translation, a kind of fluid where proposals travel between contexts. It is reasonable to assume that some perturbation will always occur on this circulation. But maybe it will be possible to manage the degree of perturbation in project translations in some profitable ways. This is the core idea of this paper and we will explore it next.

PERFORMANCE IN PROJECTS

Project Management PMI rationale (PMBOK, 2008) uses a Performance Index to measure the performance of the engagement of a resource in a task, PV/AV, where PV is the value of

Figure 1. Program of action

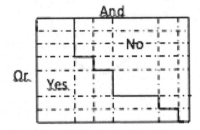

Table 1. Program of Action of the P-G (E) Translation

Work Package X		
P	G	E
PT=2	GT=3	
		AT

Box 1.

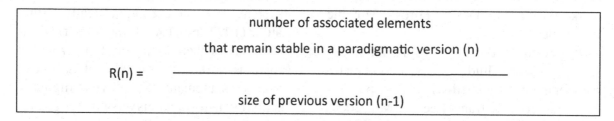

the Planned effort and AV is the Actual effort exercised to accomplish the task, both in monetary units.

Earned Value Management (EVM) integrates measurements of scope, schedule, and cost relying in a single measure unit for the project performance – cost. Earned value (EV) is defined as the budgeted cost of work performed. That means that EVM systems operate a translation of cost into cost, time into cost and scope into cost that works when integrated into the overall project. This single unit measure system is one of the strongest assets of the EVM system, for the supporters, and the weakest aspect for the detractors, see extensions EVM (2005).

Recently and motivated by misleading information provided in very specific project situations (late projects near the conclusion) there was some new developments to the EVM framework (EVM extensions) that converts project performance indicators into time with a schedule performance index – SPIt=PT/AT, where PT is the time planned to obtain EV at Actual time (AT), sometimes referred as Earned Planned Time, (Lipke, 2003), (Rodrigues, 2010).

Discussion is still alive among researchers but this approach has already provided some interesting results, allowing the correction of earlier gaps.

PERTURBATION INDEX IN PROJECTS

In Project management there are situations where bad performances could be related to bad planning but there are also bad performances situations related to bad execution. In real projects both failures of planning and execution occur. We feel the need to extend the network and consider enrolling a new actor, Good-practice.

We picture a situation where we have project planning, good-practice, and earned outcome, see Figure 2.

A focal actor P mobilizes G, the target actor, in order to attain some objectives. In the sequent chain of translations the outcome attained, E, can be not what was sought after

Figure 2. Planned (P), earned (E), and good-practice (G). Actors translating.

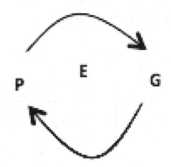

and even G, if there is a chance to learn, may change. P and G are mutually defined and both participate in E. In fact there are always uncertainties in engineering design, as well as in actor-networks.

To deal with a practical example, PT is the planned time for a work package in a Work Breakdown Structure, GT the Good Practice time for that same work package.

The schedule performance index SPIt refers to a chain of translations where only PT and AT are addressed. But we may introduce GT in the Performance Index equation by using the equation shown in Box 2.

We use "t "in SPIt to show that we are using an index depending on time.

To introduce the idea of perturbation we may observe that the allocation of most resources is not strictly fixed. We can stress resources to a certain point. For example, if we have a task that normally, in a good practice, is accomplished by 3 engineer$_x$day, maybe it would be possible to "force" it to 2 engineer$_x$day.

Table 1 is a table that intends to represent a Program of Action for the translation P-G that can be expressed as 'how long it will take Work Package X?'. Good Practice (G) states 3 engineer$_x$day for the tasks of the work package. The Project Manager is committed to make it in 2 (PT=2) and the outcome is always uncertain.

With this in mind we define a Perturbation regime index, Pr, by taking the equation shown in Box 3.

Pr can be seen as a measure of the fit between planning (PT) and practice (GT). Pr will depend mostly on the bargaining power of the project manager face to functional managers. If Pr is less than 1 we face a perturbation regime because we are stressing the resources, if it is higher than 1 we are accepting good practice durations, Figure 3.

Being bold means that in the project we consider that, due to some circumstances (routine, learning, or others), some duration should be shorter than is usually in Good Practice; being conservative means otherwise, duration would be expected to be longer. This notation gains effectiveness if circumstances and tasks are clearly identified, the reason justified and also the limits of the regime perceived.

Box 2.

$$SPIt = \frac{PT}{AT} = \frac{PT}{PT} \times \frac{PT}{AT}$$

Box 3.

$$Pr = \frac{PT}{GT}$$

Figure 3. Pr (Perturbation regime index)

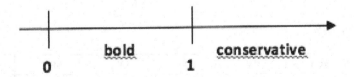

bold   conservative

0   1

Box 4.

$$\frac{GT}{AT} = GPIt$$

Using a similar notation we propose a GPIt (Good-practice Performance Index), shown in Box 4.

And then we have the equation shown in Box 5.

This conceptual evolution allows us to extend or generalize the meaning of the expressions "effort", "turbulence", or "pertur-

Box 5.

$$SPIt = Pr \times GPIt$$

bation" and "stress" used within the resource allocation process. This generalization is also aligned with ANT in the way that actors can be human and non-human (as we can stress people, we can also stress any technological device).

MANAGING PERTURBATION IN PROJECTS

The question now will be, to cope with innovative practices how much Perturbation may be allowed, while maintaining the process manageable and controllable. Of course these ideas and the meaning of the values of Pr need empirical substantiation from real projects: how much is perturbation; how much perturbation can be tolerated, etc.

We will explore six possible stylized situations defined in terms of the relative positions of PT, GT, and AT, PGE for PV>GV>EV, GPE

Table 2. Data table for PGE

t	EV	GV	PV	GT	PT	SPIt	Pr	GPIt
1.00	2.00	4.00	**6.00**	0.50	0.33	0.33	0.67	0.50
2.00	4.00	**6.00**	12.00	1.00	0.67	0.33	0.67	0.50
3.00	**6.00**	14.00	16.00	2.00	1.00	0.33	0.50	0.67

Table 3. Data table for PEG

t	EV	GV	PV	GT	PT	SPIt	Pr	GPIt
1.00	4.00	2.00	6.00	1.33	0.67	0.67	0.50	1.33
2.00	10.00	8.00	14.00	2.50	1.50	0.75	0.60	1.25
3.00	16.00	12.00	18.00	3.67	2.67	0.89	0.73	1.22

Table 4. Data table for EPG

t	EV	GV	PV	GT	PT	SPIt	Pr	GPIt
1.00	6.00	2.00	4.00	3.00	2.00	2.00	0.67	3.00
2.00	12.00	4.00	6.00	3.60	2.75	1.38	0.76	1.80
3.00	15.00	6.00	14.00	3.90	3.25	1.08	0.83	1.30

Table 5. Data table for GPE

t	EV	GV	PV	GT	PT	SPIt	Pr	GPIt
1.00	2.00	6.00	4.00	0.33	0.50	0.50	1.50	0.33
2.00	4.00	12.00	6.00	0.67	1.33	0.67	2.00	0.33
3.00	6.00	16.00	12.00	1.13	1.50	0.50	1.33	0.38

Table 6. Data table for GEP

t	EV	GV	PV	GT	PT	SPIt	Pr	GPIt
1.00	4.00	6.00	2.00	0.67	2.00	2.00	3.00	0.67
2.00	6.00	12.00	4.00	1.00	3.00	1.50	3.00	0.50
3.00	14.00	16.00	6.00	2.63	3.80	2.33	2.67	0.88

Table 7. Data table for EGP

t	EV	GV	PV	GT	PT	SPIt	Pr	GPIt
1.00	4.00	6.00	2.00	0.67	2.00	2.00	3.00	0.67
2.00	6.00	12.00	4.00	1.00	3.00	1.50	3.00	0.50
3.00	14.00	16.00	6.00	2.63	3.80	1.27	1.45	0.88

for GV>PV>EV, and so on. The descriptions are based on values depicted in data tables (see Tables 2 to 7).

PGE: Bold with Worst Outcome

In this first situation we insert a table and two graphs: Figure 4 with the cumulative values in monetary units and showing how PT (1.00) and GT (2.00) are to be found in their respective curves from the value of EV (6.00) at t=3.00. In Figure 5 we have the indexes.

For this situation we may say that SPIt looks far from favorable (0.33 for most of the time): planning was bold (Pr<1), may be unrealistic. However if we have good reasons to believe

in the planning then we may ask why it was not possible to obtain a better performance.

PEG: Bold with Unfavorable Outcome

We have a bold planning and not much favourable SPIt, although GPIt is favourable. What went wrong at the execution?

EPG: Bold with Favorable Outcome

In this situation there is a bold planning and the outcome is better. May be G is too relaxed and could be improved. It is a situation in which the good practice could easily be outperformed.

Figure 4.Values for PV, GV, and EV

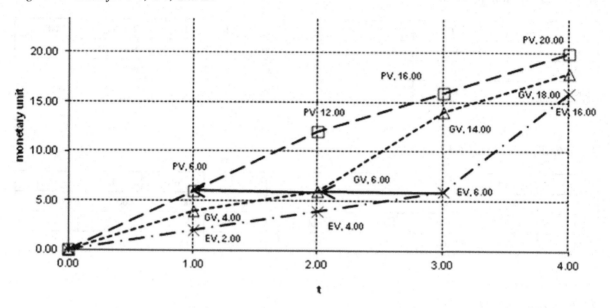

Figure 5. Indexes values (SPIt, Pr and GPIt)

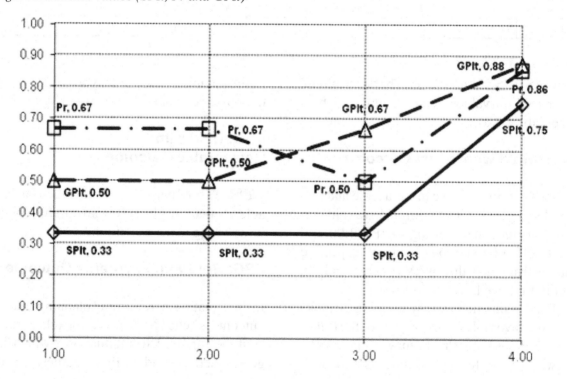

GPE: Conservative with Worst Outcome

The conservative planning amplifies GPIt. This index shows, however a bad outcome. Do the circumstances justify this bad outcome?

GEP: Conservative with Unfavorable Outcome

Now we have a conservative planning. GPIt is not much favorable but SPIt is good. May be it would be helpful to analyze the reasons for the conservative planning.

EGP: Conservative with Favorable Outcome

Here we have a conservative planning and the outcome is favorable, but GPIt shows an erratic performance. This is a direct consequence of the conservative planning. It perhaps represents a lost opportunity to do better.

COPING WITH INNOVATIVE PRACTICES

As we are interested in dealing with novel things (design) we propose paradigms by looking at what has been already made (good practices) with some possible variations. This view is justified in engineering design settings where there is a need to offer continuous innovations and practice should be included within project settings, see Figure 6. In this diagram we grafted our concept of management perturbation.

A learning cycle emerges, meaning that 'lessons learned from past project experi-

Figure 6. Managing perturbation to cope with innovative practices. Adapted from Mahmoud-Jouini (2000).

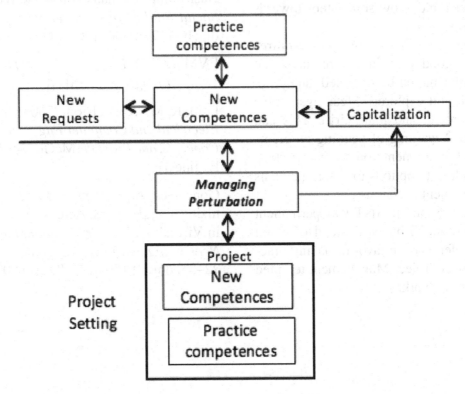

ences are incorporated in documentation and then embedded in training programs so that project managers learn from past experiences. Communication is open in such organizations, leading to a process by which project experiences are "handed down" to next generation project teams' (Barkley, 2007).

CONCLUSION

Resource allocation in engineering design projects may vary in very different ways, not only by scheduling reasons. We must allow the possibility of resource knowledge domains to be somewhat fuzzy. Perhaps the project calls for suppressing parts of some routines on normal practices or add new ones.

The use of a number to characterize translations regime enabled us the enumeration of stylized cases that we can analyze further with the lenses of ANT. This analysis can integrate Scope Management and Planning Management and allow some steps towards Risk Management.

Efficient project management requires more than good planning, it requires that relevant information be obtained, analyzed, exchanged, and explored. Meanings evolve with the project and plans drift. This can provide early warning of pending problems and impact assessments on other activities, which can lead to options for alternate plans and management decisions.

The inspiration on ANT was permanent during our research, but we decided to lighten the ANT references in favor of having practitioners from Project Management tempted to criticize our work.

REFERENCES

Akrich, M. (1992). The De-Scription of Technical Objects. In Shaping Technology/Building Society. MIT Press.

Barkley, B. T. (2007). Project Management. In *New Product Development*. The McGraw-Hill Companies.

Boudourides, M. A. (Ed.). (2001). *International Conference Spacing and Timing, Rethinking Globalization & Standardization*. Palermo, Italy: Academic Press.

Callon, M. (1986). Some elements of a sociology of translation: Domestication of the scallops and the fishermen of St Brieuc Bay. In *Power, Action and Belief: a new sociology of knowledge*. London: Routledge.

Ciborra, C. U. (2004). *Digital Technologies and the Duality of Risk*. London: CARR, LSE.

de Laet, M., & Mol. (2000). The Zimbabwe Bush Pump: Mechanics of a Fluid Technology. *Social Studies of Science*, *30*(2), 225–263. doi:10.1177/030631200030002002

EVM. (2005). *Practice Standard for Earned Value Management* (2nd ed.). PMI.

Ford, R. M., & Coulston. (2008). *Design for Electrical and Computer Engineers: Theory, Concepts, and Practice*. McGraw-Hill Science Engineering.

Gonçalves, A., & Figueiredo. (2009). Organizing Competences: Actor-Network Theory in Virtual Settings. *International Journal of Networking and Virtual Organisations*, *6*(1), 22–35. doi:10.1504/IJNVO.2009.022481

Gonçalves, A., & Figueiredo. (2010a). How to Recognize an Immutable Mobile When You Find One: Translations on Innovation and Design. *International Journal of Actor-Network Theory and Technological Innovation, 2*(2), 39–53. doi:10.4018/jantti.2010040103

Guide, P. M. B. O. K. (2008). *A Guide to the Project Management Body of Knowledge* (4th ed.). PMI.

Hanseth, O., & Braa. (2000). Who's in Control: Designers, Managers, or Technology? Infrastructure at North Hydro. In *From Control to Drift, The Dynamics of Corporate Information Infrastructures*. Oxford University Press.

Lancaster University. (n.d.). *Centre for Science Studies*. Retrieved from http://www.lancs.ac.uk/fass/centres/css/ant/ant.htm

Latour, B. (1987). *Science in Action: How to Follow Scientists and Engineers through Society*. Cambridge, MA: Harvard University Press.

Latour, B. (1990). Drawing Things Together. In M. Lynch, & S. Woolgar (Eds.), *Representation in Scientific Practice*. Cambridge, MA: MIT Press.

Latour, B., Mauguin & Teil. (1992). A Note on Socio-Technical Graphs. *Social Studies of Science, 22*(1), 33–57. doi:10.1177/030631279202022001002

Latour, B. (1993). *We Have Never Been Modern*. Brighton, UK: Harvester Wheatsheaf.

Latour, B. (1994). On Technical Mediation. *Common Knowledge, 3*(2), 29–64.

Latour, B. (1997). On Recalling ANT. In J. Law, & J. Hassard (Eds.), *Actor-network and After*. Blackwell.

Law, J. (1992). Notes on the theory of the actor-network: Ordering, strategy and heterogeneity. *Systems Practice, 5*(4), 379–393. doi:10.1007/BF01059830

Law, J., & Mol. (2000). *Situating Technoscience: An Inquiry into Spatialities*. Lancaster University.

Lipke, W. (2003, March). Schedule is Different. *The Measurable News*.

Mahmoud-Jouini, S. B. (2000). Coment passer d'une demarche réactive à une démarche proactive fondée sur des stratégies d'offres inovantes? Le cas des grandes entreprises générales de Bâtiment françaises. In *De l'idée au marché*. Vuibert.

Mol, A., & Law. (1994). Regions, networks and fluids: Anaemia and social topology. *Social Studies of Science, 24*, 641–671. doi:10.1177/030631279402400402 PMID:11639423

Monteiro, E., & Hepsø, V. (2000). Infrastructure strategy formation: seize the day at Statoil. In C. Ciborra (Ed.), *From control to drift, The dynamics of corporate information infrastructure*. Oxford University Press.

Orlikowski, W. J. (1996). Improvising Organizational Transformation over Time: A Situated Change Perspective. *Information Systems Research, 7*(1), 63–92. doi:10.1287/isre.7.1.63

Pich, M. T., Loch, & De Meyer. (2002). On Uncertainty, Ambiguity, and Complexity in Project Management. *Management Science, 48*(8), 1008–1023. doi:10.1287/mnsc.48.8.1008.163

Pons, D. (2008). Project Management for New Product Development. *Project Management Journal, 39*(2), 82–97. doi:10.1002/pmj.20052

Pons, D. J., & Raine, J. (2005). Design Mechanisms and Constrains. *Research in Engineering Design, 16,* 73–85. doi:10.1007/s00163-005-0008-9

Rodrigues, A. (2010). Effective Measurement of Time Performance using Earned Value Management – A proposed modified version for SPIU tested across various industries and project types. *PM World Today, 12*(10).

Sonnemans, P. J. M., Geudens, W. H. J., & Brombacher, A. C. (2003). Organizing product releases in time-driven development processes: A probabilistic analysis. *IMA Journal of Management Mathematics, 14,* 337–356. doi:10.1093/imaman/14.4.337

KEY TERMS AND DEFINITIONS

Actor-Network Theory (ANT): A social theory where the "social" is approached as assemblies of actors – human and non-human actants – coming together as a whole – network – by creating both material and semiotic meaning. These networks are always in the process of being re-assembled. ANT originated in the field of science and technology studies and was first developed at the *Centre de Sociologie de l'Innovation* the *École Nationale Supérieure de Mines de Paris*, in the early 1980s by Michel Callon, Bruno Latour, John Law and others. This approach was inspired by existing studies of semiotics of Algirdas Julien Greimas and the philosophy of Michel Serres.

Actors: Actants that take a specific shape by virtue of their relations with one another. They construct their own identity. Their performance in a network can be ascribed to a character, a set of competences. Relations, or interactions, amongst actors can be viewed as ´trials' leading to the construction of "meanings", both material and semiotic. A Project can be described as an actor-network where teams are also actor-networks, as well as project documents, equipment, instruments, rules, and works.

Centre of Calculation: See *Obligatory Point of Passage (OPP).*

Changes "of Regime": See *Perturbation.*

Circulation: In the context of the chapter it is a way to analyse social interactions as some kind of flow that moves around any boundaries that the analyst may try to establish between the setting – the object of the analysis – and the rest of the world.

Design-Projects: Engineering core projects where technological artifacts are created, that is, idealized and developed.

Earned Value Management: A management system that measures variances of the work done against work planned. It is in fact a calculation of actual work using budgeted references that allows to percept if the project is running late or over-run. It helps management and facilitates project control.

Good Practices: Like norms, are practices, or ways of doing things that follow patterns previously accepted as better than the alternatives.

Immutable Mobiles (IMs): Actors designed to be able to be moved – mobile – and transformed although maintained constant – immutable – while moved around settings and brought back to a *centre of calculation* to be combined with others. IMs are the entities that are to be instaurated in the class Reference, the *Mode of Existence* of Science. In a Project, functional specifications can be analysed as IMs, as Practices and Good Practices also are.

Inscription: What engineers or designers inscribe on actants, distributing competences amongst them and setting up a particular assembly.

Mode of Existence: Is a new discourse on the "social" that goes beyond the nature/society duality of modernity and where ANT, a mode of existence also, allows the mediation to other modes, namely Politics, Moral, Fiction, Law, Organization, Reference (Science), Religion, Attachment, Technology.

Obligatory Point of Passage (OPP): An actor that translates others in the first phase of a translation process. OPPs can become centres of calculation and stabilize their networks into IMs. In a successful translation the resulting network is watched as a whole and the OPP speaks on behalf of the actors. In Projects, functional specialists and project managers can be analysed as OPPs.

Perturbation: In this chapter, a concept proposed as a way to account for the risk of turbulence in a project setting when a strong actor (project manager) assigns harder conditions for a resource than the ones usually practiced. A stylized numerical description is put forward by means of a ratio of project and practice planning times for a task which allows to define changes "of regime" of turbulences.

Practices: Way of doing things, materialized in processes.

Program of Action (PA): Distribution of roles among actors on the basis of goals envisaged and competences required (from Greimas, narrative program).

Scope Management: One of the most challenging aspects in project management. Scope needs to be managed according to available resources, functionality, cost, time, organization strategy, quality, and burden crossing adaptations. Managing scope is a challenge activity in single projects, programs and portfolios.

Translation: The basic process in ANT where actors gives rise to actor-networks in interactions with one another. In doing so actors are also 'displaced' because, to some degree, they are modified. It is usually analysed as a four phase process: problematization (where OPPs are defined), 'interessement', enrolment, and mobilisation.

Turbulence: See *Perturbation*.

Work Breakdown Structure (WBS): Planning tool for division of work from the total to the parts – work package. At the top of the WBS you have the system, and at the lower levels you have work packages and single activities.

Work Package: Sets of activities that generate a deliverable output. They represent good control gates for project management.

Chapter 15
Web Tools as Actors:
The Case of Online Investing

Arthur Adamopoulos
RMIT University, Australia

Martin Dick
RMIT University, Australia

Bill Davey
RMIT University, Australia

ABSTRACT

Often actor-network theory studies find that technology has been translated through its relationships with human actors. This chapter reports on a study of online investing that found that the human actors were translated to more active and involved investors due to the changes, over time, in the online services that are available: the non-human actors. The Internet is a constantly evolving technological actor. New tools have the potential to change interactions with users. In this study, it became evident that new services had a noticeable effect on the behaviour of investors. Not only did investors report changes in their behaviour when they moved from offline to online investing, but they also reported changes in their investing strategies over time as new services became available. This study showed a new and interesting confirmation of the value of allowing non-human actors to be heard.

INTRODUCTION

This paper uses data from a study of online investing that was designed to determine key factors in investor behaviour when using online tools. An interesting phenomenon was discovered as a by-product of that study which may have implications for studies of interactions between people and the Internet. Internet based tools seem to provide such power to the technological actor that it becomes preeminent in the network.

Market investment is a very large industry, with many millions of individuals and households investing in stocks in many parts of the world. In the USA, over 20% of all households own stocks directly and that figure jumps to almost 50% when managed funds are taken

DOI: 10.4018/978-1-4666-6126-4.ch015

into account (Bucks, Kennickell, Mach, & Moore, 2009). In Hong Kong, 35% of the adult population invest directly in stocks (HKEX, 2008). Almost 50% of the adult population of Australia own shares directly and half of those monitor their investments weekly (ASX, 2004).

A significant proportion of these investors now conduct at least some of their investing activities online. In many of these markets, online brokers have taken over as the predominant channel for retail investors. Given that online trading and investing is a significant e-service industry, it is surprising that there has been relatively little research conducted in this field. Kirsty Williamson and Dimity Kingsford-Smith conducted a qualitative pilot study in 2004 that involved interviewing investors about how they seek information on the internet (Kingsford-Smith & Williamson, 2004). One of their reported observations was that many of the investors interviewed sought information and conducted activities both online and offline at the same time. They found that the primary online information source investors used was their broker's website. They reported that all their investors spoke to others about investing and sought advice from trusted family members, friends or colleagues – with few investors using online forums such as chat rooms. For some investors, there was also a community or hobby aspect to their investing activities – they were part of an investment club or often communicated with others about investing, both online and offline (Kingsford-Smith & Williamson, 2004).

How people use information systems has been a prominent research question for many years, and yet researchers have recently stressed how little we understand about system usage (Burton-Jones & Gallivan, 2007). "Despite being an important construct, IS use is still weakly conceptualized and opera-

tionalized as frequency, duration, or variety of system functions used" (Barki, Titah, & Boffo, 2007). "Researchers must also consider the nature, extent, quality, and appropriateness of the system use" (DeLone & McLean, 2003). Burton-Jones has proposed that system usage should be investigated by taking three elements into consideration: the System, the User and the Task. He has classified system usage measurements from 'Very Lean' to 'Very Rich' (Burton-Jones & Straub, 2006).

A great deal of research has investigated why people use information systems. The research reported here was looking at the way they are used; the nature of the interactions between the technology and the person. This study, of the nature and practice of use of technology rather than the reasons for its adoption, is particularly apt for the application of actor-network theory and allows investigation of the 'very rich' nature of technology use.

An extensive study using an Actor-Network Theory approach was conducted. The results of the wider study are reported elsewhere. The purpose of this paper is to present an unusual finding that emerged within the study. The contention presented here is that an enduring network of mostly non-human actors has, in effect, produced an ongoing translation of humans.

METHOD

Tatnall (2009), following Law (1991), identifies Actor-Network Theory as considering social and technical determinism to be flawed and proposes a social and technical account in which nothing is purely social and nothing is purely technical. Tatnall suggests that Actor-Network Theory allows us to look at non-human entities such as computers, computer programs and the Internet. All these non-human entities are to be considered in the same way as a human entity; existing through the interactions with other entities forming a network. Actor-Network Theory requires us to interview human actors but also to study in detail the non-human actors so we can determine the nature of the network.

The first stage of this process was to identify a number of people and interview them to determine the non-human actors involved in online investing. A convenience sample of Australian residents was chosen through a viral marketing campaign. A diverse group of males and females in a number of age groups starting at age 30 was recruited. These people were deliberately chosen to have diverging backgrounds, but all were using tools that allowed investment in the Australian Stock Exchange (ASX). Occupations included information technology, accounting, finance, and education with three interviewees being retired. The interviewees also fell into the categories of short term traders and long term investors. A short term trader is one who buys and sells a stock within a short period of time: months, weeks, days or even hours. Long term investors generally buy a stock with a view to benefiting from the dividend income arising from that stock over a number of years.

The interviews were conducted face-to-face with most interviews lasting about an hour. Interviews were recorded and then transcribed. From the transcriptions, a number of non-human actors were identified and for each of these, the nature of the interactions between them and the human actors. The purpose of the interviews was to determine the nature of the translation(s) of the online services available to investors. The choice of the word 'translation' derives from Callon, who defines it as "the methods by which an actor enrols others" (Callon, 1986).

This outcome should be differentiated from assessing the services in any way as the Actor-Network Theory is to understand the nature of the translations:

A faithful application of ANT is not concerned with assessing what is dysfunctional behaviour and what is not. Rather, the concern is with understanding in some detail how and why translation processes evolve in certain ways. (Mähring, Holmström, Keil, & Montealegre, 2004)

For this paper it was decided to report the outcome using an early 'definition' of translation that considers a translation to be in four parts:

1. **Problematization:** An actor (the initiator) identifies the interests of other actors and defines problems and solutions that will require the initiator to be a part of the network.
2. **Interessement Stage:** Actors commit to the problematization and an alliance of the actors commences the network.
3. **Enrollment:** Involves a set of strategies through which initiators seek to convince other actors to embrace the underlying ideas of the growing actor-network, and to be an active part of the whole project.
4. **Mobilization:** Initiators seek to secure continued support to the underlying ideas from the enrolled actors. With allies mobilized, an actor-network achieves stability (Callon, 1986; Mähring et al., 2004).

The outcome of a translation, as different actors compete in 'trials of strength' (Latour, 1987), will be weak or strong translations.

A strong translation will be a durable actor-network with intentions, goals and beliefs. The existence of a booming online investor community hints that such a network exists.

RESULTS

Identified actors included the human actors: investor, family and friends of the investor and professionals involved in the investment industry (such as brokers). The non-human actors include the stock exchange, information sources such as television and print media, and a rich variety of applications roughly described as 'Internet Services'.

We found the investor to be not a single human actor but a range of types of humans. For our purposes, those people who trade frequently were called 'short term traders'. Others, sometimes the same people but in a different aspect of their behaviour, had a portfolio that was occasionally added to but very seldom sold. These were called 'long term investors'. These two types of investor, even when they were the same people, responded to questions in different ways depending upon whether the topic was their short or long term portfolio. In some of the interviews, the questions were asked twice: once for their short-term portfolio and once for their long-term portfolio.

The complex non-human actor list is what we have collated together, black boxed, and called 'Internet Services'. This set of actors includes: online brokers, online banks, third party research services, chat rooms, general financial Websites, blogs, e-mail, and search engines such as Google. The relationships between the human and non-human actors are complex and a model of them has been presented elsewhere. As an approximation we can say that the relationships can be categorised in the following terms:

- **Investment Approach:** The relationship between the actors is often characterized by the investment approach of the investor. Some investors, for example, are heavily focused on charting price movements of stocks. These investors may have strong relationships with charting software packages and online providers of historical price data. Other investors may use the 'fundamentals' approach where they look into the financials of a company before investing. These investors may have relationships with an online stock recommendation newsletter, online financial data providers and the stock's website. Many investors described using a mix of the two approaches while others said they used charting only for their short term trading and fundamentals only for their long term investing portfolios.

- **Investment Goals:** The differences between human actors can often be characterized by looking at the goals of the investor. These goals can have a profound effect on the relationship between investor and the tools/services that they use. One major factor is the investment time frame. Short term traders are usually interested in identifying a stock to purchase either by looking at chart movements or by getting a 'hot tip'. Those that work on 'tips' either source these tips from trusted networks of friends and family, or may also source them from publicly available online sources such as investment chat rooms or blogs. Others may be looking to invest long term in a particular sector such as mining or oil stocks. These investors may use their online broker or even Google to identify the best and largest companies in the sector.

- **Influences:** The relationship between an investor and the other people around them often determines the way in which Internet Services are used. Some investors, particularly chartists, express the desire to only invest by their own thinking and not be influenced by any external information, or "noise". They may make heavy use of particular online services but not communicate with any other human actors. Other investors are the complete opposite and operate solely on the basis of investing on recommendations from trusted friends and family members. These investors have strong relationships with other human actors but then also use online services to verify information they may have received.

All of this so far presents a picture, however complex, of a static set of actors and relationships between them. On further investigation it was identified that the online services available to investors are constantly changing over time. Some investors described how their investing activities had changed from before and after starting to invest online. Many said they were more actively involved in investing after going online and that they preferred online brokers to human brokers, primarily for cost reasons but also for convenience and availability. Others described how their investing had changed over the years from when they began, even when they had only ever invested online. Some of these changes were attributed to changing risk profiles or past profits or losses, but some of the changes were due to the changing online services that they gained access to. One investor described how he initially only invested in steel companies because that was the industry he worked in and understood. The Internet and the information services available on it allowed him to learn about other industries to the point where he felt he understood them enough to invest in them.

This led to a different line of questioning. Investors were asked 'would you be investing this way if the online services were not available'? Investors commonly reported that the presence of the online tools and the immediacy of the information and services available motivated them to trade online and to trade more often than they otherwise might. Many also said that they gained a sense of control over their investing by having information available at their 'fingertips', rather than having to go through a human broker. When asked if they would still invest in stocks if the online tools were not available, most said they would definitely be less interested in stock market trading, either still investing but on a smaller scale or not investing at all.

As in many cases with actor-network theory investigations, this occurrence of relationships change over time is not unusual. What became starkly apparent however is that the non-human actor –in this case the services available on the Internet, was changing and causing a translation in the investor: the human actor. Following Latour (1996), Tatnall (2002) points out that actor-network theory sees the world as being composed of hybrid entities containing both human and non-human elements. Changes in the network result from decisions being made by actors, and these decisions involve the exercise of power. The possession of power does not create change as actors must be also persuaded to perform actions for change to occur. This often leads investigators using actor-network theory to concentrate on the way in which human actors persuade each other. In this case we have found that the non-human actor, Internet Services, has been most persuasive in changing the nature of the human actor.

To identify this Translation of a human actor the four stages of translation can be enunciated as follows:

- **Problematization:** The data seems to indicate that a significant actor that might be characterized as the initiator in the Australian investment market is the provision of services by Banks. The ASX is completely electronic in Australia. Although most traditional broking firms offered online reports, it was not until the banks offered low cost online broking that this form of investment practice became common. The Banking industry problematization could be expressed in terms such as 'people with an online bank account should see stocks as another financial instrument that banking manages online'.

- **Interessement Stage:** The Banks were able to easily explain the problematization in terms that the ASX would like, as the ASX is principally interested in turnover, which is where their income derives. They were already in possession of software that included bank accounts and security. A third major actor was the Internet. This actor had access to large numbers of people and saw online banking as a long term ally.
- **Enrollment:** Enrollment for this network was a simply understood process, as each of the actors (banks, ASX, Internet) are all to make financial gains directly from the problematization. The Internet was able to recruit human actors into the network by providing them access to transactional services (via the Banks) and information services (via other Internet Services from other actors) at a cheaper cost than previously available and also provide information more easily and quickly than previously. This combination provided a compelling case for investors to join the network for their investing activities.
- **Mobilization:** After a very short time the initiators were able to gain support from a large range of non-human actors and their human allies. The investment industry is large and contains a plethora of services. Each service could be recruited with the promise of the number of investors already in the cadre. As investors used these

services over a period of time, they integrated them into their standard operating practices, strengthening the stability of the network. As many of the investors we interviewed stated, they would not be investing in stocks if it wasn't for the online services available to them. The non-human actors also encourage the human actors to be more active in their investing and therefore their usage, by providing information in a timely manner and giving the human actors a sense of control. As new services are constantly becoming available, the network is in a constant state of change but at the same time becoming more stable rather than less.

CONCLUSION

This study has shown online investing to be an innovation that has become translated in the Australian context in a very special way. The strongest actor in this case has been the provision of low cost online broking by the banking sector. In this case the technological actor in the form of online investing systems has been the agent of mobilization. It became clear to us in the analysis of interviews that the technological actor was driving the human actors. The most convincing evidence of this phenomenon came from interviewees who told stories of changing their online behaviour radically as the result of new tools becoming available. The use of Actor-Network Theory to examine the finance industry seems apt. The next stage of our research will be to look at the issue of calculative agencies in the online investing environment (David & Halbert, 2013).

The case of online investors seems to be a powerful case in which to study the interactions between people and the Internet. In this case we found an interesting confirmation of the power of allowing non-human actors to be heard. Because the non-human actor changed so much over the course of the investigation it was obvious that the human actors were changing as a result. Often actor-network theory studies conclude that technology has been translated through its relationships with the human actors. In our case we found that the human actor has been translated to a more active and involved investor due to the change in the online services that are available: the non-human actors.

REFERENCES

ASX. (2004). *Share Ownership Study 2004*. Australian Stock Exchange.

Barki, H., Titah, R., & Boffo, C. (2007). Information System Use–Related Activity: An Expanded Behavioral Conceptualization of Individual-Level Information System Use. *Information Systems Research, 18*(2), 173–192. doi:10.1287/isre.1070.0122

Bucks, B. K., Kennickell, A. B., Mach, T. L., & Moore, K. B. (2009). *Changes in U.S. Family Finances from 2004 to 2007: Evidence from the Survey of Consumer Finances*. US Federal Reserve.

Burton-Jones, A., & Gallivan, M. (2007). Toward a deeper understanding of system usage in organizations: A multilevel perspective. *Management Information Systems Quarterly, 31*(4), 657–679.

Burton-Jones, A., & Straub, D. W. (2006). Reconceptualizing System Usage: An Approach and Empirical Test. *Information Systems Research, 17*, 228. doi:10.1287/isre.1060.0096

Callon, M. (1986). Some Elements of a Sociology of Translation: Domestication of the Scallops and the Fishermen of St Brieuc Bay. In J. Law (Ed.), *Power, Action & Belief: A New Sociology of Knowledge?* (pp. 196–229). London: Routledge & Kegan Paul.

David, L., & Halbert, L. (2013). Finance Capital, Actor-Network Theory and the Struggle Over Calculative Agencies in the Business Property Markets of Mexico City Metropolitan Region. *Regional Studies*, 1–14.

DeLone, W. H., & McLean, E. R. (2003). The DeLone and McLean model of information systems success: A ten-year update. *Journal of Management Information Systems, 19*, 9.

HKEX. (2008). *Retail Investor Survey 2007*. Hong Kong Exchanges and Clearing Ltd.

Kingsford-Smith, D., & Williamson, K. (2004). How Do Online Investors Seek Information, And What Does This Mean for Regulation?. *The Journal of Information, Law and Technology, (2)*.

Latour, B. (1987). *Science in Action: How to Follow Engineers and Scientists Through Society*. Milton Keynes, UK: Open University Press.

Latour, B. (1996). *Aramis or the Love of Technology*. Cambridge, MA: Harvard University Press.

Law, J. (Ed.). (1991). *A Sociology of Monsters: Essays on Power, Technology and Domination*. London: Routledge.

Mähring, M., Holmström, J., Keil, M., & Montealegre, R. (2004). Trojan actor-networks and swift translation: Bringing actor-network theory to IT project escalation studies. *Information Technology & People, 17*(2), 210–238. doi:10.1108/09593840410542510

Tatnall, A. (2002). Socio-Technical Processes in the Development of Electronic Commerce Curricula. In A. Wenn (Ed.), *Skilling the e-Business Professional* (pp. 9–21). Melbourne, Australia: Heidelberg Press.

Tatnall, A. (2009). Actor-Network Theory Applied to Information Systems Research. In M. Khosrow-Pour (Ed.), *Encyclopedia of Information Science and Technology* (2nd ed., Vol. 1, pp. 20–24). Hershey, PA: Information Science Reference.

KEY TERMS AND DEFINITIONS

Internet Services: All information sources and services available on the Internet at a particular point in time.

Investment Approach: The method used to invest - charts, fundamental analysis or word of mouth stock tips.

Investment Goals: The focus of a portfolio - by time frame (short or long term), or by industry (biotech stocks) or by sector (blue chips).

Long Term Investor: Individual who buys stocks to hold for years and profit from dividends as well as a gradual rise in the stock price.

Portfolio: A portion of an individual's funds invested with a particular goal. Individuals may have multiple portfolios at the same time.

Short Term Trader: Individual who seeks to profit from short term (from minutes to weeks) price fluctuations in stocks.

Related References

To continue our tradition of advancing information science and technology research, we have compiled a list of recommended IGI Global readings. These references will provide additional information and guidance to further enrich your knowledge and assist you with your own research and future publications.

Acilar, A., & Karamasa, Ç. (2013). Factors Affecting E-Commerce Adoption by Small Businesses in a Developing Country: A Case Study of a Small Hotel. In S. Chhabra (Ed.), *ICT Influences on Human Development, Interaction, and Collaboration* (pp. 174–184). Hershey, PA: Information Science Reference.

Adami, I., Antona, M., & Stephanidis, C. (2014). Ambient Assisted Living for People with Motor Impairments. In G. Kouroupetroglou (Ed.), *Disability Informatics and Web Accessibility for Motor Limitations* (pp. 76–104). Hershey, PA: Medical Information Science Reference.

Adamopoulos, A., Dick, M., & Davey, B. (2012). Actor-Network Theory and the Online Investor. [IJANTTI]. *International Journal of Actor-Network Theory and Technological Innovation*, 4(2), 25–31. doi:10.4018/jantti.2012040103

Adapa, S., & Valenzuela, F. (2014). Case Study on Customer's Ambidextrous Nature of Trust in Internet Banking: Australian Context. In V. Jham, & S. Puri (Eds.), *Cases on Consumer-Centric Marketing Management* (pp. 206–229). Hershey, PA: Business Science Reference.

Ahamer, G. (2014). GISS and GISP Facilitate Higher Education and Cooperative Learning Design. In S. Mukerji, & P. Tripathi (Eds.), *Handbook of Research on Transnational Higher Education* (pp. 1–21). Hershey, PA: Information Science Reference.

Al-Shqairat, Z. I., & Altarawneh, I. I. (2011). The Role of Partnership in E-Government Readiness: The Knowledge Stations (KSs) Initiative in Jordan. [IJTHI]. *International Journal of Technology and Human Interaction*, 7(3), 16–34. doi:10.4018/jthi.2011070102

Al-Weshah, G., Al-Hyari, K., Abu-Elsamen, A., & Al-Nsour, M. (2013). Electronic Networks and Gaining Market Share: Opportunities and Challenges. In S. Chhabra (Ed.), *ICT Influences on Human Development, Interaction, and Collaboration* (pp. 142–157). Hershey, PA: Information Science Reference.

Alawneh, A., Al-Refai, H., & Batiha, K. (2011). E-Business Adoption by Jordanian Banks: An Exploratory Study of the Key Factors and Performance Indicators. In A. Tatnall (Ed.), *Actor-Network Theory and Technology Innovation: Advancements and New Concepts* (pp. 113–128). Hershey, PA: Information Science Reference. doi:10.4018/978-1-60960-587-2.ch412

Albarelli, A., Bergamasco, F., & Torsello, A. (2014). Learning Computer Vision through the Development of a Camera-Trackable Game Controller. In F. Cipolla-Ficarra (Ed.), *Advanced Research and Trends in New Technologies, Software, Human-Computer Interaction, and Communicability* (pp. 154–163). Hershey, PA: Information Science Reference.

Albrecht, K. (2014). Microchip-Induced Tumors in Laboratory Rodents and Dogs: A Review of the Literature 1990–2006. In M. Michael, & K. Michael (Eds.), *Uberveillance and the Social Implications of Microchip Implants: Emerging Technologies* (pp. 281–317). Hershey, PA: Information Science Reference.

Ali, Y. U. K. S. E. L. K. (2014). Using Information Retrieval for Interaction with Mobile Devices. In K. Rızvanoğlu, & G. Çetin (Eds.) Research and Design Innovations for Mobile User Experience (pp. 286-302). Hershey, PA: Information Science Reference. doi: doi:10.4018/978-1-4666-4446-5.ch015

Ali Yüksel, K. (2014). Gestural Interaction with Mobile Devices Based on Magnetic Field. In K. Rızvanoğlu, & G. Çetin (Eds.), *Research and Design Innovations for Mobile User Experience* (pp. 203–222). Hershey, PA: Information Science Reference.

Anders, B. A. (2014). More than Just Data: The Importance of Motivation, Examples, and Feedback in Comprehending and Retaining Digital Information. In S. Hai-Jew (Ed.), *Packaging Digital Information for Enhanced Learning and Analysis: Data Visualization, Spatialization, and Multidimensionality* (pp. 37–46). Hershey, PA: Information Science Reference.

Andrade, A. D. (2012). From Intermediary to Mediator and Vice Versa: On Agency and Intentionality of a Mundane Sociotechnical System. In A. Tatnall (Ed.), *Social Influences on Information and Communication Technology Innovations* (pp. 195–204). Hershey, PA: Information Science Reference. doi:10.4018/978-1-4666-1559-5.ch014

Angulo, J. (2014). Users as Prosumers of PETs: The Challenge of Involving Users in the Creation of Privacy Enhancing Technologies. In M. Pańkowska (Ed.), *Frameworks of IT Prosumption for Business Development* (pp. 178–199). Hershey, PA: Business Science Reference.

Ariani, Y., & Yuliar, S. (2011). Translating Biofuel, Discounting Farmers: The Search for Alternative Energy in Indonesia. In A. Tatnall (Ed.), *Actor-Network Theory and Technology Innovation: Advancements and New Concepts* (pp. 68–79). Hershey, PA: Information Science Reference.

Arslan, P. (2014). Collaborative Participation in Personalized Health through Mobile Diaries. In K. Rızvanoğlu, & G. Çetin (Eds.), *Research and Design Innovations for Mobile User Experience* (pp. 150–181). Hershey, PA: Information Science Reference.

Augusto, A. B., & Correia, M. E. (2014). A Mobile-Based Attribute Aggregation Architecture for User-Centric Identity Management. In A. Ruiz-Martinez, R. Marin-Lopez, & F. Pereniguez-Garcia (Eds.), *Architectures and Protocols for Secure Information Technology Infrastructures* (pp. 266–287). Hershey, PA: Information Science Reference.

Ávila, I., Menezes, E., & Braga, A. M. (2014). Strategy to Support the Memorization of Iconic Passwords. In K. Blashki, & P. Isaias (Eds.), *Emerging Research and Trends in Interactivity and the Human-Computer Interface* (pp. 239–259). Hershey, PA: Information Science Reference.

Awwal, M. A. (2012). Influence of Age and Genders on the Relationship between Computer Self-Efficacy and Information Privacy Concerns. [IJTHI]. *International Journal of Technology and Human Interaction*, 8(1), 14–37. doi:10.4018/jthi.2012010102

Baladrón, C., Aguiar, J. M., Calavia, L., Carro, B., & Sánchez-Esguevillas, A. (2014). Learning on the Move in the Web 2.0: New Initiatives in M-Learning. In H. Lim, & F. Sudweeks (Eds.), *Innovative Methods and Technologies for Electronic Discourse Analysis* (pp. 437–458). Hershey, PA: Information Science Reference.

Baldauf, M., Fröhlich, P., Buchta, J., & Stürmer, T. (2013). From Touchpad to Smart Lens: A Comparative Study on Smartphone Interaction with Public Displays. [IJMHCI]. *International Journal of Mobile Human Computer Interaction*, 5(2), 1–20. doi:10.4018/jmhci.2013040101

Bali, R. K., & Wickramasinghe, N. (2012). RAD and Other Innovative Approaches to Facilitate Superior Project Management. In A. Tatnall (Ed.), *Social Influences on Information and Communication Technology Innovations* (pp. 149–155). Hershey, PA: Information Science Reference. doi:10.4018/978-1-4666-1559-5.ch010

Barazzetti, L. (2014). Multi-Photo Fusion through Projective Geometry. In F. Cipolla-Ficarra (Ed.), *Advanced Research and Trends in New Technologies, Software, Human-Computer Interaction, and Communicability* (pp. 164–173). Hershey, PA: Information Science Reference.

Barazzetti, L., & Scaioni, M. (2014). Reality-Based 3D Modelling from Images and Laser Scans: Combining Accuracy and Automation. In F. Cipolla-Ficarra (Ed.), *Advanced Research and Trends in New Technologies, Software, Human-Computer Interaction, and Communicability* (pp. 294–306). Hershey, PA: Information Science Reference.

Barrantes, D., & Maurizia D´Antoni,. (2014). Special Educational Needs Workshop Online: Play Activity – Homework in the Jungle. In F. Cipolla-Ficarra (Ed.), *Advanced Research and Trends in New Technologies, Software, Human-Computer Interaction, and Communicability* (pp. 174-182). Hershey, PA: Information Science Reference. doi:10.4018/978-1-4666-4490-8.ch017

Baruh, L., & Popescu, M. (2014). Trapped in My Mobility: How a Principle of "Control over Communicative Interaction" Can Guide Privacy by Design in Mobile Ecosystems. In K. Rızvanoğlu, & G. Çetin (Eds.), *Research and Design Innovations for Mobile User Experience* (pp. 223–244). Hershey, PA: Information Science Reference.

Basham, R. (2014). Surveilling the Elderly: Emerging Demographic Needs and Social Implications of RFID Chip Technology Use. In M. Michael, & K. Michael (Eds.), *Uberveillance and the Social Implications of Microchip Implants: Emerging Technologies* (pp. 169–185). Hershey, PA: Information Science Reference.

Bedrina, T., Parodi, A., Clematis, A., & Quarati, A. (2014). Mashing-Up Weather Networks Data to Support Hydro-Meteorological Research. In F. Cipolla-Ficarra (Ed.), *Advanced Research and Trends in New Technologies, Software, Human-Computer Interaction, and Communicability* (pp. 245–254). Hershey, PA: Information Science Reference.

Bergamasco, F., Albarelli, A., & Torsello, A. (2014). A Practical Setup for Projection-Based Augmented Maps. In F. Cipolla-Ficarra (Ed.), *Advanced Research and Trends in New Technologies, Software, Human-Computer Interaction, and Communicability* (pp. 13–22). Hershey, PA: Information Science Reference.

Bergmann, N. W. (2014). Ubiquitous Computing for Independent Living. In I. Association (Ed.), *Assistive Technologies: Concepts, Methodologies, Tools, and Applications* (pp. 679–692). Hershey, PA: Information Science Reference.

Berler, A., & Apostolakis, I. (2014). Normalizing Cross-Border Healthcare in Europe via New E-Prescription Paradigms. In C. El Morr (Ed.), *Research Perspectives on the Role of Informatics in Health Policy and Management* (pp. 168–208). Hershey, PA: Medical Information Science Reference.

Bezuayehu, L., Stilan, E., & Peesapati, S. T. (2014). Icon Metaphors for Global Cultures. In K. Blashki, & P. Isaias (Eds.), *Emerging Research and Trends in Interactivity and the Human-Computer Interface* (pp. 34–53). Hershey, PA: Information Science Reference.

Bielenia-Grajewska, M. (2011). Actor-Network-Theory in Medical e-Communication – The Role of Websites in Creating and Maintaining Healthcare Corporate Online Identity. [IJANTTI]. *International Journal of Actor-Network Theory and Technological Innovation*, *3*(1), 39–53. doi:10.4018/jantti.2011010104

Bielenia-Grajewska, M. (2011). A Potential Application of Actor Network Theory in Organizational Studies: The Company as an Ecosystem and its Power Relations from the ANT Perspective. In A. Tatnall (Ed.), *Actor-Network Theory and Technology Innovation: Advancements and New Concepts* (pp. 247–258). Hershey, PA: Information Science Reference.

Bielenia-Grajewska, M. (2013). Actor-Network-Theory in Medical E-Communication: The Role of Websites in Creating and Maintaining Healthcare Corporate Online Identity. In A. Tatnall (Ed.), *Social and Professional Applications of Actor-Network Theory for Technology Development* (pp. 156–172). Hershey, PA: Information Science Reference.

Bingley, S., & Burgess, S. (2011). Using I-D maps to Represent the Adoption of Internet Applications by Local Cricket Clubs. In A. Tatnall (Ed.), *Actor-Network Theory and Technology Innovation: Advancements and New Concepts* (pp. 80–94). Hershey, PA: Information Science Reference.

Biswas, P. (2014). A Brief Survey on User Modelling in Human Computer Interaction. In I. Association (Ed.), *Assistive Technologies: Concepts, Methodologies, Tools, and Applications* (pp. 102–119). Hershey, PA: Information Science Reference.

Bloom, L., & Dole, S. (2014). Virtual School of the Smokies. In S. Mukerji, & P. Tripathi (Eds.), *Handbook of Research on Transnational Higher Education* (pp. 674–689). Hershey, PA: Information Science Reference.

Bowe, B. J., Blom, R., & Freedman, E. (2013). Negotiating Boundaries between Control and Dissent: Free Speech, Business, and Repressitarian Governments. In J. Lannon, & E. Halpin (Eds.), *Human Rights and Information Communication Technologies: Trends and Consequences of Use* (pp. 36–55). Hershey, PA: Information Science Reference.

Bowers, L., Rebola, C. B., & Vela, P. (2014). Communication Technologies for Older Adults in Retirement Communities. In F. Cipolla-Ficarra (Ed.), *Advanced Research and Trends in New Technologies, Software, Human-Computer Interaction, and Communicability* (pp. 491–501). Hershey, PA: Information Science Reference.

Buchel, O., & Sedig, K. (2014). From Data-Centered to Activity-Centered Geospatial Visualizations. In M. Huang, & W. Huang (Eds.), *Innovative Approaches of Data Visualization and Visual Analytics* (pp. 266–287). Hershey, PA: Information Science Reference.

Burke, M. E., & Speed, C. (2014). Knowledge Recovery: Applications of Technology and Memory. In M. Michael, & K. Michael (Eds.), *Uberveillance and the Social Implications of Microchip Implants: Emerging Technologies* (pp. 133–142). Hershey, PA: Information Science Reference.

Burlea, A. S. (2014). The Challenges of the Prosumer as Entrepreneur in IT. In M. Pańkowska (Ed.), *Frameworks of IT Prosumption for Business Development* (pp. 1–15). Hershey, PA: Business Science Reference.

Cabitza, F., & Gesso, I. (2014). Reporting a User Study on a Visual Editor to Compose Rules in Active Documents. In K. Blashki, & P. Isaias (Eds.) *Emerging Research and Trends in Interactivity and the Human-Computer Interface* (pp. 182-203). Hershey, PA: Information Science Reference. doi: doi:10.4018/978-1-4666-4623-0.ch009

Cagliero, L., & Mahoto, N. A. (2014). Visualization of High-Level Associations from Twitter Data. In S. Hai-Jew (Ed.), *Packaging Digital Information for Enhanced Learning and Analysis: Data Visualization, Spatialization, and Multidimensionality* (pp. 164–183). Hershey, PA: Information Science Reference.

Çalışkan, V., Öztürk, Ö., & Rızvanoğlu, K. (2014). Mobile Accessibility in Touchscreen Devices: Implications from a Pilot Study with Blind Users on iOS Applications in iPhone and iPad. In K. Rızvanoğlu, & G. Çetin (Eds.), *Research and Design Innovations for Mobile User Experience* (pp. 182–202). Hershey, PA: Information Science Reference.

Carroll, N., Richardson, I., & Whelan, E. (2012). Service Science: An Actor-Network Theory Approach. [IJANTTI]. *International Journal of Actor-Network Theory and Technological Innovation*, 4(3), 51–69. doi:10.4018/jantti.2012070105

Cecez-Kecmanovic, D., & Nagm, F. (2011). Have You Taken Your Guys on the Journey?: An ANT Account of IS Project Evaluation. In A. Tatnall (Ed.), *Actor-Network Theory and Technology Innovation: Advancements and New Concepts* (pp. 1–19). Hershey, PA: Information Science Reference.

Celotto, E., Ellero, A., & Ferretti, P. (2014). Rough Set Analysis and Short-Medium Term Tourist Services Demand Forecasting. In F. Cipolla-Ficarra (Ed.), *Advanced Research and Trends in New Technologies, Software, Human-Computer Interaction, and Communicability* (pp. 341–349). Hershey, PA: Information Science Reference.

Chan, S., Fisher, K., & Sauer, P. (2014). Student Development of E-Workbooks: A Case for Situated-Technology Enhanced Learning (STEL) Using Net Tablets. In D. McConatha, C. Penny, J. Schugar, & D. Bolton (Eds.), *Mobile Pedagogy and Perspectives on Teaching and Learning* (pp. 20–40). Hershey, PA: Information Science Reference.

Chao, L., Wen, Y., Chen, P., Lin, C., Lin, S., Guo, C., & Wang, W. (2012). The Development and Learning Effectiveness of a Teaching Module for the Algal Fuel Cell: A Renewable and Sustainable Battery. [IJTHI]. *International Journal of Technology and Human Interaction, 8*(4), 1–15. doi:10.4018/jthi.2012100101

Chen, W., Juang, Y., Chang, S., & Wang, P. (2012). Informal Education of Energy Conservation: Theory, Promotion, and Policy Implication. [IJTHI]. *International Journal of Technology and Human Interaction, 8*(4), 16–44. doi:10.4018/jthi.2012100102

Chia-Wen, T., Pei-Di, S., & Yen-Ting, L. (2013). A Quasi-Experiment on Computer Multimedia Integration into AIDS Education: A Study of Four Senior High Schools in Chennai, India. In S. Chhabra (Ed.), *ICT Influences on Human Development, Interaction, and Collaboration* (pp. 260–272). Hershey, PA: Information Science Reference.

Chiu, M. (2012). Gaps Between Valuing and Purchasing Green-Technology Products: Product and Gender Differences. [IJTHI]. *International Journal of Technology and Human Interaction, 8*(3), 54–68. doi:10.4018/jthi.2012070106

Cho, B., & Woodward, L. (2014). New Demands of Reading in the Mobile Internet Age. In D. McConatha, C. Penny, J. Schugar, & D. Bolton (Eds.), *Mobile Pedagogy and Perspectives on Teaching and Learning* (pp. 187–204). Hershey, PA: Information Science Reference.

Christidis, K., Papailiou, N., Apostolou, D., & Mentzas, G. (2013). Semantic Interfaces for Personal and Social Knowledge Work. In J. Wang (Ed.), *Intelligence Methods and Systems Advancements for Knowledge-Based Business* (pp. 213–230). Hershey, PA: Information Science Reference.

Cipolla-Ficarra, F. V. (2014). Negative Exponent Fraction: A Strategy for a New Virtual Image into the Financial Sector. In F. Cipolla-Ficarra (Ed.), *Advanced Research and Trends in New Technologies, Software, Human-Computer Interaction, and Communicability* (pp. 255–268). Hershey, PA: Information Science Reference.

Cipolla-Ficarra, F. V. (2014). Software and Emerging Technologies for Education, Culture, Entertainment, and Commerce. In F. Cipolla-Ficarra (Ed.), *Advanced Research and Trends in New Technologies, Software, Human-Computer Interaction, and Communicability* (pp. 1–12). Hershey, PA: Information Science Reference.

Cipolla-Ficarra, F. V. (2014). The Excellence of the Video Games: Past and Present. In F. Cipolla-Ficarra (Ed.), *Advanced Research and Trends in New Technologies, Software, Human-Computer Interaction, and Communicability* (pp. 511–520). Hershey, PA: Information Science Reference.

Cipolla-Ficarra, F. V., & Alma, J. (2014). Banking Online: Design for a New Credibility. In F. Cipolla-Ficarra (Ed.), *Advanced Research and Trends in New Technologies, Software, Human-Computer Interaction, and Communicability* (pp. 71–82). Hershey, PA: Information Science Reference.

Cipolla-Ficarra, F. V., & Alma, J. (2014). Static Graphics for Dynamic Information. In F. Cipolla-Ficarra (Ed.), *Advanced Research and Trends in New Technologies, Software, Human-Computer Interaction, and Communicability* (pp. 230–244). Hershey, PA: Information Science Reference.

Cipolla-Ficarra, F. V., Alma, J., & Carré, J. (2014). Human Factors in Computer Science, New Technologies, and Scientific Information. In F. Cipolla-Ficarra (Ed.), *Advanced Research and Trends in New Technologies, Software, Human-Computer Interaction, and Communicability* (pp. 480–490). Hershey, PA: Information Science Reference.

Cipolla-Ficarra, F. V., Alma, J., Carré, J., & Cipolla-Ficarra, M. (2014). Design and Behaviour Computer Animation for Children. In F. Cipolla-Ficarra (Ed.), *Advanced Research and Trends in New Technologies, Software, Human-Computer Interaction, and Communicability* (pp. 401–412). Hershey, PA: Information Science Reference.

Cipolla-Ficarra, F. V., & Cipolla-Ficarra, M. (2014). Universality and Communicability in Computer Animation. In F. Cipolla-Ficarra (Ed.), *Advanced Research and Trends in New Technologies, Software, Human-Computer Interaction, and Communicability* (pp. 131–142). Hershey, PA: Information Science Reference.

Cipolla-Ficarra, F. V., Cipolla-Ficarra, M., & Alma, J. (2014). Knowledge and Background of the Multimedia Culture: A Study of the Spatio-Temporal Context in Claymation and Computer Animation for Children and Adults. In F. Cipolla-Ficarra (Ed.), *Advanced Research and Trends in New Technologies, Software, Human-Computer Interaction, and Communicability* (pp. 452–465). Hershey, PA: Information Science Reference.

Cipolla-Ficarra, F. V., Cipolla-Ficarra, M., Alma, J., & Quiroga, A. (2014). Web Divide and Paper Unite: Towards a Model of the Local Tourist Information for All. In F. Cipolla-Ficarra (Ed.), *Advanced Research and Trends in New Technologies, Software, Human-Computer Interaction, and Communicability* (pp. 536–543). Hershey, PA: Information Science Reference.

Cipolla-Ficarra, F. V., & Ficarra, V. M. (2014). Anti-Models for Universitary Education: Analysis of the Catalans Cases in Information and Communication Technologies. In F. Cipolla-Ficarra (Ed.), *Advanced Research and Trends in New Technologies, Software, Human-Computer Interaction, and Communicability* (pp. 43–60). Hershey, PA: Information Science Reference.

Cipolla-Ficarra, F. V., & Ficarra, V. M. (2014). Art, Future, and New Technologies: Research or Business? In F. Cipolla-Ficarra (Ed.), *Advanced Research and Trends in New Technologies, Software, Human-Computer Interaction, and Communicability* (pp. 280–293). Hershey, PA: Information Science Reference.

Cipolla-Ficarra, F. V., Quiroga, A., & Alma, J. (2014). Towards a Cyber-Destructors Assessment Method. In F. Cipolla-Ficarra (Ed.), *Advanced Research and Trends in New Technologies, Software, Human-Computer Interaction, and Communicability* (pp. 431–440). Hershey, PA: Information Science Reference.

Cipolla-Ficarra, F. V., Quiroga, A., & Carré, J. (2014). Storyboard and Computer Animation for Children: Communicability Evaluation. In F. Cipolla-Ficarra (Ed.), *Advanced Research and Trends in New Technologies, Software, Human-Computer Interaction, and Communicability* (pp. 203–219). Hershey, PA: Information Science Reference.

Cipolla-Ficarra, F. V., Quiroga, A., & Carré, J. (2014). Web Attacks and the ASCII Files. In F. Cipolla-Ficarra (Ed.), *Advanced Research and Trends in New Technologies, Software, Human-Computer Interaction, and Communicability* (pp. 595–605). Hershey, PA: Information Science Reference.

Cipolla-Ficarra, F. V., Quiroga, A., Carré, J., Alma, J., & Cipolla-Ficarra, M. (2014). Museum Information and Communicability Evaluation. In F. Cipolla-Ficarra (Ed.), *Advanced Research and Trends in New Technologies, Software, Human-Computer Interaction, and Communicability* (pp. 319–340). Hershey, PA: Information Science Reference.

Clay, A., Lombardo, J., Couture, N., & Conan, J. (2014). Bi-Manual 3D Painting: An Interaction Paradigm for Augmented Reality Live Performance. In F. Cipolla-Ficarra (Ed.), *Advanced Research and Trends in New Technologies, Software, Human-Computer Interaction, and Communicability* (pp. 423–430). Hershey, PA: Information Science Reference.

Clemmons, K., Nolen, A., & Hayn, J. A. (2014). Constructing Community in Higher Education Regardless of Proximity: Re-Imagining the Teacher Education Experience within Social Networking Technology. In S. Mukerji, & P. Tripathi (Eds.), *Handbook of Research on Transnational Higher Education* (pp. 713–729). Hershey, PA: Information Science Reference.

Clusella, M. M., & Mitre, M. G. (2014). "E-Culture System": A New Infonomic and Symbionomic Technical Resource to Serve the Intercultural Communication. In F. Cipolla-Ficarra (Ed.), *Advanced Research and Trends in New Technologies, Software, Human-Computer Interaction, and Communicability* (pp. 466–479). Hershey, PA: Information Science Reference.

Cole, F. T. (2012). Negotiating the Socio-Material in and about Information Systems: An Approach to Native Methods. In A. Tatnall (Ed.), *Social Influences on Information and Communication Technology Innovations* (pp. 173–182). Hershey, PA: Information Science Reference. doi:10.4018/978-1-4666-1559-5.ch012

Colman, J., & Gnanayutham, P. (2014). Assistive Technologies for Brain-Injured Gamers. In G. Kouroupetroglou (Ed.), *Assistive Technologies and Computer Access for Motor Disabilities* (pp. 28–56). Hershey, PA: Medical Information Science Reference.

Cook, R. G., & Sutton, R. (2014). Administrators' Assessments of Online Courses and Student Retention in Higher Education: Lessons Learned. In S. Mukerji, & P. Tripathi (Eds.), *Handbook of Research on Transnational Higher Education* (pp. 138–150). Hershey, PA: Information Science Reference.

Cordella, A. (2012). Information Infrastructure: An Actor-Network Perspective. In A. Tatnall (Ed.), *Social Influences on Information and Communication Technology Innovations* (pp. 20–39). Hershey, PA: Information Science Reference. doi:10.4018/978-1-4666-1559-5.ch002

Costa, D., & Duarte, C. (2014). Improving Interaction with TV-Based Applications through Adaptive Multimodal Fission. In K. Blashki, & P. Isaias (Eds.), *Emerging Research and Trends in Interactivity and the Human-Computer Interface* (pp. 54–73). Hershey, PA: Information Science Reference.

Costello, R. (2014). Evaluating E-Learning from an End User Perspective. In M. Pańkowska (Ed.), *Frameworks of IT Prosumption for Business Development* (pp. 259–283). Hershey, PA: Business Science Reference.

Costello, R. (2014). Improving IT Market Development through IT Solutions for Prosumers. In M. Pańkowska (Ed.), *Frameworks of IT Prosumption for Business Development* (pp. 16–30). Hershey, PA: Business Science Reference.

Cucchiara, S., Ligorio, M. B., & Fujita, N. (2014). Understanding Online Discourse Strategies for Knowledge Building Through Social Network Analysis. In H. Lim, & F. Sudweeks (Eds.), *Innovative Methods and Technologies for Electronic Discourse Analysis* (pp. 42–62). Hershey, PA: Information Science Reference.

Cucchiarini, C., & Strik, H. (2014). Second Language Learners' Spoken Discourse: Practice and Corrective Feedback through Automatic Speech Recognition. In H. Lim, & F. Sudweeks (Eds.), *Innovative Methods and Technologies for Electronic Discourse Analysis* (pp. 169–189). Hershey, PA: Information Science Reference.

Dafoulas, G. A., & Saleeb, N. (2014). 3D Assistive Technologies and Advantageous Themes for Collaboration and Blended Learning of Users with Disabilities. In I. Association (Ed.), *Assistive Technologies: Concepts, Methodologies, Tools, and Applications* (pp. 421–453). Hershey, PA: Information Science Reference.

Damásio, M. J., Henriques, S., Teixeira-Botelho, I., & Dias, P. (2014). Mobile Internet in Portugal: Adoption Patterns and User Experiences. In K. Rızvanoğlu, & G. Çetin (Eds.), *Research and Design Innovations for Mobile User Experience* (pp. 94–113). Hershey, PA: Information Science Reference.

Daradoumis, T., & Lafuente, M. M. (2014). Studying the Suitability of Discourse Analysis Methods for Emotion Detection and Interpretation in Computer-Mediated Educational Discourse. In H. Lim, & F. Sudweeks (Eds.), *Innovative Methods and Technologies for Electronic Discourse Analysis* (pp. 119–143). Hershey, PA: Information Science Reference.

Davis, B., & Mason, P. (2014). Positioning Goes to Work: Computer-Aided Identification of Stance Shifts and Semantic Themes in Electronic Discourse Analysis. In H. Lim, & F. Sudweeks (Eds.), *Innovative Methods and Technologies for Electronic Discourse Analysis* (pp. 394–413). Hershey, PA: Information Science Reference.

De Luise, D. L., & Hisgen, D. (2014). MLW and Bilingualism: Case Study and Critical Evaluation. In F. Cipolla-Ficarra (Ed.), *Advanced Research and Trends in New Technologies, Software, Human-Computer Interaction, and Communicability* (pp. 555–586). Hershey, PA: Information Science Reference. doi:10.4018/978-1-4666-6042-7.ch001

De Pace, C., & Stasolla, F. (2014). Promoting Environmental Control, Social Interaction, and Leisure/Academy Engagement Among People with Severe/Profound Multiple Disabilities Through Assistive Technology. In G. Kouroupetroglou (Ed.), *Assistive Technologies and Computer Access for Motor Disabilities* (pp. 285–319). Hershey, PA: Medical Information Science Reference.

De Vincentis, S. (2011). Complexifying the 'Visualised' Curriculum with Actor-Network Theory. [IJANTTI]. *International Journal of Actor-Network Theory and Technological Innovation*, *3*(2), 32–45. doi:10.4018/jantti.2011040103

De Vincentis, S. (2013). Complexifying the 'Visualised' Curriculum with Actor-Network Theory. In A. Tatnall (Ed.), *Social and Professional Applications of Actor-Network Theory for Technology Development* (pp. 31–44). Hershey, PA: Information Science Reference.

Deering, P., Tatnall, A., & Burgess, S. (2012). Adoption of ICT in Rural Medical General Practices in Australia: An Actor-Network Study. In A. Tatnall (Ed.), *Social Influences on Information and Communication Technology Innovations* (pp. 40–51). Hershey, PA: Information Science Reference. doi:10.4018/978-1-4666-1559-5.ch003

Douai, A. (2013). "In YouTube We Trust": Video Exchange and Arab Human Rights. In J. Lannon, & E. Halpin (Eds.), *Human Rights and Information Communication Technologies: Trends and Consequences of Use* (pp. 57–71). Hershey, PA: Information Science Reference.

Douglas, G., Morton, H., & Jack, M. (2012). Remote Channel Customer Contact Strategies for Complaint Update Messages. [IJTHI]. *International Journal of Technology and Human Interaction*, *8*(2), 43–55. doi:10.4018/jthi.2012040103

Driouchi, A. (2013). ICTs and Coordination for Poverty Alleviation. In ICTs for Health, Education, and Socioeconomic Policies: Regional Cases (pp. 165-189). Hershey, PA: Information Science Reference. doi: doi:10.4018/978-1-4666-3643-9.ch008

Driouchi, A. (2013). ICTs and Socioeconomic Performance with Focus on ICTs and Health. In ICTs for Health, Education, and Socioeconomic Policies: Regional Cases (pp. 104-125). Hershey, PA: Information Science Reference. doi: doi:10.4018/978-1-4666-3643-9.ch005

Driouchi, A. (2013). ICTs, Youth, Education, Health, and Prospects of Further Coordination. In ICTs for Health, Education, and Socioeconomic Policies: Regional Cases (pp. 126-145). Hershey, PA: Information Science Reference. doi: doi:10.4018/978-1-4666-3643-9.ch006

Driouchi, A. (2013). Need of a Balance between Fragmented and Coordinated Decision-Making. In ICTs for Health, Education, and Socioeconomic Policies: Regional Cases (pp. 1-24). Hershey, PA: Information Science Reference. doi: doi:10.4018/978-1-4666-3643-9.ch001

Driouchi, A. (2013). Risk Factors, Health, Education, and Poverty: Can ICTs Help1? In ICTs for Health, Education, and Socioeconomic Policies: Regional Cases (pp. 252-277). Hershey, PA: Information Science Reference. doi: doi:10.4018/978-1-4666-3643-9.ch012

Driouchi, A. (2013). Social Deficits, Social Cohesion, and Prospects from ICTs. In ICTs for Health, Education, and Socioeconomic Policies: Regional Cases (pp. 230-251). Hershey, PA: Information Science Reference. doi: doi:10.4018/978-1-4666-3643-9.ch011

Driouchi, A. (2013). Socioeconomic Reforms, Human Development, and the Millennium Development Goals with ICTs for Coordination. In ICTs for Health, Education, and Socioeconomic Policies: Regional Cases (pp. 211-229). Hershey, PA: Information Science Reference. doi: doi:10.4018/978-1-4666-3643-9.ch010

Driouchi, A. (2013). The Triple Helix as a Model for Coordination with Potential for ICTs. In ICTs for Health, Education, and Socioeconomic Policies: Regional Cases (pp. 191-210). Hershey, PA: Information Science Reference. doi: doi:10.4018/978-1-4666-3643-9.ch009

Driouchi, A. (2013). Women Empowerment and ICTs in Developing Economies. In ICTs for Health, Education, and Socioeconomic Policies: Regional Cases (pp. 146-164). Hershey, PA: Information Science Reference. doi: doi:10.4018/978-1-4666-3643-9.ch007

Duarte, C., Ribeiro, A., & Nunes, R. (2014). Studying Natural Interaction in Multimodal, Multi-Surface, Multiuser Scenarios. In K. Blashki, & P. Isaias (Eds.), *Emerging Research and Trends in Interactivity and the Human-Computer Interface* (pp. 160–181). Hershey, PA: Information Science Reference.

Dueck, J., & Rempel, M. (2013). Human Rights and Technology: Lessons from Alice in Wonderland. In J. Lannon, & E. Halpin (Eds.), *Human Rights and Information Communication Technologies: Trends and Consequences of Use* (pp. 1–20). Hershey, PA: Information Science Reference.

Durán, E. B., Álvarez, M. M., & Únzaga, S. I. (2014). Generic Model of a Multi-Agent System to Assist Ubiquitous Learning. In F. Cipolla-Ficarra (Ed.), *Advanced Research and Trends in New Technologies, Software, Human-Computer Interaction, and Communicability* (pp. 544–554). Hershey, PA: Information Science Reference.

Edenius, M., & Rämö, H. (2011). An Office on the Go: Professional Workers, Smartphones and the Return of Place. [IJTHI]. *International Journal of Technology and Human Interaction, 7*(1), 37–55. doi:10.4018/jthi.2011010103

Eke, D. O. (2011). ICT Integration in Nigeria: The Socio-Cultural Constraints. [IJTHI]. *International Journal of Technology and Human Interaction, 7*(2), 21–27. doi:10.4018/jthi.2011040103

Elbanna, A. R. (2011). The Theoretical and Analytical Inclusion of Actor Network Theory and its Implication on ICT Research. In A. Tatnall (Ed.), *Actor-Network Theory and Technology Innovation: Advancements and New Concepts* (pp. 130–142). Hershey, PA: Information Science Reference.

Fidler, C. S., Kanaan, R. K., & Rogerson, S. (2011). Barriers to e-Government Implementation in Jordan: The Role of Wasta. [IJTHI]. *International Journal of Technology and Human Interaction, 7*(2), 9–20. doi:10.4018/jthi.2011040102

Filho, J. R. (2013). ICT and Human Rights in Brazil: From Military to Digital Dictatorship. In J. Lannon, & E. Halpin (Eds.), *Human Rights and Information Communication Technologies: Trends and Consequences of Use* (pp. 86–99). Hershey, PA: Information Science Reference.

Forbes, D. (2014). Listening and Learning through ICT with Digital Kids: Dynamics of Interaction, Power, and Mutual Learning between Student Teachers and Children in Online Discussion. In K. Sullivan, P. Czigler, & J. Sullivan Hellgren (Eds.), *Cases on Professional Distance Education Degree Programs and Practices: Successes, Challenges, and Issues* (pp. 149–176). Hershey, PA: Information Science Reference.

Freeman, I., & Freeman, A. (2014). Capacity Building for Different Abilities Using ICT. In I. Association (Ed.), *Assistive Technologies: Concepts, Methodologies, Tools, and Applications* (pp. 261–276). Hershey, PA: Information Science Reference.

Frigo, C. A., & Pavan, E. E. (2014). Prosthetic and Orthotic Devices. In I. Association (Ed.), *Assistive Technologies: Concepts, Methodologies, Tools, and Applications* (pp. 549–613). Hershey, PA: Information Science Reference.

Fuhrer, C., & Cucchi, A. (2012). Relations Between Social Capital and Use of ICT: A Social Network Analysis Approach. [IJTHI]. *International Journal of Technology and Human Interaction, 8*(2), 15–42. doi:10.4018/jthi.2012040102

Garbin, B., Staunton, K., & Burdon, M. (2014). Tracking Legislative Developments in Relation to "Do Not Track" Initiatives. In M. Michael, & K. Michael (Eds.), *Uberveillance and the Social Implications of Microchip Implants: Emerging Technologies* (pp. 235–259). Hershey, PA: Information Science Reference.

Gibson, R. (2014). Utilizing Augmented Reality in Library Information Search. In S. Hai-Jew (Ed.), *Packaging Digital Information for Enhanced Learning and Analysis: Data Visualization, Spatialization, and Multidimensionality* (pp. 93–102). Hershey, PA: Information Science Reference.

Giraldo, F. D., Villegas, M. L., & Collazos, C. A. (2014). The Use of HCI Approaches into Distributed CSCL Activities Applied to Software Engineering Courses. In I. Management Association (Ed.), *Software Design and Development: Concepts, Methodologies, Tools, and Applications* (pp. 2033-2050). Hershey, PA: Information Science Reference. doi: doi:10.4018/978-1-4666-4301-7.ch094

Goh, K. N., Chen, Y. Y., & Chow, C. H. (2014). Location-Based Data Visualisation Tool for Tuberculosis and Dengue: A Case Study in Malaysia. In K. Blashki, & P. Isaias (Eds.), *Emerging Research and Trends in Interactivity and the Human-Computer Interface* (pp. 260–282). Hershey, PA: Information Science Reference.

Gomez, R. (2013). Success Factors in Public Access Computing for Development. In S. Chhabra (Ed.), *ICT Influences on Human Development, Interaction, and Collaboration* (pp. 97–116). Hershey, PA: Information Science Reference.

Gomez, R., & Camacho, K. (2013). Users of ICT at Public Access Centers: Age, Education, Gender, and Income Differences in Users. In S. Chhabra (Ed.), *ICT Influences on Human Development, Interaction, and Collaboration* (pp. 1–21). Hershey, PA: Information Science Reference.

Gonçalves, F. A., & Figueiredo, J. (2012). How to Recognize an Immutable Mobile When You Find One: Translations on Innovation and Design. In A. Tatnall (Ed.), *Social Influences on Information and Communication Technology Innovations* (pp. 92–106). Hershey, PA: Information Science Reference. doi:10.4018/978-1-4666-1559-5.ch006

Gonçalves, F. A., & Figueiredo, J. (2012). Negotiating Meaning: An ANT Approach to the Building of Innovations. In A. Tatnall (Ed.), *Social Influences on Information and Communication Technology Innovations* (pp. 117–131). Hershey, PA: Information Science Reference. doi:10.4018/978-1-4666-1559-5.ch008

Groba, B., Pousada, T., & Nieto, L. (2014). Assistive Technologies, Tools and Resources for the Access and Use of Information and Communication Technologies by People with Disabilities. In I. Association (Ed.), *Assistive Technologies: Concepts, Methodologies, Tools, and Applications* (pp. 246–260). Hershey, PA: Information Science Reference.

Gu, L., Aiken, M., Wang, J., & Wibowo, K. (2011). The Influence of Information Control upon On-line Shopping Behavior. [IJTHI]. *International Journal of Technology and Human Interaction*, 7(1), 56–66. doi:10.4018/jthi.2011010104

Guimarães da Silva, C. (2014). An Information Visualization-Based Approach for Exploring Databases: A Case Study for Learning Management Systems. In M. Huang, & W. Huang (Eds.), *Innovative Approaches of Data Visualization and Visual Analytics* (pp. 288–315). Hershey, PA: Information Science Reference.

Guo, Y. (2014). The Axiomatic Usability Evaluation Method. In K. Blashki, & P. Isaias (Eds.), *Emerging Research and Trends in Interactivity and the Human-Computer Interface* (pp. 353–373). Hershey, PA: Information Science Reference.

Hai-Jew, S. (2014). Multimodal Mapping of a University's Formal and Informal Online Brand: Using NodeXL to Extract Social Network Data in Tweets, Digital Contents, and Relational Ties. In S. Hai-Jew (Ed.), *Packaging Digital Information for Enhanced Learning and Analysis: Data Visualization, Spatialization, and Multidimensionality* (pp. 120–162). Hershey, PA: Information Science Reference.

Hai-Jew, S. (2014). Structuring an Emergent and Transdisciplinary Online Curriculum: A One Health Case. In S. Hai-Jew (Ed.), *Packaging Digital Information for Enhanced Learning and Analysis: Data Visualization, Spatialization, and Multidimensionality* (pp. 299–331). Hershey, PA: Information Science Reference.

Hayes, A. (2014). Uberveillance: Where Wear and Educative Arrangement. In M. Michael, & K. Michael (Eds.), *Uberveillance and the Social Implications of Microchip Implants: Emerging Technologies* (pp. 46–62). Hershey, PA: Information Science Reference.

Heikkinen, M. V., & Töyli, J. (2013). Modeling Intention to Use Novel Mobile Peer-To-Peer Services. In I. Lee (Ed.), *Mobile Applications and Knowledge Advancements in E-Business* (pp. 26–41). Hershey, PA: Business Science Reference.

Heimgärtner, R. (2014). Intercultural User Interface Design. In K. Blashki, & P. Isaias (Eds.), *Emerging Research and Trends in Interactivity and the Human-Computer Interface* (pp. 1–33). Hershey, PA: Information Science Reference.

Hersh, M. A., & Leporini, B. (2014). An Overview of Accessibility and Usability of Educational Games. In I. Association (Ed.), *Assistive Technologies: Concepts, Methodologies, Tools, and Applications* (pp. 63–101). Hershey, PA: Information Science Reference.

Hodge, B. (2014). Critical Electronic Discourse Analysis: Social and Cultural Research in the Electronic Age. In H. Lim, & F. Sudweeks (Eds.), *Innovative Methods and Technologies for Electronic Discourse Analysis* (pp. 191–209). Hershey, PA: Information Science Reference.

Hsiao, S., Chen, D., Yang, C., Huang, H., Lu, Y., & Huang, H. et al. (2013). Chemical-Free and Reusable Cellular Analysis: Electrochemical Impedance Spectroscopy with a Transparent ITO Culture Chip. [IJTHI]. *International Journal of Technology and Human Interaction*, 8(3), 1–9. doi:10.4018/jthi.2012070101

Hsu, M., Yang, C., Wang, C., & Lin, Y. (2013). Simulation-Aided Optimal Microfluidic Sorting for Monodispersed Microparticles. [IJTHI]. *International Journal of Technology and Human Interaction*, 8(3), 10–18. doi:10.4018/jthi.2012070102

Huang, C., Liang, C., & Lin, E. (2014). A Study on Emotion Releasing Effect with Music and Color. In F. Cipolla-Ficarra (Ed.), *Advanced Research and Trends in New Technologies, Software, Human-Computer Interaction, and Communicability* (pp. 23–31). Hershey, PA: Information Science Reference.

Huang, C., & Lin, E. (2014). Using Emotion Map System to Implement the Generative Chinese Style Music with Wu Xing Theory. In F. Cipolla-Ficarra (Ed.), *Advanced Research and Trends in New Technologies, Software, Human-Computer Interaction, and Communicability* (pp. 183–192). Hershey, PA: Information Science Reference.

Ihamäki, P., & Luimula, M. (2014). Players' Experience in a Sport Geocaching Game. In K. Blashki, & P. Isaias (Eds.), *Emerging Research and Trends in Interactivity and the Human-Computer Interface* (pp. 127–143). Hershey, PA: Information Science Reference.

Ionescu, A. (2013). ICTs and Gender-Based Rights. In J. Lannon, & E. Halpin (Eds.), *Human Rights and Information Communication Technologies: Trends and Consequences of Use* (pp. 214–234). Hershey, PA: Information Science Reference.

Issa, T., & Isaias, P. (2014). Promoting Human-Computer Interaction and Usability Guidelines and Principles Through Reflective Journal Assessment. In K. Blashki, & P. Isaias (Eds.), *Emerging Research and Trends in Interactivity and the Human-Computer Interface* (pp. 375–394). Hershey, PA: Information Science Reference.

Iyamu, T. (2012). Theoretical Analysis of Strategic Implementation of Enterprise Architecture. In A. Tatnall (Ed.), *Social Influences on Information and Communication Technology Innovations* (pp. 132–148). Hershey, PA: Information Science Reference. doi:10.4018/978-1-4666-1559-5.ch009

Iyamu, T., & Roode, D. (2012). The Use of Structuration Theory and Actor Network Theory for Analysis: Case Study of a Financial Institution in South Africa. In A. Tatnall (Ed.), *Social Influences on Information and Communication Technology Innovations* (pp. 1–19). Hershey, PA: Information Science Reference. doi:10.4018/978-1-4666-1559-5.ch001

Jacob, A. F., da Mata, E. C., Santana, Á. L., Francês, C. R., Costa, J. C., & Barros, F. D. (2014). Adapting Chatterbots' Interaction for Use in Children's Education. In K. Blashki, & P. Isaias (Eds.), *Emerging Research and Trends in Interactivity and the Human-Computer Interface* (pp. 413–428). Hershey, PA: Information Science Reference.

Jan, Y., Lin, M., Shiao, K., Wei, C., Huang, L., & Sung, Q. (2013). Development of an Evaluation Instrument for Green Building Literacy among College Students in Taiwan. [IJTHI]. *International Journal of Technology and Human Interaction*, *8*(3), 31–45. doi:10.4018/jthi.2012070104

Jonsson, K., Holmström, J., Lyytinen, K., & Nilsson, A. (2012). Desituating Context in Ubiquitous Computing: Exploring Strategies for the Use of Remote Diagnostic Systems for Maintenance Work. In A. Tatnall (Ed.), *Social Influences on Information and Communication Technology Innovations* (pp. 156–172). Hershey, PA: Information Science Reference. doi:10.4018/978-1-4666-1559-5.ch011

Joshi, D., & Rath, R. (2014). Experimental Research Approaches for Mobile UX in Emerging Markets. In K. Rızvanoğlu, & G. Çetin (Eds.), *Research and Design Innovations for Mobile User Experience* (pp. 128–138). Hershey, PA: Information Science Reference.

Kaba, B., & Osei-Bryson, K. (2012). An Empirical Investigation of External Factors Influencing Mobile Technology Use in Canada: A Preliminary Study. [IJTHI]. *International Journal of Technology and Human Interaction*, *8*(2), 1–14. doi:10.4018/jthi.2012040101

Kaczmarek, T., & Węckowski, D. G. (2014). Harvesting Deep Web Data through Produser Involvement. In M. Pańkowska (Ed.), *Frameworks of IT Prosumption for Business Development* (pp. 200–221). Hershey, PA: Business Science Reference.

Kallergi, A., & Verbeek, F. J. (2014). Playful Interfaces for Scientific Image Data: A Case for Storytelling. In K. Blashki, & P. Isaias (Eds.), *Emerging Research and Trends in Interactivity and the Human-Computer Interface* (pp. 471–489). Hershey, PA: Information Science Reference.

Kammüller, F., Probst, C. W., & Raimondi, F. (2014). Application of Verification Techniques to Security: Model Checking Insider Attacks. In F. Cipolla-Ficarra (Ed.), *Advanced Research and Trends in New Technologies, Software, Human-Computer Interaction, and Communicability* (pp. 61–70). Hershey, PA: Information Science Reference.

Karakatič, S., Podgorelec, V., & Heričko, M. (2014). Building Recommendation Service with Social Networks and Semantic Databases. In F. Cipolla-Ficarra (Ed.), *Advanced Research and Trends in New Technologies, Software, Human-Computer Interaction, and Communicability* (pp. 83–92). Hershey, PA: Information Science Reference.

Karp, J. (2014). Global Tracking Systems in the Australian Interstate Trucking Industry. In M. Michael, & K. Michael (Eds.), *Uberveillance and the Social Implications of Microchip Implants: Emerging Technologies* (pp. 226–234). Hershey, PA: Information Science Reference.

Kasimin, H., & Ibrahim, H. (2011). Managing Multi-Organizational Interaction Issues: A Case Study of Information Technology Transfer in Public Sector of Malaysia. In A. Tatnall (Ed.), *Actor-Network Theory and Technology Innovation: Advancements and New Concepts* (pp. 192–206). Hershey, PA: Information Science Reference.

Kavianpour, S., Ismail, Z., & Shanmugam, B. (2014). Differences between Role of Strong Ties and Weak Ties in Information Diffusion on Social Network Sites. In F. Cipolla-Ficarra (Ed.), *Advanced Research and Trends in New Technologies, Software, Human-Computer Interaction, and Communicability* (pp. 307–311). Hershey, PA: Information Science Reference.

Kawlra, A. (2013). From Rural Outsourcing to Rural Opportunities: Developing an ICT Mediated Distributed Production Enterprise in Tamil Nadu, India. In S. Chhabra (Ed.), *ICT Influences on Human Development, Interaction, and Collaboration* (pp. 158–173). Hershey, PA: Information Science Reference.

Keane, K., & Nisi, V. (2014). Experience Prototyping: Gathering Rich Understandings to Guide Design. In K. Blashki, & P. Isaias (Eds.), *Emerging Research and Trends in Interactivity and the Human-Computer Interface* (pp. 224–237). Hershey, PA: Information Science Reference.

Kendall, R. (2014). Podcasting and Pedagogy. In D. McConatha, C. Penny, J. Schugar, & D. Bolton (Eds.), *Mobile Pedagogy and Perspectives on Teaching and Learning* (pp. 41–57). Hershey, PA: Information Science Reference.

Kennan, M. A., Cecez-Kecmanovic, D., & Underwood, J. (2012). Having a Say: Voices for all the Actors in ANT Research? In A. Tatnall (Ed.), *Social Influences on Information and Communication Technology Innovations* (pp. 52–67). Hershey, PA: Information Science Reference. doi:10.4018/978-1-4666-1559-5.ch004

Kennedy-Clark, S., & Thompson, K. (2014). Using Multimodal Discourse Analysis to Identify Patterns of Problem Solving Processes in a Computer-Supported Collaborative Environment. In H. Lim, & F. Sudweeks (Eds.), *Innovative Methods and Technologies for Electronic Discourse Analysis* (pp. 94–118). Hershey, PA: Information Science Reference.

Kontogeorgakopoulos, A., Wechsler, R., & Keay-Bright, W. (2014). Camera-Based Motion Tracking and Performing Arts for Persons with Motor Disabilities and Autism. In G. Kouroupetroglou (Ed.), *Disability Informatics and Web Accessibility for Motor Limitations* (pp. 294–322). Hershey, PA: Medical Information Science Reference.

Krajnc, M., Podgorelec, V., & Heričko, M. (2014). Context-Aware Mobile System for Public Bus Transportation. In F. Cipolla-Ficarra (Ed.), *Advanced Research and Trends in New Technologies, Software, Human-Computer Interaction, and Communicability* (pp. 102–111). Hershey, PA: Information Science Reference.

Kratky, A., & Ho, D. (2014). Story Objects: An Interactive Installation to Excavate Immigrant History and Identity through Evocative Everyday Objects. In F. Cipolla-Ficarra (Ed.), *Advanced Research and Trends in New Technologies, Software, Human-Computer Interaction, and Communicability* (pp. 377–387). Hershey, PA: Information Science Reference.

Kripanont, N., & Tatnall, A. (2011). Modelling the Adoption and Use of Internet Technologies in Higher Education in Thailand. In A. Tatnall (Ed.), *Actor-Network Theory and Technology Innovation: Advancements and New Concepts* (pp. 95–112). Hershey, PA: Information Science Reference.

Krishnaswamy, K., & Oates, T. (2014). Pathway to Independence: Past, Present, and Beyond via Robotics. In G. Kouroupetroglou (Ed.), *Disability Informatics and Web Accessibility for Motor Limitations* (pp. 153–201). Hershey, PA: Medical Information Science Reference.

Kumar, M., & Sareen, M. (2013). Impact of Technology-Related Environment Issues on Trust in B2B E-Commerce. In S. Chhabra (Ed.), *ICT Influences on Human Development, Interaction, and Collaboration* (pp. 22–42). Hershey, PA: Information Science Reference.

Kurosu, M. (2014). Concept of User Experience and Issues to be Discussed. In M. Pańkowska (Ed.), *Frameworks of IT Prosumption for Business Development* (pp. 31–47). Hershey, PA: Business Science Reference.

Kymäläinen, T., & Siltanen, S. (2014). Co-Designing Novel Interior Design Services that Utilise Augmented Reality: A Case Study. In F. Cipolla-Ficarra (Ed.), *Advanced Research and Trends in New Technologies, Software, Human-Computer Interaction, and Communicability* (pp. 269–279). Hershey, PA: Information Science Reference.

Laarni, J., & Aaltonen, I. (2014). Anticipation Dialogue Method in Participatory Design. In K. Blashki, & P. Isaias (Eds.), *Emerging Research and Trends in Interactivity and the Human-Computer Interface* (pp. 315–330). Hershey, PA: Information Science Reference.

Laksov, K. B., Silén, C., & Boman, L. E. (2014). Implementation of Scholarship of Teaching and Learning through an On-Line Masters Program. In K. Sullivan, P. Czigler, & J. Sullivan Hellgren (Eds.), *Cases on Professional Distance Education Degree Programs and Practices: Successes, Challenges, and Issues* (pp. 258–295). Hershey, PA: Information Science Reference.

Lee, H., & Baek, E. (2012). Facilitating Deep Learning in a Learning Community. [IJTHI]. *International Journal of Technology and Human Interaction*, 8(1), 1–13. doi:10.4018/jthi.2012010101

Leng, H. K., & Leng, D. (2014). Marketing Higher Educational Institutions on Social Network Sites. In S. Mukerji, & P. Tripathi (Eds.), *Handbook of Research on Transnational Higher Education* (pp. 175–188). Hershey, PA: Information Science Reference.

Libreri, C., & Graffigna, G. (2014). How Web 2.0 Shapes Patient Knowledge Sharing: The Case of Diabetes in Italy. In C. El Morr (Ed.), *Research Perspectives on the Role of Informatics in Health Policy and Management* (pp. 238–260). Hershey, PA: Medical Information Science Reference.

Lin, C., Chu, L., & Hsu, H. (2013). Study on the Performance and Exhaust Emissions of Motorcycle Engine Fuelled with Hydrogen-Gasoline Compound Fuel. [IJTHI]. *International Journal of Technology and Human Interaction*, 8(3), 69–81. doi:10.4018/jthi.2012070107

Lin, T., Wu, Z., Tang, N., & Wu, S. (2013). Exploring the Effects of Display Characteristics on Presence and Emotional Responses of Game Players. [IJTHI]. *International Journal of Technology and Human Interaction*, 9(1), 50–63. doi:10.4018/jthi.2013010104

Linden, L., & Saunders, C. (2011). Linux Kernel Developers Embracing Authors Embracing Licenses. In A. Tatnall (Ed.), *Actor-Network Theory and Technology Innovation: Advancements and New Concepts* (pp. 143–161). Hershey, PA: Information Science Reference.

Lipschutz, R. D., & Hester, R. J. (2014). We Are the Borg! Human Assimilation into Cellular Society. In M. Michael, & K. Michael (Eds.), *Uberveillance and the Social Implications of Microchip Implants: Emerging Technologies* (pp. 366–407). Hershey, PA: Information Science Reference.

Loureiro, I. F., Leão, C. P., Costa, F., Teixeira, J., & Arezes, P. M. (2014). ETdAnalyser: A Model-Based Architecture for Ergonomic Decision Intervention. In K. Blashki, & P. Isaias (Eds.), *Emerging Research and Trends in Interactivity and the Human-Computer Interface* (pp. 284–300). Hershey, PA: Information Science Reference.

Luczak, H., Schlick, C. M., Jochems, N., Vetter, S., & Kausch, B. (2014). Touch Screens for the Elderly: Some Models and Methods, Prototypical Development and Experimental Evaluation of Human-Computer Interaction Concepts for the Elderly. In I. Association (Ed.), *Assistive Technologies: Concepts, Methodologies, Tools, and Applications* (pp. 377–396). Hershey, PA: Information Science Reference.

Luor, T., Lu, H., Johanson, R. E., & Yu, H. (2012). Minding the Gap Between First and Continued Usage of a Corporate E-Learning English-language Program. [IJTHI]. *International Journal of Technology and Human Interaction, 8*(1), 55–74. doi:10.4018/jthi.2012010104

Luzón, M. (2014). Engaging in Scientific Controversies in Science Blogs: The Expression of Allegiance and Ideological Commitment. In H. Lim, & F. Sudweeks (Eds.), *Innovative Methods and Technologies for Electronic Discourse Analysis* (pp. 235–259). Hershey, PA: Information Science Reference.

Magjuka, R. J., & Liu, X. (2014). A Case Study of Online MBA Courses: Online Facilitation, Case-Based Learning, and Virtual Team. In K. Sullivan, P. Czigler, & J. Sullivan Hellgren (Eds.), *Cases on Professional Distance Education Degree Programs and Practices: Successes, Challenges, and Issues* (pp. 200–232). Hershey, PA: Information Science Reference.

Malhotra, L., & Bansal, A. J. (2014). Prediction of Change-Prone Classes Using Machine Learning and Statistical Techniques. In F. Cipolla-Ficarra (Ed.), *Advanced Research and Trends in New Technologies, Software, Human-Computer Interaction, and Communicability* (pp. 193–202). Hershey, PA: Information Science Reference.

Mandasari, V., & Theng, L. B. (2014). Enhancing the Acquisition of Social Skills through the Interactivity of Multimedia. In K. Blashki, & P. Isaias (Eds.), *Emerging Research and Trends in Interactivity and the Human-Computer Interface* (pp. 95–126). Hershey, PA: Information Science Reference.

Mann, S. (2014). Veillance: Beyond Surveillance, Dataveillance, Uberveillance, and the Hypocrisy of One-Sided Watching. In M. Michael, & K. Michael (Eds.), *Uberveillance and the Social Implications of Microchip Implants: Emerging Technologies* (pp. 32–45). Hershey, PA: Information Science Reference.

Manning, K., Wong, L., & Tatnall, A. (2012). Aspects of e-Learning in a University. In A. Tatnall (Ed.), *Social Influences on Information and Communication Technology Innovations* (pp. 219–229). Hershey, PA: Information Science Reference. doi:10.4018/978-1-4666-1559-5.ch016

Marache-Francisco, C., & Brangier, E. (2014). The Gamification Experience: UXD with a Gamification Background. In K. Blashki, & P. Isaias (Eds.), *Emerging Research and Trends in Interactivity and the Human-Computer Interface* (pp. 205–223). Hershey, PA: Information Science Reference.

Marcengo, A., & Rapp, A. (2014). Visualization of Human Behavior Data: The Quantified Self. In M. Huang, & W. Huang (Eds.), *Innovative Approaches of Data Visualization and Visual Analytics* (pp. 236–265). Hershey, PA: Information Science Reference.

Marcus, A. (2014). The Driving Machine: Combining Information Design/Visualization with Persuasion Design to Change Behavior. In K. Rızvanoğlu, & G. Çetin (Eds.), *Research and Design Innovations for Mobile User Experience* (pp. 1–21). Hershey, PA: Information Science Reference.

Marcus, A. (2014). The Travel Machine: Combining Information Design/Visualization with Persuasion Design to Change Behavior. In K. Rızvanoğlu, & G. Çetin (Eds.), *Research and Design Innovations for Mobile User Experience* (pp. 22–46). Hershey, PA: Information Science Reference.

Martin, J., & McKay, E. (2013). Mental Health, Post-Secondary Education, and Information Communications Technology. In J. Lannon, & E. Halpin (Eds.), *Human Rights and Information Communication Technologies: Trends and Consequences of Use* (pp. 196–213). Hershey, PA: Information Science Reference. doi:10.4018/978-1-4666-4422-9.ch063

Mayes, P. (2014). Interactive Advertising: Displays of Identity and Stance on YouTube. In H. Lim, & F. Sudweeks (Eds.), *Innovative Methods and Technologies for Electronic Discourse Analysis* (pp. 260–284). Hershey, PA: Information Science Reference.

McAuley, J., O'Connor, A., & Lewis, D. (2014). Community Management Matters: Advanced Visual Analytics for Online Community Managers. In M. Huang, & W. Huang (Eds.), *Innovative Approaches of Data Visualization and Visual Analytics* (pp. 349–384). Hershey, PA: Information Science Reference.

McDonald, A., & Helmer, S. (2011). A Comparative Case Study of Indonesian and UK Organisational Culture Differences in IS Project Management. [IJTHI]. *International Journal of Technology and Human Interaction, 7*(2), 28–37. doi:10.4018/jthi.2011040104

McGee, E. M. (2014). Neuroethics and Implanted Brain Machine Interfaces. In M. Michael, & K. Michael (Eds.), *Uberveillance and the Social Implications of Microchip Implants: Emerging Technologies* (pp. 351–365). Hershey, PA: Information Science Reference.

McLoughlin, C. (2014). Open, Flexible and Participatory Pedagogy in the Era of Globalisation: Technology, Open Education and International E-Learning. In V. Wang (Ed.), *International Education and the Next-Generation Workforce: Competition in the Global Economy* (pp. 224–239). Hershey, PA: Information Science Reference. doi:10.4018/978-1-4666-4979-8.ch002

Meawad, F., & Ahmed, G. (2014). Designing Browser-Style Augmented Reality Applications. In K. Rızvanoğlu, & G. Çetin (Eds.), *Research and Design Innovations for Mobile User Experience* (pp. 245–262). Hershey, PA: Information Science Reference.

Meissonierm, R., Bourdon, I., Amabile, S., & Boudrandi, S. (2012). Toward an Enacted Approach to Understanding OSS Developer's Motivations. [IJTHI]. *International Journal of Technology and Human Interaction, 8*(1), 38–54. doi:10.4018/jthi.2012010103

Mencía, B. L., Pardo, D. D., Trapote, A. H., & Gómez, L. A. (2014). Embodied Conversational Agents in Interactive Applications for Children with Special Educational Needs. In I. Association (Ed.), *Assistive Technologies: Concepts, Methodologies, Tools, and Applications* (pp. 811–840). Hershey, PA: Information Science Reference.

Menelas, B. J. (2014). Virtual Reality Technologies (Visual, Haptics, and Audio) in Large Datasets Analysis. In M. Huang, & W. Huang (Eds.), *Innovative Approaches of Data Visualization and Visual Analytics* (pp. 111–132). Hershey, PA: Information Science Reference.

Meredith, J., & Potter, J. (2014). Conversation Analysis and Electronic Interactions: Methodological, Analytic and Technical Considerations. In H. Lim, & F. Sudweeks (Eds.), *Innovative Methods and Technologies for Electronic Discourse Analysis* (pp. 370–393). Hershey, PA: Information Science Reference.

Mesquita, A. (2013). Human-Information Interaction and Technical Communication: Concepts and Frameworks. [IJTHI]. *International Journal of Technology and Human Interaction, 9*(1), 96–98. doi:10.4018/jthi.2013010107

Michael, K. (2014). Towards the Blanket Coverage DNA Profiling and Sampling of Citizens in England, Wales, and Northern Ireland. In M. Michael, & K. Michael (Eds.), *Uberveillance and the Social Implications of Microchip Implants: Emerging Technologies* (pp. 187–207). Hershey, PA: Information Science Reference.

Michael, M. G. (2014). Introduction: On the "Birth" of Uberveillance. In M. Michael, & K. Michael (Eds.), *Uberveillance and the Social Implications of Microchip Implants: Emerging Technologies* (pp. 1–31). Hershey, PA: Information Science Reference.

Millán-Calenti, J. C., & Maseda, A. (2014). Telegerontology®: A New Technological Resource for Elderly Support. In I. Association (Ed.), *Assistive Technologies: Concepts, Methodologies, Tools, and Applications* (pp. 705–719). Hershey, PA: Information Science Reference.

Molla, A., & Peszynski, K. (2013). E-Business in Agribusiness: Investigating the E-Readiness of Australian Horticulture Firms. In S. Chhabra (Ed.), *ICT Influences on Human Development, Interaction, and Collaboration* (pp. 78–96). Hershey, PA: Information Science Reference.

Montañola-Sales, C., Rubio-Campillo, X., Cela-Espin, J. M., Casanovas-Garcia, J., & Kaplan-Marcusan, A. (2014). Overview on Agent-Based Social Modelling and the Use of Formal Languages. In P. Fonseca i Casas (Ed.), *Formal Languages for Computer Simulation: Transdisciplinary Models and Applications* (pp. 333–377). Hershey, PA: Information Science Reference.

Morikawa, C., & Lyons, M. J. (2014). Design and Evaluation of Vision-Based Head and Face Tracking Interfaces for Assistive Input. In G. Kouroupetroglou (Ed.), *Assistive Technologies and Computer Access for Motor Disabilities* (pp. 180–205). Hershey, PA: Medical Information Science Reference.

Morueta, R. T., Gómez, J. I., & Gómez, Á. H. (2012). B-Learning at Universities in Andalusia (Spain): From Traditional to Student-Centred Learning. [IJTHI]. *International Journal of Technology and Human Interaction, 8*(2), 56–76. doi:10.4018/jthi.2012040104

Mosindi, O., & Sice, P. (2011). An Exploratory Theoretical Framework for Understanding Information Behaviour. [IJTHI]. *International Journal of Technology and Human Interaction, 7*(2), 1–8. doi:10.4018/jthi.2011040101

Muhammad, I., Teoh, S. Y., & Wickramasinghe, N. (2012). Why Using Actor Network Theory (ANT) Can Help to Understand the Personally Controlled Electronic Health Record (PCEHR) in Australia. [IJANTTI]. *International Journal of Actor-Network Theory and Technological Innovation, 4*(2), 44–60. doi:10.4018/jantti.2012040105

Murphy, J., Lee, R., & Swinger, E. (2011). Student Perceptions and Adoption of University Smart Card Systems. [IJTHI]. *International Journal of Technology and Human Interaction, 7*(3), 1–15. doi:10.4018/jthi.2011070101

Mushi, R. T., & Chilimo, W. (2013). Contribution of Information and Communication Technologies to Malaria Control in Tanzania. In S. Chhabra (Ed.), *ICT Influences on Human Development, Interaction, and Collaboration* (pp. 132–141). Hershey, PA: Information Science Reference.

Naidoo, R. (2012). A Socio-Technical Account of an Internet-Based Self-Service Technology Implementation: Why Call-Centres Sometimes 'Prevail' in a Multi-Channel Context? In A. Tatnall (Ed.), *Social Influences on Information and Communication Technology Innovations* (pp. 68–91). Hershey, PA: Information Science Reference. doi:10.4018/978-1-4666-1559-5.ch005

Naidoo, V., & Naidoo, Y. (2014). Home Telecare, Medical Implant, and Mobile Technology: Evolutions in Geriatric Care. In C. El Morr (Ed.), *Research Perspectives on the Role of Informatics in Health Policy and Management* (pp. 222–237). Hershey, PA: Medical Information Science Reference.

Nieva, E. G., Peralta, M. F., & Beltramone, D. A. (2014). Home Automation by Brain-Computer Interface. In F. Cipolla-Ficarra (Ed.), *Advanced Research and Trends in New Technologies, Software, Human-Computer Interaction, and Communicability* (pp. 502–510). Hershey, PA: Information Science Reference.

Nisi, V., Nunes, N., Isarankura, K., & Forlizzi, J. (2014). "Cozinha da Madeira": A Sustainable Tourism Service Concept for Madeira Island. In F. Cipolla-Ficarra (Ed.), *Advanced Research and Trends in New Technologies, Software, Human-Computer Interaction, and Communicability* (pp. 364–376). Hershey, PA: Information Science Reference.

Notley, T., & Hankey, S. (2013). Human Rights Defenders and the Right to Digital Privacy and Security. In J. Lannon, & E. Halpin (Eds.), *Human Rights and Information Communication Technologies: Trends and Consequences of Use* (pp. 157–175). Hershey, PA: Information Science Reference.

Novak, D., & Verber, D. (2014). New Generation of Artificial Intelligence for Real-Time Strategy Games. In F. Cipolla-Ficarra (Ed.), *Advanced Research and Trends in New Technologies, Software, Human-Computer Interaction, and Communicability* (pp. 220–229). Hershey, PA: Information Science Reference.

Ntoa, S., Margetis, G., Antona, M., & Stephanidis, C. (2014). Scanning-Based Interaction Techniques for Motor Impaired Users. In G. Kouroupetroglou (Ed.), *Assistive Technologies and Computer Access for Motor Disabilities* (pp. 57–89). Hershey, PA: Medical Information Science Reference.

Olaniyan, M., & Graham, D. (2014). Media Streaming for Technological Innovation in Higher Education. In S. Mukerji, & P. Tripathi (Eds.), *Handbook of Research on Transnational Higher Education* (pp. 691–712). Hershey, PA: Information Science Reference.

Oye, N. D., Iahad, N. A., & Rahim, N. Z. (2013). An Application of the UTAUT Model for Understanding Acceptance and Use of ICT by Nigerian University Academicians. In S. Chhabra (Ed.), *ICT Influences on Human Development, Interaction, and Collaboration* (pp. 214–229). Hershey, PA: Information Science Reference.

Palmer, D., Warren, I., & Miller, P. (2014). ID Scanners and Überveillance in the Night-Time Economy: Crime Prevention or Invasion of Privacy? In M. Michael, & K. Michael (Eds.), *Uberveillance and the Social Implications of Microchip Implants: Emerging Technologies* (pp. 208–225). Hershey, PA: Information Science Reference.

Pańkowska, M. (2014). Information Technology Prosumption Acceptance by Business Information System Consultants. In M. Pańkowska (Ed.), *Frameworks of IT Prosumption for Business Development* (pp. 119–141). Hershey, PA: Business Science Reference.

Passera, S., & Haapio, H. (2014). The Quest for Clarity: How Visualization Improves the Usability and User Experience of Contracts. In M. Huang, & W. Huang (Eds.), *Innovative Approaches of Data Visualization and Visual Analytics* (pp. 191–217). Hershey, PA: Information Science Reference.

Peevers, G., Douglas, G., & Jack, M. A. (2011). Multimedia Technology in the Financial Services Sector: Customer Satisfaction with Alternatives to Face-to-Face Interaction in Mortgage Sales. [IJTHI]. *International Journal of Technology and Human Interaction*, 7(4), 17–30. doi:10.4018/jthi.2011100102

Peevers, G., Williams, R., Douglas, G., & Jack, M. A. (2013). Usability Study of Fingerprint and Palmvein Biometric Technologies at the ATM. [IJTHI]. *International Journal of Technology and Human Interaction*, 9(1), 78–95. doi:10.4018/jthi.2013010106

Perakslis, C. (2014). Willingness to Adopt RFID Implants: Do Personality Factors Play a Role in the Acceptance of Uberveillance? In M. Michael, & K. Michael (Eds.), *Uberveillance and the Social Implications of Microchip Implants: Emerging Technologies* (pp. 144–168). Hershey, PA: Information Science Reference.

Pereira, R., Hornung, H., & Baranauskas, M. C. (2014). Cognitive Authority Revisited in Web Social Interaction. In M. Pańkowska (Ed.), *Frameworks of IT Prosumption for Business Development* (pp. 142–157). Hershey, PA: Business Science Reference.

Pfeiffer, T. (2014). Gaze-Based Assistive Technologies. In G. Kouroupetroglou (Ed.), *Assistive Technologies and Computer Access for Motor Disabilities* (pp. 90–109). Hershey, PA: Medical Information Science Reference.

Pino, A. (2014). Augmentative and Alternative Communication Systems for the Motor Disabled. In G. Kouroupetroglou (Ed.), *Disability Informatics and Web Accessibility for Motor Limitations* (pp. 105–152). Hershey, PA: Medical Information Science Reference.

Pino, A. (2014). Free Assistive Technology Software for Persons with Motor Disabilities. In G. Kouroupetroglou (Ed.), *Assistive Technologies and Computer Access for Motor Disabilities* (pp. 110–152). Hershey, PA: Medical Information Science Reference.

Podgorelec, V., & Dobrina, M. (2014). Educational Social Networks as a Means for Better Social Cohesion in Education. In F. Cipolla-Ficarra (Ed.), *Advanced Research and Trends in New Technologies, Software, Human-Computer Interaction, and Communicability* (pp. 112–120). Hershey, PA: Information Science Reference.

Podgorelec, V., & Grašič, B. (2014). Semantic Web Services-Based Knowledge Management Framework. In F. Cipolla-Ficarra (Ed.), *Advanced Research and Trends in New Technologies, Software, Human-Computer Interaction, and Communicability* (pp. 121–130). Hershey, PA: Information Science Reference.

Popescu, D. A., & Nicolae, D. (2014). Degree of Similarity of Web Applications. In F. Cipolla-Ficarra (Ed.), *Advanced Research and Trends in New Technologies, Software, Human-Computer Interaction, and Communicability* (pp. 312–318). Hershey, PA: Information Science Reference.

Previtali, M., Scaioni, M., Barazzetti, L., Brumana, R., & Oreni, D. (2014). An Algorithm for Occlusion-Free Texture Mapping from Oriented Images. In F. Cipolla-Ficarra (Ed.), *Advanced Research and Trends in New Technologies, Software, Human-Computer Interaction, and Communicability* (pp. 32–42). Hershey, PA: Information Science Reference.

Prietch, S. S., & Filgueiras, L. V. (2014). Developing Emotion-Libras 2.0: An Instrument to Measure the Emotional Quality of Deaf Persons while Using Technology. In K. Blashki, & P. Isaias (Eds.), *Emerging Research and Trends in Interactivity and the Human-Computer Interface* (pp. 74–94). Hershey, PA: Information Science Reference.

Puls, S., & Wörn, H. (2014). Seamless Interfacing: Situation Awareness through Action Recognition and Spatio-Temporal Reasoning. In K. Blashki, & P. Isaias (Eds.), *Emerging Research and Trends in Interactivity and the Human-Computer Interface* (pp. 144–159). Hershey, PA: Information Science Reference.

Quinlan, A. (2012). Imagining a Feminist Actor-Network Theory. [IJANTTI]. *International Journal of Actor-Network Theory and Technological Innovation*, 4(2), 1–9. doi:10.4018/jantti.2012040101

Quinlan, A., Quinlan, E., & Nelson, D. (2011). Performing Actor-Network Theory in the Post-Secondary Classroom. [IJANTTI]. *International Journal of Actor-Network Theory and Technological Innovation*, 3(4), 1–10. doi:10.4018/jantti.2011100101

Quinlan, A., Quinlan, E., & Nelson, D. (2013). Performing Actor-Network Theory in the Post-Secondary Classroom. In A. Tatnall (Ed.), *Social and Professional Applications of Actor-Network Theory for Technology Development* (pp. 56–66). Hershey, PA: Information Science Reference.

Rahman, H., & Ramos, I. (2013). Empowerment of SMEs Through Open Innovation Strategies: Life Cycle of Technology Management. In S. Chhabra (Ed.), *ICT Influences on Human Development, Interaction, and Collaboration* (pp. 185–202). Hershey, PA: Information Science Reference.

Rambaree, K. (2014). Computer-Aided Deductive Critical Discourse Analysis of a Case Study from Mauritius with ATLAS-ti 6.2. In H. Lim, & F. Sudweeks (Eds.), *Innovative Methods and Technologies for Electronic Discourse Analysis* (pp. 346–368). Hershey, PA: Information Science Reference.

Ramos, I., & Fernandes, J. (2013). Web-Based Intellectual Property Marketplace: A Survey of Current Practices. In S. Chhabra (Ed.), *ICT Influences on Human Development, Interaction, and Collaboration* (pp. 203–213). Hershey, PA: Information Science Reference.

Ranganathan, K. (2013). Leapfrogging the Digital Divide: Myth or Reality for Emerging Regions? In S. Chhabra (Ed.), *ICT Influences on Human Development, Interaction, and Collaboration* (pp. 230–242). Hershey, PA: Information Science Reference.

Ratten, V. (2013). Social Cognitive Theory in Mobile Banking Innovations. In I. Lee (Ed.), *Mobile Applications and Knowledge Advancements in E-Business* (pp. 42–55). Hershey, PA: Business Science Reference.

Rayna, T., Striukova, L., & Landau, S. (2011). The Critical Role of Market Segmentation: Evidence from the Audio Player Market. In A. Tatnall (Ed.), *Actor-Network Theory and Technology Innovation: Advancements and New Concepts* (pp. 162–177). Hershey, PA: Information Science Reference.

Rebedea, T., Trausan-Matu, S., & Chiru, C. (2014). Inter-Animation between Utterances in Collaborative Chat Conversations. In H. Lim, & F. Sudweeks (Eds.), *Innovative Methods and Technologies for Electronic Discourse Analysis* (pp. 63–93). Hershey, PA: Information Science Reference.

Reio, T. G. Jr, & Whitehead, C. L. (2014). Using Technology to Address Workforce Readiness Skills. In V. Wang (Ed.), *International Education and the Next-Generation Workforce: Competition in the Global Economy* (pp. 154–169). Hershey, PA: Information Science Reference.

Renner, J., & Click, I. (2014). The Writing on My Wall: Freedom of Expression, First Amendment, and Social Media – New Faculty Rights Concerns. In S. Mukerji, & P. Tripathi (Eds.), *Handbook of Research on Transnational Higher Education* (pp. 350–367). Hershey, PA: Information Science Reference.

Riveiro, M. (2014). The Importance of Visualization and Interaction in the Anomaly Detection Process. In M. Huang, & W. Huang (Eds.), *Innovative Approaches of Data Visualization and Visual Analytics* (pp. 133–150). Hershey, PA: Information Science Reference.

Rivera, L. A., & DeSouza, G. N. (2014). Haptic and Gesture-Based Assistive Technologies for People with Motor Disabilities. In G. Kouroupetroglou (Ed.), *Assistive Technologies and Computer Access for Motor Disabilities* (pp. 1–27). Hershey, PA: Medical Information Science Reference.

Rogers, L. S. (2014). Communities of Communication: Using Social Media as Medium for Supporting Teacher Interpersonal Development. In D. McConatha, C. Penny, J. Schugar, & D. Bolton (Eds.), *Mobile Pedagogy and Perspectives on Teaching and Learning* (pp. 58–73). Hershey, PA: Information Science Reference.

Rowan, L., & Bigum, C. (2011). Reassembling the Problem of the Under-Representation of Girls in IT Courses. In A. Tatnall (Ed.), *Actor-Network Theory and Technology Innovation: Advancements and New Concepts* (pp. 208–222). Hershey, PA: Information Science Reference.

Ryokai, K., & Agogino, A. (2013). Off the Paved Paths: Exploring Nature with a Mobile Augmented Reality Learning Tool. [IJMHCI]. *International Journal of Mobile Human Computer Interaction*, 5(2), 21–49. doi:10.4018/jmhci.2013040102

Salazar, L. H., Lacerda, T., Nunes, J. V., & Gresse von Wangenheim, C. (2013). A Systematic Literature Review on Usability Heuristics for Mobile Phones. [IJMHCI]. *International Journal of Mobile Human Computer Interaction*, 5(2), 50–61. doi:10.4018/jmhci.2013040103

Saleh, A. I. (2013). A Novel Strategy for Managing User's Locations in PCS Networks Based on a Novel Hot Spots Topology. In S. Chhabra (Ed.), *ICT Influences on Human Development, Interaction, and Collaboration* (pp. 43–77). Hershey, PA: Information Science Reference.

Samanta, S. K., Woods, J., & Ghanbari, M. (2011). Automatic Language Translation: An Enhancement to the Mobile Messaging Services. [IJTHI]. *International Journal of Technology and Human Interaction*, 7(1), 1–18. doi:10.4018/jthi.2011010101

Schleicher, R., Westermann, T., Weiss, B., Wechsung, I., & Möller, S. (2014). Research on Mobile HCI: Taken Out of Context? In K. Rızvanoğlu, & G. Çetin (Eds.), *Research and Design Innovations for Mobile User Experience* (pp. 76–93). Hershey, PA: Information Science Reference.

Schultz, R. A. (2014). Causes of Conflict with the Ecosystem. In Technology versus Ecology: Human Superiority and the Ongoing Conflict with Nature (pp. 159-178). Hershey, PA: Information Science Reference. doi: doi:10.4018/978-1-4666-4586-8.ch009

Schultz, R. A. (2014). Civilization. In Technology versus Ecology: Human Superiority and the Ongoing Conflict with Nature (pp. 125-144). Hershey, PA: Information Science Reference. doi: doi:10.4018/978-1-4666-4586-8.ch007

Schultz, R. A. (2014). More about Modern Technology. In Technology versus Ecology: Human Superiority and the Ongoing Conflict with Nature (pp. 145-158). Hershey, PA: Information Science Reference. doi: doi:10.4018/978-1-4666-4586-8.ch008

Schultz, R. A. (2014). Science, Modern Technology, and Corporations. In Technology versus Ecology: Human Superiority and the Ongoing Conflict with Nature (pp. 46-65). Hershey, PA: Information Science Reference. doi: doi:10.4018/978-1-4666-4586-8.ch003

Schultz, R. A. (2014). The Role of Science and Technology. In Technology versus Ecology: Human Superiority and the Ongoing Conflict with Nature (pp. 213-230). Hershey, PA: Information Science Reference. doi: doi:10.4018/978-1-4666-4586-8.ch011

Sharifzadeh, M., Zamani, G., Karami, E. H., Khalili, D., & Tatnall, A. (2012). The Iranian Wheat Growers' Climate Information Use: An Actor-Network Theory Perspective. [IJANTTI]. *International Journal of Actor-Network Theory and Technological Innovation*, 4(4), 1–22. doi:10.4018/jantti.2012100101

Sharma, D. K., & Sharma, A. K. (2013). Search Engine: A Backbone for Information Extraction in ICT Scenario. In S. Chhabra (Ed.), *ICT Influences on Human Development, Interaction, and Collaboration* (pp. 117–131). Hershey, PA: Information Science Reference.

Shen, J., & Eder, L. B. (2011). An Examination of Factors Associated with User Acceptance of Social Shopping Websites. [IJTHI]. *International Journal of Technology and Human Interaction, 7*(1), 19–36. doi:10.4018/jthi.2011010102

Shih, Y. (2014). Weaving Web 2.0 and Facial Expression Recognition into the 3D Virtual English Classroom. In S. Hai-Jew (Ed.), *Packaging Digital Information for Enhanced Learning and Analysis: Data Visualization, Spatialization, and Multidimensionality* (pp. 71–92). Hershey, PA: Information Science Reference.

Shih, Y., & Leonard, M. (2014). Immersion and Interaction via Avatars within Google Street View: Opening Possibilities beyond Traditional Cultural Learning. In S. Hai-Jew (Ed.), *Packaging Digital Information for Enhanced Learning and Analysis: Data Visualization, Spatialization, and Multidimensionality* (pp. 266–281). Hershey, PA: Information Science Reference. doi:10.4018/978-1-4666-4979-8.ch037

Shing, O. C., Phooi, S. K., & Li-Minn, A. (2014). The Realization of Customer Satisfaction with Technology Integrations. In V. Jham, & S. Puri (Eds.), *Cases on Consumer-Centric Marketing Management* (pp. 39–52). Hershey, PA: Business Science Reference.

Sikorski, M. (2014). Evolution of End User Participation in IT Projects. In M. Pańkowska (Ed.), *Frameworks of IT Prosumption for Business Development* (pp. 48–63). Hershey, PA: Business Science Reference.

Singer, A. E. (2013). Corporate Moral Agency and Artificial Intelligence. [IJSODIT]. *International Journal of Social and Organizational Dynamics in IT, 3*(1), 1–13. doi:10.4018/ijsodit.2013010101

Smith, C. (2014). (Re)Engineering Cultural Heritage Contexts using Creative Human Computer Interaction Techniques and Mixed Reality Methodologies. In F. Cipolla-Ficarra (Ed.), *Advanced Research and Trends in New Technologies, Software, Human-Computer Interaction, and Communicability* (pp. 441–451). Hershey, PA: Information Science Reference.

Sørensen, C. G., & Sørensen, M. G. (2014). The Conceptual Pond: A Persuasive Tool for Quantifiable Qualitative Assessment. In K. Blashki, & P. Isaias (Eds.), *Emerging Research and Trends in Interactivity and the Human-Computer Interface* (pp. 449–469). Hershey, PA: Information Science Reference.

Sosa, M. D. (2014). Revision of the Groupware Users Interface Development Methods. In F. Cipolla-Ficarra (Ed.), *Advanced Research and Trends in New Technologies, Software, Human-Computer Interaction, and Communicability* (pp. 587–594). Hershey, PA: Information Science Reference.

Souto, V. T. (2014). A Framework for Designing Interactive Digital Learning Environments for Young People. In K. Blashki, & P. Isaias (Eds.), *Emerging Research and Trends in Interactivity and the Human-Computer Interface* (pp. 429–447). Hershey, PA: Information Science Reference.

Spichkova, M. (2014). Design of Formal Languages and Interfaces: "Formal" Does Not Mean "Unreadable". In K. Blashki, & P. Isaias (Eds.), *Emerging Research and Trends in Interactivity and the Human-Computer Interface* (pp. 301–314). Hershey, PA: Information Science Reference.

Spöhrer, M. (2013). Murphy's Law in Action: The Formation of the Film Production Network of Paul Lazarus' Barbarosa (1982)- An Actor-Network-Theory Case Study. [IJAN-TTI]. *International Journal of Actor-Network Theory and Technological Innovation*, 5(1), 19–39. doi:10.4018/jantti.2013010102

Stebick, D. M., & Paxton, M. L. (2014). Bridging the Gap: 21ST Century Media Meets Theoretical Pedagogical Literacy Practices. In H. Lim, & F. Sudweeks (Eds.), *Innovative Methods and Technologies for Electronic Discourse Analysis* (pp. 1–18). Hershey, PA: Information Science Reference.

Stinson, J., & Gill, N. (2014). Internet-Based Chronic Disease Self-Management for Youth. In I. Association (Ed.), *Assistive Technologies: Concepts, Methodologies, Tools, and Applications* (pp. 224–245). Hershey, PA: Information Science Reference.

Stoicovy, C. E. (2014). Culturally Responsive Online Learning for Asian/Pacific Islanders in a Pacific Island University. In K. Sullivan, P. Czigler, & J. Sullivan Hellgren (Eds.), *Cases on Professional Distance Education Degree Programs and Practices: Successes, Challenges, and Issues* (pp. 177–199). Hershey, PA: Information Science Reference.

Strömberg-Jakka, M. (2013). Social Assistance via the Internet: The Case of Finland in the European Context. In J. Lannon, & E. Halpin (Eds.), *Human Rights and Information Communication Technologies: Trends and Consequences of Use* (pp. 177–195). Hershey, PA: Information Science Reference.

Suki, N. M., Ramayah, T., Ming, M. K., & Suki, N. M. (2011). Factors Enhancing Employed Job Seekers Intentions to Use Social Networking Sites as a Job Search Tool. [IJTHI]. *International Journal of Technology and Human Interaction*, 7(2), 38–54. doi:10.4018/jthi.2011040105

Sullivan, K. P. (2014). Studying Professional Degrees via the Internet: Challenges, Issues, and Relevance from the Student's Perspective. In K. Sullivan, P. Czigler, & J. Sullivan Hellgren (Eds.), *Cases on Professional Distance Education Degree Programs and Practices: Successes, Challenges, and Issues* (pp. 1–27). Hershey, PA: Information Science Reference.

Sun, X., Golightly, D., Cranwell, J., Bedwell, B., & Sharples, S. (2013). Participant Experiences of Mobile Device-Based Diary Studies. [IJMHCI]. *International Journal of Mobile Human Computer Interaction*, 5(2), 62–83. doi:10.4018/jmhci.2013040104

Tatnall, A. (2011). Innovation Translation, Innovation Diffusion and the Technology Acceptance Model: Comparing Three Different Approaches to Theorising Technological Innovation. In A. Tatnall (Ed.), *Actor-Network Theory and Technology Innovation: Advancements and New Concepts* (pp. 52–66). Hershey, PA: Information Science Reference.

Tatnall, A. (2012). On Actors, Networks, Hybrids, Black Boxes and Contesting Programming Languages. In A. Tatnall (Ed.), *Social Influences on Information and Communication Technology Innovations* (pp. 183–194). Hershey, PA: Information Science Reference. doi:10.4018/978-1-4666-1559-5.ch013

Taylor, J. M., Santuzzi, A. M., & Cogburn, D. L. (2013). Trust and Member Satisfaction in a Developing Virtual Organization: The Roles of Leader Contact and Experience with Technology. [IJSODIT]. *International Journal of Social and Organizational Dynamics in IT*, 3(1), 32–46. doi:10.4018/ijsodit.2013010103

Tennyson, R. D. (2014). Computer Interventions for Children with Disabilities: Review of Research and Practice. In I. Association (Ed.), *Assistive Technologies: Concepts, Methodologies, Tools, and Applications* (pp. 841–864). Hershey, PA: Information Science Reference.

Tiwary, U. S., & Siddiqui, T. J. (2014). Working Together with Computers: Towards a General Framework for Collaborative Human Computer Interaction. In I. Association (Ed.), *Assistive Technologies: Concepts, Methodologies, Tools, and Applications* (pp. 141–162). Hershey, PA: Information Science Reference.

Truong, Y. (2011). Antecedents of Consumer Acceptance of Mobile Television Advertising. [IJTHI]. *International Journal of Technology and Human Interaction, 7*(3), 70–83. doi:10.4018/jthi.2011070105

Tsai, C. (2011). How Much Can Computers and Internet Help?: A Long-Term Study of Web-Mediated Problem-Based Learning and Self-Regulated Learning. [IJTHI]. *International Journal of Technology and Human Interaction, 7*(1), 67–81. doi:10.4018/jthi.2011010105

Tsai, W. (2013). An Investigation on Undergraduate's Bio-Energy Engineering Education Program at the Taiwan Technical University. [IJTHI]. *International Journal of Technology and Human Interaction, 8*(3), 46–53. doi:10.4018/jthi.2012070105

Tummons, J. (2011). Actors, Networks and Assessment: An Actor-Network Critique of Quality Assurance in Higher Education In England. In A. Tatnall (Ed.), *Actor-Network Theory and Technology Innovation: Advancements and New Concepts* (pp. 178–191). Hershey, PA: Information Science Reference.

Uden, L., & Francis, J. (2011). Service Innovation Using Actor Network Theory. In A. Tatnall (Ed.), *Actor-Network Theory and Technology Innovation: Advancements and New Concepts* (pp. 20–40). Hershey, PA: Information Science Reference.

Underwood, J., & McCabe, B. (2012). Using Actor-Network Theory to Research the Adoption of Inter-Organizational Information Systems. In K. Vaidya (Ed.), *Inter-Organizational Information Systems and Business Management: Theories for Researchers* (pp. 83–98). Hershey, PA: Business Science Reference.

Valeria, N., & Theng, L. B. (2014). An Affective Computer-Mediated Learning for Persons with Motor Impairments. In G. Kouroupetroglou (Ed.), *Disability Informatics and Web Accessibility for Motor Limitations* (pp. 323–369). Hershey, PA: Medical Information Science Reference.

Varnava-Marouchou, D., & Minott, M. A. (2014). Experiences of an Online Doctoral Course in Teacher Education. In K. Sullivan, P. Czigler, & J. Sullivan Hellgren (Eds.), *Cases on Professional Distance Education Degree Programs and Practices: Successes, Challenges, and Issues* (pp. 28–48). Hershey, PA: Information Science Reference.

Veen, W., van Staalduinen, J., & Hennis, T. (2011). Informal Self-regulated Learning in Corporate Organizations. In G. Dettori, & D. Persico (Eds.), *Fostering Self-Regulated Learning through ICT* (pp. 364–379). Hershey, PA: Information Science Reference.

Verber, D. (2014). Cloud-Assisted Services for Mobile Applications: CLASS-MA. In F. Cipolla-Ficarra (Ed.), *Advanced Research and Trends in New Technologies, Software, Human-Computer Interaction, and Communicability* (pp. 93–101). Hershey, PA: Information Science Reference.

Wagner, K. K., & Dole, S. (2014). Social Media in Higher Education: Using Wiki for Online Gifted Education Courses. In S. Mukerji, & P. Tripathi (Eds.), *Handbook of Research on Transnational Higher Education* (pp. 730–750). Hershey, PA: Information Science Reference.

Walck, P. E., & Kalyango, Y. (2014). Journalism and Media: From Mellowed Pedagogy to New Mobile Learning Tools. In D. McConatha, C. Penny, J. Schugar, & D. Bolton (Eds.), *Mobile Pedagogy and Perspectives on Teaching and Learning* (pp. 221–237). Hershey, PA: Information Science Reference.

Wallgren, L. G., Leijon, S., & Andersson, K. M. (2011). IT Managers' Narratives on Subordinates' Motivation at Work: A Case Study. [IJTHI]. *International Journal of Technology and Human Interaction*, 7(3), 35–49. doi:10.4018/jthi.2011070103

Walton, A. L., DeVaney, S. A., & Sandall, D. L. (2011). Graduate Students' Perceptions of Privacy and Closed Circuit Television Systems in Public Settings. [IJTHI]. *International Journal of Technology and Human Interaction*, 7(3), 50–69. doi:10.4018/jthi.2011070104

Wang, H. (2014). A Guide to Assistive Technology for Teachers in Special Education. In I. Association (Ed.), *Assistive Technologies: Concepts, Methodologies, Tools, and Applications* (pp. 12–25). Hershey, PA: Information Science Reference.

Wang, H., & Bowerman, J. (2014). The Impact of Visual Complexity on Children's Learning Websites in Relation to Aesthetic Preference and Learning Motivation. In K. Blashki, & P. Isaias (Eds.), *Emerging Research and Trends in Interactivity and the Human-Computer Interface* (pp. 395–412). Hershey, PA: Information Science Reference.

Wang, S., Ku, C., & Chu, C. (2013). Sustainable Campus Project: Potential for Energy Conservation and Carbon Reduction Education in Taiwan. [IJTHI]. *International Journal of Technology and Human Interaction*, 8(3), 19–30. doi:10.4018/jthi.2012070103

Warwick, K., & Gasson, M. N. (2014). Practical Experimentation with Human Implants. In M. Michael, & K. Michael (Eds.), *Uberveillance and the Social Implications of Microchip Implants: Emerging Technologies* (pp. 64–132). Hershey, PA: Information Science Reference.

Weber, R. H. (2013). ICT Policies Favouring Human Rights. In J. Lannon, & E. Halpin (Eds.), *Human Rights and Information Communication Technologies: Trends and Consequences of Use* (pp. 21–35). Hershey, PA: Information Science Reference.

Welch, K. C., Lahiri, U., Sarkar, N., Warren, Z., Stone, W., & Liu, C. (2014). Affect-Sensitive Computing and Autism. In I. Association (Ed.), *Assistive Technologies: Concepts, Methodologies, Tools, and Applications* (pp. 865–883). Hershey, PA: Information Science Reference.

Whyte, K. P., List, M., Stone, J. V., Grooms, D., Gasteyer, S., & Thompson, P. B. et al. (2014). Uberveillance, Standards, and Anticipation: A Case Study on Nanobiosensors in U.S. Cattle. In M. Michael, & K. Michael (Eds.), *Uberveillance and the Social Implications of Microchip Implants: Emerging Technologies* (pp. 260–279). Hershey, PA: Information Science Reference.

Wickramasinghe, N., & Bali, R. K. (2011). The Need for Rich Theory to Realize the Vision of Healthcare Network Centric Operations: The Case for Combining ANT and Social Network Analysis. In A. Tatnall (Ed.), *Actor-Network Theory and Technology Innovation: Advancements and New Concepts* (pp. 41–51). Hershey, PA: Information Science Reference.

Wickramasinghe, N., Bali, R. K., & Goldberg, S. (2011). Using S'ANT for Facilitating Superior Understanding of Key Factors in the Design of a Chronic Disease Self-Management Model. In A. Tatnall (Ed.), *Actor-Network Theory and Technology Innovation: Advancements and New Concepts* (pp. 223–233). Hershey, PA: Information Science Reference.

Wickramasinghe, N., Tatnall, A., & Bali, R. K. (2012). Using Actor-Network Theory to Facilitate a Superior Understanding of Knowledge Creation and Knowledge Transfer. In A. Tatnall (Ed.), *Social Influences on Information and Communication Technology Innovations* (pp. 205–218). Hershey, PA: Information Science Reference. doi:10.4018/978-1-4666-1559-5.ch015

Wiebelhaus-Brahm, E. (2013). Truth-Seeking at a Distance: Engaging Diaspora Populations in Transitional Justice Processes. In J. Lannon, & E. Halpin (Eds.), *Human Rights and Information Communication Technologies: Trends and Consequences of Use* (pp. 72–85). Hershey, PA: Information Science Reference.

Wilson, C., & Dunn, A. (2013). Contingency and Hybridity in the Study of Digital Advocacy Networks: Implications of the Egyptian Protest Movement. In J. Lannon, & E. Halpin (Eds.), *Human Rights and Information Communication Technologies: Trends and Consequences of Use* (pp. 100–121). Hershey, PA: Information Science Reference.

Wobbrock, J. O. (2014). Improving Pointing in Graphical User Interfaces for People with Motor Impairments Through Ability-Based Design. In G. Kouroupetroglou (Ed.), *Assistive Technologies and Computer Access for Motor Disabilities* (pp. 206–253). Hershey, PA: Medical Information Science Reference.

Wong, C. Y., Seet, G., Sim, S. K., & Pang, W. C. (2014). A Hierarchically Structured Collective of Coordinating Mobile Robots Supervised by a Single Human. In I. Management Association (Ed.), Software Design and Development: Concepts, Methodologies, Tools, and Applications (pp. 1142-1164). Hershey, PA: Information Science Reference. doi: doi:10.4018/978-1-4666-4301-7.ch056

Yamazaki, T. (2014). Assistive Technologies in Smart Homes. In I. Association (Ed.), *Assistive Technologies: Concepts, Methodologies, Tools, and Applications* (pp. 663–678). Hershey, PA: Information Science Reference.

Yang, Y., Wang, X., & Li, L. (2013). Use Mobile Devices to Wirelessly Operate Computers. [IJTHI]. *International Journal of Technology and Human Interaction, 9*(1), 64–77. doi:10.4018/jthi.2013010105

Zahabi, L. (2014). Visualizing Information-Triage: A Speculative and Metaphoric Interface for Making Sense of Online Searching. In M. Huang, & W. Huang (Eds.), *Innovative Approaches of Data Visualization and Visual Analytics* (pp. 316–338). Hershey, PA: Information Science Reference.

Zammar, N. (2012). Social Network Sites: The Science of Building and Maintaining Online Communities, a Perspective from Actor-Network Theory. In A. Tatnall (Ed.), *Social Influences on Information and Communication Technology Innovations* (pp. 107–116). Hershey, PA: Information Science Reference. doi:10.4018/978-1-4666-1559-5.ch007

Zanolla, S., Canazza, S., Rodà, A., & Foresti, G. L. (2014). Learning by Means of an Interactive Multimodal Environment. In F. Cipolla-Ficarra (Ed.), *Advanced Research and Trends in New Technologies, Software, Human-Computer Interaction, and Communicability* (pp. 143–153). Hershey, PA: Information Science Reference.

Zelenkauskaite, A. (2014). Analyzing Blending Social and Mass Media Audiences through the Lens of Computer-Mediated Discourse. In H. Lim, & F. Sudweeks (Eds.), *Innovative Methods and Technologies for Electronic Discourse Analysis* (pp. 304–326). Hershey, PA: Information Science Reference.

Ziegler, J., Döring, R., Pfeffer, J., & Urbas, L. (2014). Hand Gesture Recognition as Means for Mobile Human Computer Interaction in Adverse Working Environments. In K. Blashki, & P. Isaias (Eds.), *Emerging Research and Trends in Interactivity and the Human-Computer Interface* (pp. 331–352). Hershey, PA: Information Science Reference.

Compilation of References

Aanestad, M., & Hanseth, O. (2000). *Implementing open network technologies in complex work practices: A case from telemedicine.* Paper presented at the IFIP 8.2. Aalborg, Denmark.

Aaronson, J. W., Murphy-Cullen, C. L., Chop, W. M., & Frey, R. D. (2001). Electronic medical records: the family practice resident perspective. *Family Medicine, 33*(2), 128–132. PMID:11271741

Aarts, J., Doorewaard, H., & Berg, M. (2004). Understanding Implementation: The Case of a Computerized Physician Order Entry System in a Large Dutch University Medical Center. *Journal of the American Medical Informatics Association, 11*, 207–216. doi:10.1197/jamia. M1372 PMID:14764612

Aarts, J., & Koppel, R. (2009). Implementation of computerized physician order entry in seven countries. *Health Affairs (Project Hope), 28*(2), 404–414. doi:10.1377/hlthaff.28.2.404 PMID:19275996

Abbott, T. (2005). *The adoption of eHealth in Australia.* Paper presented at the Health Informatics Conference Melbourne 2005. Melbourne, Australia.

Abernethy, M. A., Horne, M., Lillis, A. M., Malina, M. A., & Selto, F. H. (2005). A multi-method approach to building causal performance maps from expert knowledge. *Management Accounting Research, 16*, 135–155. doi:10.1016/j.mar.2005.03.003

Acquisti, A., & Gross, R. (2006). *Imagined Communities: Awareness, information sharing, and privacy on the Facebook.* Paper presented at the 2006 privacy enhancing technologies (PET) Workshop. Cambridge, UK. Retrieved from http://privacy.cs.Cmu.edu/dataprivacy/projects/facebook/facebook2.pdf retrieved 2/2/2011

Adam, T. (2010). *Determining an e-learning model for students with learning disabilities: An analysis of web-based technologies and curriculum.* (PhD Dissertation). Victoria University, Melbourne, Australia.

Adam, T., & Tatnall, A. (2007). Building a virtual knowledge community of schools for children with special needs. Prague, Czech Republic: Information Technologies for Education and Training (iTET), Charles University.

Adam, T., & Tatnall, A. (2008b). Using ICT to Improve the Education of Students with Learning Disabilities. In *Learning to Live in the Knowledge Society.* Springer.

Adam, T., & Tatnall, A. (2010). Use of ICT to Assist Students with Learning Difficulties: An Actor-Network Analysis. In *Key Competencies in the Knowledge Society.* Springer.

Adamopoulos, A., Dick, M., & Davey, W. (2012). Actor-Network Theory and the Online Investor. *International Journal of Actor-Network Theory and Technological Innovation, 4*(2). doi:10.4018/jantti.2012040103

Adam, T. (2011). A Petri Net Model for Analysing E-Learning and Learning Difficulties. *International Journal of Actor-Network Theory and Technological Innovation, 3*(4), 11–21. doi:10.4018/jantti.2011100102

Adam, T., & Tatnall, A. (2008a). ICT and Inclusion: Students with Special Needs. In *The New 21st Century Workplace.* Heidelberg Press.

Adam, T., & Tatnall, A. (2012). School Children with Learning Disabilities: An Actor-Network Analysis of the Use of ICT to Enhance Self-Esteem and Improve Learning Outcomes. *International Journal of Actor-Network Theory and Technological Innovation, 4*(2), 10–24. doi:10.4018/jantti.2012040102

Agarwal, R., & Prasad, J. (1997). The Role of Innovation Characteristics and Perceived Voluntariness in the Acceptance of Information Technologies. *Decision Sciences*, *28*(3), 557–582. doi:10.1111/j.1540-5915.1997.tb01322.x

Agran, M. (1997). Teaching Self-Instructional Skills to Persons with Mental Retardation: A Descriptive and Experimental Analysis. *Education and Training of the Mentally Retarded*, *21*, 273–281.

AIHW. (2007). *National Indicators for Monitoring Diabetes: Report of the Diabetes Indicators Review Subcommittee of the National Diabetes Data Working*. AIHW.

AIHW. (2008). *Diabetes: Australian Facts 2008*. Canberra, Australia: Australian Institute of Health and Welfare.

Aikat, D. D. (2001). Pioneers of the early digital era: Innovative ideas that shaped computing in 1833-1945. *International Journal of Research into New Media Technologies*, *7*(4), 52–81. doi:10.1177/135485650100700404

Akhlaghpour, S., Wu, J., Lapointe, L., & Pinsonneault, A. (2009). Re-examining the Status of IT in IT Research-An Update on Orlikowski and Iacono (2001). In *Proceedings of AMCIS 2009*. AMCIS.

Akrich, M. (1992). The De-Scription of Technical Objects. In W. Bijker, & J. Law (Eds.), *Shaping Technology, Building Society: Studies in Sociotechnical Change*. Cambridge, MA: MIT Press.

Akrich, M., & Latour, B. (1992). A summary of a convenient vocabulary for the semiotics of human and nonhuman assemblies. In W. Bijker, & J. Law (Eds.), *Shaping Technology, Building Society: Studies in Sociotechnical Change*. Cambridge, MA: MIT Press.

Al-Azmi, S., Al-Enezi, N., & Chowdhury, R. (2009). Users' attitudes to an electronic medical record system and its correlates: A multivariate analysis. *The HIM Journal*, *38*(2), 33–40. PMID:19546486

Alcouffe, S., Berland, N., & Levant, Y. (2008). Actor-networks and the diffusion of management accounting innovations: A comparative study. *Management Accounting Research*, *19*, 1–17. doi:10.1016/j.mar.2007.04.001

Al-Hajri, S., & Tatnall, A. (2011). *A Study of Adoption of Internet Banking in Oman in the Early 2000s, Re-Interpreted Using Innovation Translation Informed by Actor-Network Theory*. Paper presented at the International Information Systems Conference (iiSC-2011). Muscat, Oman.

Ali, S., & Green, P. (2007). IT governance mechanisms in public sector organisations: An Australian context. *Journal of Global Information Management*, *15*(4), 41–63. doi:10.4018/jgim.2007100103

America Institute of Medicine. (2001). *Crossing the Quality Chasm - A New Health System for the 21st Century Committee on Quality of Health Care*. National Academy Press.

Ammenwerth, E., Iller, C., & Mahler, C. (2006). IT-adoption and the interaction of task, technology and individuals: A fit framework and a case study. *BMC Medical Informatics and Decision Making*, 6. PMID:16451720

André, B., Ringdal, G. I., Loge, J. H., Rannestad, T., Laerum, H., & Kaasa, S. (2008). Experiences with the implementation of computerized tools in health care units: A review article. *International Journal of Human-Computer Interaction*, *24*(8), 753–775. doi:10.1080/10447310802205768

Ankolekar, A., Krotzsch, M., & Tran Tand Vrandecic, D. (2007). The Two Cultures, Mashing up Web 2.0 and Semantic Web. In Proceedings of IW3C2, (pp. 825 – 834). IW3C2.

Aral, S., & Weill, P. (2007). IT Assets, Organizational Capabilities, and Firm Performance: How Resource Allocations and Organizational Differences Explain Performance Variation. *Organization Science*, *18*(5), 763–780. doi:10.1287/orsc.1070.0306

Ariani, Y., & Yuliar, S. (2009). Opening the Indonesian Bio-Fuel Box: How Scientists Modulate the Social. *International Journal of Actor-Network Theory and Technological Innovation*, *1*(2).

Armstrong, B. K. et al. (2007). Challenges in Health and Health Care For Australia. *The Medical Journal of Australia*, *187*(9), 485–489. PMID:17979607

ArticlesBase.com – McMillan. (2009). *Two-tier health care*. Retrieved August 26, 2011, from http://www.articles-base.com/health-articles/twotier-health-care-1020480.html

Artikov, I., Hoffman, S. J., & Lynne, G. D., PytlikZillg, L. M., Hu, Q., Tomkins, A. J., ... Waltman, W. J. (2006). Understanding the influence of climate forecasts on farmer decisions as pllanedbehavior. *Journal of Applied Meteorology and Climatology, 45*, 1202–1214. doi:10.1175/JAM2415.1

Ashish, K. (2009). Use of electronic health records in U.S. hospitals. *Medical Benefits*. doi:101056NEJMsa0900592, 25

ASX. (2004). *Share Ownership Study 2004*. Australian Stock Exchange.

Atkins, C., & Sampson, J. (2002). Critical appraisal guidelines for single case study research. In *Proceedings of the Xth European Conference on Information Systems (ECIS)*. Gdansk, Poland: ECIS.

Australian Curriculum Assessment and Reporting Authority. (2010a). *My School*. Retrieved November 2010, from http://www.myschool.edu.au/SchoolSearch.aspx

Australian Curriculum Assessment and Reporting Authority. (2010b). *Naplan - National Assessment Program*. Retrieved Jan 2012, from http://www.naplan.edu.au/home_page.html

Australian Education Union. (2010). *Australian Education Union Submission to the Senate Education Employment and Workplace Relations Committee into the Administration and Reporting of NAPLAN Testing*. Melbourne, Australia: Australian Education Union.

Australian Education Union. (2011). *Stop League Tables*. Retrieved Jan 2012, from http://www.aeufederal.org.au/LT/index2.html

Australian Government – Department of Health and Ageing. (2011). *Private health insurance*. Retrieved August 26, 2011, from http://www.health.gov.au/internet/main/publishing.nsf/Content/health-privatehealth-consumers-glossary.htm

Australian Schools Directory. (2011). *Australian Schools Directory: The Only Guide to all Australian Primary and Secondary Schools*. Retrieved from http://www.australianschoolsdirectory.com.au/

Avgerou, C. (2008). Information systems in developing countries: A critical research review. *Journal of Information Technology, 23*, 133–146. doi:10.1057/palgrave.jit.2000136

Avison, D., & Fitzgerald, G. (2006). *Information Systems Development Methodologies, Techniques & Tools* (4th ed.). London: McGraw-Hill.

Baccarini, D., Salm, G., & Love, P. E. D. (2004). Management of risk in information technology projects. *Industrial Management & Data Systems, 104*(4), 286–295. doi:10.1108/02635570410530702

Bahensky, J. A., Jaana, M., & Ward, M. M. (2008). Health care information technology in rural America: Electronic medical record adoption status in meeting the national agenda. *The Journal of Rural Health, 24*(2), 101–105. doi:10.1111/j.1748-0361.2008.00145.x PMID:18397442

Bakhshaie, A. (2008). *Testing an actor-network theory model of innovation adoption with econometric methods*. (Thesis). Virginia Polytechnic Institute and State University, Blacksburg, VA.

Bali, R. K., & Wickramasinghe, N. (2010). Critical Success factors in the Management of Projects Using Innovative Approaches. *International Journal of Electronic Healthcare, 2*(3), 33–39.

Bannister, F. (2001). Dismantling the silos: Extracting new value from IT investments in public administration. *Information Systems Journal, 11*(1), 65–84. doi:10.1046/j.1365-2575.2001.00094.x

Barki, H., Titah, R., & Boffo, C. (2007). Information System Use–Related Activity: An Expanded Behavioral Conceptualization of Individual-Level Information System Use. *Information Systems Research, 18*(2), 173–192. doi:10.1287/isre.1070.0122

Barkley, B. T. (2007). Project Management. In *New Product Development*. The McGraw-Hill Companies.

Barmer, G. E. K. Krankenkasse. (2011). *Elektronische Gesundheitskarte*. Retrieved July 29, 2011, from https://www.barmer-gek.de/barmer/web/Portale/Versicherte/Leistungen-Services/Leistungen-Beitraege/Multilexikon_20Leistungen/Alle_20Eintr_C3_A4ge/Gesundheitskarte.html?w-cm=CenterColumn_tdocid

Basch, P. (2005). Electronic Health Records and the National Health Information Network: Affordable, Adoptable, and Ready for Prime Time? *Annals of Internal Medicine*, *143*(3), 227–228. doi:10.7326/0003-4819-143-3-200508020-00009 PMID:16061921

Bassey, M. (1999). *Case Study Research in Educational Settings*. Open University Press.

Bates. (2005). Physicians And Ambulatory Electronic Health Records. *Health Aff, 24*, 1180 - 1189.

Bath, P. (2008). Health informatics: Current issues and challenges. *Journal of Information Science*, *34*(4), 501. doi:10.1177/0165551508092267

Baum, D., Spann, M., Füller, J., & Pedit, T. (2013). *Social Media Campaigns for New Product Introductions*. ECIS 2013 Completed Research.

Benatar, S. R. (1993). Reforming the South African health care system. *Physician Executive*, *19*(5), 51–57. PMID:10130951

Benjamin, J. (1988). *The bonds of love: Psychoanalysis, feminism, and the problem of domination*. New York: Random House.

Bergeron, F., & Raymond, L. (1992). Planning of information systems to gain a competitive edge. *Journal of Small Business Management*, *30*(1), 21–26.

Berg, M. (2001). Implementing information systems in health care organizations: Myths and challenges. *International Journal of Medical Informatics*, *64*(2-3), 143–156. doi:10.1016/S1386-5056(01)00200-3 PMID:11734382

Berg, M., & Timmermans, S. (2000). Orders and their others: On the constitution of universalities in medical work. *Configurations*, *8*, 31–61. doi:10.1353/con.2000.0001

Bernstein, M. L., McCreless, T., & Côté, M. J. (2007). Five constants of information technology adoption in healthcare. *Hospital Topics*, *85*(1), 17–25. doi:10.3200/HTPS.85.1.17-26 PMID:17405421

Berry, F. S., Choi, S. O., Goa, W. X., Jang, H., Kwon, M., & Word, J. (2004). Three traditions of network research: What the public management research agenda can learn from other research communities. *Public Administration Review*, *64*(5), 539–552. doi:10.1111/j.1540-6210.2004.00402.x

Bichard, M. (2000). Creativity, leadership and change. *Public Money and Management*, *20*(2), 41–46. doi:10.1111/1467-9302.00210

Bielenia-Grajewska, M. (2009). Actor-Network Theory in Intercultural Communication – Translation through the Prism of Innovation, Technology, Networks and Semiotics. *International Journal of Actor-Network Theory and Technological Innovation*, *1*(4), 53–69. doi:10.4018/jantti.2009062304

Bielenia-Grajewska, M. (2011). Actor-Network-Theory in Medical e-Communication – The Role of Websites in Creating and Maintaining Healthcare Corporate Online Identity. *International Journal of Actor-Network Theory and Technological Innovation*, *3*(1), 39–53. doi:10.4018/jantti.2011010104

Bielski, L. (2008). Eight tech innovations that took banking into the 21st century. *ABA Banking Journal*, *100*(11), 74–86.

Bigum, C. (1998). Solutions in Search of Educational Problems: Speaking for Computers in Schools. *Educational Policy*, *12*(5), 586–601. doi:10.1177/0895904898012005007

Binder, J., Howes, A., & Sutcliffe, A. (2009). *The Problem of Conflicting Social Spheres: Effects of Network Structure on Experienced Tension in Social Network Sites*. Paper presented at CHI 2009. Boston, MA.

Bistričić, A. (2006). Project information system. *Tourism and Hospitality Management*, *12*(2), 213–224.

Bloomfield, B. P., Coombs, R., Cooper, D. J., & Rea, D. (1992). Machines and manoeuvres: Responsibility accounting and construction of hospital information systems. *Accounting Management and Information Technologies, 2,* 197–219. doi:10.1016/0959-8022(92)90009-H

BMG – Bundesministerium fuer Gesundheit. (2009). *Daten des Gesundheitswesens 2009.* Retrieved August 26, 2011, from http://www.bmg.bund.de/uploads/publications/BMG-G-09030-Daten-des-Gesundheitswesens_200907.pdf

Bolter, J. D., & Macintyre, B. (2007). Is It Live or Is It AR? *IEEE Internet Computing.* Retrieved from http://www.spectrum.ieee.org/aug07/5377

Boonstra, A., & Broekhuis, M. (2010). Barriers to the acceptance of electronic medical records by physicians from systematic review to taxonomy and interventions. *BMC Health Services Research, 10*(1), 231. doi:10.1186/1472-6963-10-231 PMID:20691097

Bossen, C. (2007). Test the artefact – Develop the organization: The implementation of an electronic medication plan. *International Journal of Medical Informatics, 76*(1), 13–21. doi:10.1016/j.ijmedinf.2006.01.001 PMID:16455299

Bots, P. W. G. (2007). Analysis of multi-actor policy contexts using perception graphs. In *Proceedings of IEEE/WIC/ACM International Conference on Intelligent Agent Technology* (pp. 160-167). IEEE.

Bots, P. W. G. (2008). *Analyzing actor networks while assuming frame rationality.* Paper presented at the conference on Networks in Political Science (NIPS). Cambridge, MA.

Bots, P. W. G., Van Twist, M. J. W., & Van Duin, J. H. R. (2000). Automatic pattern detection in stakeholder networks. In J.F. Nunamaker & R.H. Sprague (Eds.), *Proceedings 33rd Hawaii International Conference on System Sciences.* Los Alamitos, CA: IEEE Press.

Bots, P. W. G., van Twist, M. J. W., & van Duin, R. (1999). Designing a power tool for policy analysts: Dynamic actor network analysis. In *Proceedings of Thirty–Second Annual Hawaii International Conference on System Sciences.* IEEE.

Boudourides, M. A. (Ed.). (2001). *International Conference Spacing and Timing, Rethinking Globalization & Standardization.* Palermo, Italy: Academic Press.

Boulos, M. G., & Wheeler, S. (2007). The emerging Web 2.0 social software: An enabling suite of sociable technologies in health and health care education. *Health Information and Libraries Journal, 24,* 2–23. doi:10.1111/j.1471-1842.2007.00701.x PMID:17331140

Bowker, G. C., & Star, S. L. (2000). *Sorting things out – Classification and its consequences.* Cambridge, MA: MIT Press.

Boyatzis, R. (1998). *Transforming Qualitative Information Thematic analysis And Code Development.* Sage Publications.

Boyd, D., & Heer, J. (2006). Profiles as Conversation: Networked Identity Performance on Friendster. In *Proceedings of Thirty-Ninth Hawai'i International Conference on System Sciences,* (pp. 59–69). Los Alamitos, CA: IEEE Press.

Boyd, D. (2007). Why Youth (Heart) Social Network Sites: The Role of Networked Publics in Teenage Social Life. In D. Buckingham (Ed.), *MacArthur Foundation Series on Digital Learning – Youth, Identity, and Digital Media Volume.* Cambridge, MA: MIT Press.

Boyd, D., & Ellison, N. B. (2007). Social Network Sites: Definition, History, and Scholarship. *Journal of Computer-Mediated Communication, 13*(1), 210–230. doi:10.1111/j.1083-6101.2007.00393.x

Braa, J., Monteiro, E., & Sahay, S. (2004). Networks of action: Sustainable health information systems across developing countries. *Management Information Systems Quarterly, 28*(3), 337–362.

Brandtzæg, P.B., & Heim, J. (2011). A typology of social networking sites users. *Int. J. Web Based Communities, 7*(1).

Brennan, S., & Taylor-Butts, A. (2008). *Sexual assault in Canada (Catalogue 85F0033M - No. 19).* Ottawa, Canada: Canadian Centre for Justice Statistics Profile Series, Statistics Canada.

Brey, P. (1997). Philosophy of technology meets social constructivism. *Society of Philosophy & Technology, 2*, 3–4.

Britt, H., Miller, G. C., Charles, J., Pan, Y., Valenti, L., Henderson, J., & Knox, S. (2007). *General Practice Activity in Australia 2005-06 (Cat. no. GEP 16)*. Canberra, Australia: AIHW.

Brooks, A. (2007). Feminist standpoint epistemology: Building knowledge and empowerment through women's lived experience. In S. Nagy Hesse-Biber, & P. Leavy (Eds.), *Feminist research practice: A primer* (pp. 53–82). Thousand Oaks, CA: Sage. doi:10.4135/9781412984270. n3

Brynjolfsson, E. (1993). The productivity paradox of information technology. *Communications of the ACM, 36*, 66–77. doi:10.1145/163298.163309

Bucks, B. K., Kennickell, A. B., Mach, T. L., & Moore, K. B. (2009). *Changes in U.S. Family Finances from 2004 to 2007: Evidence from the Survey of Consumer Finances.* US Federal Reserve.

Bulgren, J. (1998). Effectiveness of a concept teaching routine in enhancing the performance of LD students in secondary-level mainstream classes. *Learning Disability Quarterly, 11*.

Bumiller, K. (2008). *In an abusive state: How neoliberalism appropriated the feminist movement against sexual violence.* Durham, NC: Duke University Press.

Burton-Jones, A., & Gallivan, M. (2007). Toward a deeper understanding of system usage in organizations: A multilevel perspective. *Management Information Systems Quarterly, 31*(4), 657–679.

Burton-Jones, A., & Straub, D. W. (2006). Reconceptualizing System Usage: An Approach and Empirical Test. *Information Systems Research, 17*, 228. doi:10.1287/ isre.1060.0096

Busch, K. V. (1997). Applying Actor Network Theory to Curricula Change in Medical Schools: Policy Strategies for Initiating and Sustaining Change. In *Proceedings of Midwest Research-to-Practice Conference in Adult, Continuing and Community Education Conference.* Michigan State University.

Butler, J. (2004). *Undoing Gender.* New York: Routledge.

Callon, M. (1986). Some elements of the sociology of translation: Domestication of the scallops and the fisherman of St Brieuc Bay. In Power, Action and Belief: A New Sociology of Knowledge (pp. 196-223). Routledge.

Callon, M. (1986). Some elements of a sociology of translation: Domestication of the scallops and the fisherman of St. Brieue Bay. In J. Law (Ed.), *Power, action, and belief: A new sociology of knowledge?* (pp. 196–223). London: Routledge.

Callon, M. (1986a). The sociology of an actor-network: The case of the electric vehicle. In M. Callon, J. Law, & A. Rips (Eds.), *Mapping the dynamics of science and technology* (pp. 19–34). Basingstoke, UK: Macmillan.

Callon, M. (1991). Techno-economic Networks and Irreversibility. In J. Law (Ed.), *A Sociology of Monsters? Essays on Power, Technology and Domination.* London: Routledge.

Callon, M. (1999). Actor-Network Theory - The Market Test. In *Actor Network Theory and After.* Oxford, UK: Blackwell Publishers.

Callon, M., Courtial, J. P., Turner, W. A., & Bauin, S. (1983). From translations to problematic networks: An introduction to co-word analysis. *Social Sciences Information. Information Sur les Sciences Sociales, 22*(2), 191–235. doi:10.1177/053901883022002003

Callon, M., & Latour, B. (1981). Unscrewing the Big Leviathan: How actors macrostructure reality and how sociologists help them to do so. In K. D. Knorr-Cetina, & A. V. Cicoure (Eds.), *Advances in Social Theory and Methodology: Toward an Integration of Micro- and Macro-Sociologies.* Routledge and Kegan Paul.

Callon, M., & Latour, B. (1992). Don't throw the baby out with the Bath School! A reply to Collins and Yearley. In A. Pickering (Ed.), *Science as Practice and Culture.* Chicago: Chicago University Press.

Callon, M., & Law, J. (1982). Of interests and their transformation: Enrolment and counter-enrolment. *Social Studies of Science, 12*(4), 615–625. doi:10.1177/030631282012004006

Callon, M., & Law, J. (1995). Agency and the Hybrid Collectif. *The South Atlantic Quarterly, 94,* 481–507.

Callon, M., & Law, J. (2005). On Qualculation, Agency and Otherness. *Society and Space, 23,* 717–733.

Callon, M., & Muniesa, F. (2005). Economic markets as calculative collective devices. *Organization Studies, 26*(8), 1229–1250. doi:10.1177/0170840605056393

Cameron, T. A. (2005). Updating subjective risks in the presence of conflicting information: An application to climate change. *Journal of Risk and Uncertainty, 30*(1), 63–97. doi:10.1007/s11166-005-5833-8

Cannon, M. F., & Tanner, M. D. (2005). *Healthy Competition: What's Holding Back Health Care and How to Free IT.* Cato Institute.

Carberry, P., Hammer, G. L., Meinke, H., & Bange, M. (2000). The potential value of seasonal climate forecasting in managing cropping systems. In G. L. Hammer, N. Nicholls, & C. Mitchell (Eds.), *Applications of seasonal climate forecasting in agricultural and natural ecosystems: The Australian experience.* Dordrecht, The Netherlands: Kluwer Academic Publishers. doi:10.1007/978-94-015-9351-9_12

Car, J., Anandan, C., Black, A., Cresswell, K., Pagliari, C., & McKinstry, B. et al. (2008). *The Impact of eHealth on the Quality & Safety of Healthcare: A Systematic Overview and Synthesis of the Literature.* NHS Connecting for Health Evaluation Programme.

Carmen Lemos, M., Finan, T. J., Fox, R. W., Nelson, D. R., & Tucker, J. (2002). The use of seasonal climate forecasting in policy making: Lessons from North East Brazil. *Climatic Change, 55,* 479–507. doi:10.1023/A:1020785826029

Carr, N. (2003). IT Doesn't Matter. *Harvard Business Review, 81*(5), 41–49. PMID:12747161

Carr, N. (2004). *Does IT matter? Information technology and the corrosion of competitive advantage.* Harvard Business School Press.

Carroll, N. (2012). *Service science: An empirical study on the socio-technical dynamics of public sector service network innovation.* (PhD Thesis). University of Limerick. Retrieved from https://www.academia.edu/2259268/Service_Science_An_Empirical_Study_on_the_Socio-Technical_Dynamics_of_Public_Sector_Service_Network_Innovation

Carroll, N., Whelan, E., & Richardson, I. (2010). Applying Social Network Analysis to Discover Service Innovation within Agile Service Networks. *Journal of Service Science, 2*(4), 225–244. doi:10.1287/serv.2.4.225

Carroll, N., Whelan, E., & Richardson, I. (2012). Service Science – An Actor Network Theory Approach. *International Journal of Actor-Network Theory and Technological Innovation, 4*(3), 51–69. doi:10.4018/jantti.2012070105

Cash, J. (2008). Technology can make or break the hospital-physician relationship. *Healthcare Financial Management, 62*(12), 104–109. PMID:19069330

Casper, M., & Clarke, A. (1998). Making the pap smear into the 'right tool' for the job: Cervical cancer screening in the USA, circa 1940-95. *Social Studies of Science, 28*(2), 255–290. doi:10.1177/030631298028002003 PMID:11620085

Catwell, L., & Sheikh, A. (2009). Evaluating eHealth Interventions: The Need for Continuous Systemic Evaluation. *PLoS Medicine, 6*(8), e1000126. doi:10.1371/journal.pmed.1000126 PMID:19688038

Cecez-Kecmanovic, D., & Nagm, F. (2009). Have you Taken your Guys on the Journey? An ANT Account of IS Project Evaluation. *International Journal of Actor-Network Theory and Technological Innovation, 1*(1), 1–22. doi:10.4018/jantti.2009010101

Chan, L., & Dally, K. (2000). Review of the Literature. In *Mapping the Territory. Primary Students with Learning Difficulties: Literature and Numeracy.* Canberra, Australia: Department of Education, Training and Youth Affairs.

Chen, D. Q., Mocker, M., Preston, D. S., & Teubner, A. (2010). Information systems strategy: Reconceptualization, measurement, and implications. *Management Information Systems Quarterly, 34*(2), 233–259.

Chiu, C.-M., Hsu, M.-H., & Wang, E. T. D. (2006). Understanding knowledge sharing in virtual communities: An integration of social capital and social cognitive theories. *Decision Support Systems, 42*(3), 1872–1888. doi:10.1016/j.dss.2006.04.001

Cho, S., Mathiassen, L., & Nilsson, A. (2008). Contextual dynamics during health information systems implementation: An event-based actor-network approach. *European Journal of Information Systems, 17*, 614–630. doi:10.1057/ejis.2008.49

Chun, H., Kwak, H., Eom, Y. H., Ahn, Y. Y., Moon, S., & Jeong, H. (2008). *Comparison of Online Social Relations in Terms of Volume vs. Interaction: A Case Study of Cyworld.* Paper presented at IMC'08. Vouliagmeni, Greece.

Ciborra, C. U. (2004). *Digital Technologies and the Duality of Risk.* London: CARR, LSE.

Ciborra, C. U. (2002). *The Labyrinths of Information: Challenging the Wisdom of Systems.* Oxford University Press.

Ciborra, C. U., & Hanseth, O. (1998). From Tool to Gestell Agendas for Managing the Information Infrastructure. *Information Technology & People, 11*(4), 305–327. doi:10.1108/09593849810246129

Ciborra, C. U., & Lanzara, G. F. (1994). Formative Contexts and Information Technology: Understanding the dynamics of innovation in organisations. *Accting., Mgmt., &. Info. Tech., 4*, 61–86.

Clancy, T. (2004). *IT development projects.* Retrieved from www.standishgroup.com/msie.htm

Clarke, A., & Montini, T. (1993). The many faces of RU486: Tales of situated knowledges and technological contestations. *Science, Technology & Human Values, 18*(1), 42–78. doi:10.1177/016224399301800104 PMID:11652075

Cohen, N. (2009, March 27). When stars Twitter, a ghost may be lurking. *The New York Times.*

Cohn, K., Berman, J., Chaiken, B., Green, D., Green, M., Morrison, D., & Scherger, J. (2009). Engaging physicians to adopt healthcare information technology. *Journal of Healthcare Management, 54*(5), 291. PMID:19831114

Coiera, E. (2004). Four rules for the reinvention of health care. *BMJ (Clinical Research Ed.), 328*(7449), 1197–1199. doi:10.1136/bmj.328.7449.1197 PMID:15142933

Commonwealth Fund. (2010). *International Profiles of Health Care Systems.* Retrieved August 26, 2011, from http://www.commonwealthfund.org/~/media/Files/%20Publications/Fund%20Report/2010/Jun/1417_Squires_Intl_Profiles_622.pdf

Commonwealth of Australia. (1998). Literacy for All: The Challenge for Australian Schools. Commonwealth Literacy Policies for Australian Schools. *Australian Schooling Monograph Series No. 1/1998.* Retrieved from http://www.dest.gov.au/archive/schools/literacy&numeracy/publications/lit4all.htm

Conry-Murray. (2009, March 21). *Can enterprise social networking pay off?* Retrieved from http://internetrevolution.com/document.asp?Docid=173854

Constant, D., Sproull, L., & Kiesler, S. (1996). The kindness of strangers: The usefulness of electronic weak ties for technical advice. *Organization Science, 7*, 119–135. doi:10.1287/orsc.7.2.119

Cooper, D. R., & Schindler, P. S. (2003). *Business Research Methods* (8th ed.). McGraw-Hill.

Cordella, A., & Shaikh, M. (2006). *From Epistemology to Ontology: Challenging the Constructed Truth of ANT* (Working Paper). London: Department of Information Systems, London School of Economics.

Cordella, A. (2010). Information Infrastructure: an Actor-Network Perspective. *International Journal of Actor-Network Theory and Technological Innovation, 2*(1), 27–53. doi:10.4018/jantti.2010071602

Cordella, A., & Iannacci, F. (2010). Information systems in the public sector: The e-Government enactment framework. *The Journal of Strategic Information Systems, 19*(1), 52–66. doi:10.1016/j.jsis.2010.01.001

Cornford, T., & Smithson, S. (1996). *Project Research in Information Systems: A Student's Guide*. Macmillan.

Costello, G. I., & Tuchen, J. H. (1998). A comparative study of business to consumer electronic healthcare within the Australian insurance sector. *Journal of Information Technology, 13*, 153–167. doi:10.1080/026839698344800

Cresswell, K. M., Worth, A., & Sheikh, A. (2010). Actor-Network Theory and its role in understanding the implementation of information technology developments in healthcare. *BMC Medical Informatics and Decision Making, 10*, 67. doi:10.1186/1472-6947-10-67 PMID:21040575

Cresswell, K., Worth, A., & Sheikh, A. (2011). Implementing and adopting electronic health record systems: How actor-network theory can support evaluation. *Clinical Governance: An International Journal, 16*(4), 320–336. doi:10.1108/14777271111175369

Crowston, K., & Williams, M. (2000). Reproduced and emergent genres of communication on the World-Wide Web. *The Information Society, 16*(3), 201–215. doi:10.1080/019722400050133652

Culnan, M., & Armstrong, P. K. (1997). *Information Privacy Concerns, Procedural Fairness and Impersonal Trust: An Empirical Investigation*. Retrieved from http://citeseer.ist.psu.edu/viewdoc/summary?doi=10.1.1.40.5243

Cummings, J., Massey, A. P., & Ramesh, V. (2009). Web 2.0 Proclivity: Understanding How Personal Use Influences Organisational Adoption.[GIGDOC.]. *Proceedings of GIGDOC, 09*, 257–263.

Cussins, C. (1998). Ontological Choreography Agency for Women Patients in an Infertility Clinic. In M. Berg, & A. Mol (Eds.), *Differences in Medicine: Unravelling Practices, Techniques and Bodies*. Durham, NC: Duke University Press.

DA. (2007). *Diabetes Facts*. New South Wales, Australia: Diabetes Australia.

Darke, P., Shanks, G., & Broadbent, M. (1998). Successfully Completing Case Study Research: Combining Rigor, Relevance and Pragmatism. *Information Systems Journal, 8*, 273–289. doi:10.1046/j.1365-2575.1998.00040.x

Darking, M. L., & Whitley, E. A. (2007). Towards an Understanding of Floss: Infrastructures, Materiality and the Digital Business Ecosystem. *Science Studies, 20*(2), 13–33.

Davey, W., & Tatnall, A. (2012). Using ANT to Guide Technological Adoption: the Case of School Management Software. *International Journal of Actor-Network Theory and Technological Innovation, 4*(4), 47–61. doi:10.4018/jantti.2012100103

Davey, W., & Tatnall, A. (2013). Social Technologies in Education: An Actor-Network Analysis. In *Open and Social Technologies for Networked Learning*. Heidelberg, Germany: Springer. doi:10.1007/978-3-642-37285-8_17

David, L., & Halbert, L. (2013). Finance Capital, Actor-Network Theory and the Struggle Over Calculative Agencies in the Business Property Markets of Mexico City Metropolitan Region. *Regional Studies*, 1–14.

Davidson, E., & Heslinga, D. (2006). Bridging the IT - Adoption Gap for Small Physician Practices: An Action Research Study on Electronic Health Records. *Information Systems Management, 24*(1), 15. doi:10.1080/10580530601036786

Davis, P. (2010). *Development of a Framework for the Assessment of the Role and Impact of Technology on the Public Procurement Process: An Irish Health Sector Study*. (Doctoral Thesis). Dublin Institute of Technology, Dublin, Ireland. Retrieved on 28/10/2011 from website: http://arrow.dit.ie/cgi/viewcontent.cgi?article=1026&context=engdoc&sei-redir=1#search=%22ireland%20must%20examine%20technology%20public%20sector%22

Davis, A. (1983). *Women, race & class*. New York: Vintage Books.

Davis, F. (1986). *A Technology Acceptance Model for Empirically Testing New End-User Information Systems: Theory and Results*. Boston, MA: MIT.

Davis, F. (1989). Perceived Usefulness, Perceived Ease of Use, and User Acceptance of Information Technology. *Management Information Systems Quarterly, 13*(3), 319–340. doi:10.2307/249008

de Beauvoir, S. (1957). *The second sex*. New York: Vintage Books.

de Laet, M., & Mol, A. (2000). The Zimbabwe bush pump: Mechanics of a fluid technology. *Social Studies of Science, 30,* 225–263. doi:10.1177/030631200030002002

Deering, P., Tatnall, A., & Burgess, S. (2010). Adoption of ICT in Rural Medical General Practices in Australia - An Actor-Network Study. *International Journal of Actor-Network Theory and Technological Innovation, 2*(1), 54–69. doi:10.4018/jantti.2010071603

DEETYA. (1999). *Students with Learning Difficulties. 2010.* Retrieved from http://www.dest.gov.au/archive/sectors/school_education/publications.../swd_rtf.htm

DeLone, W. H., & McLean, E. R. (2003). The DeLone and McLean model of information systems success: A ten-year update. *Journal of Management Information Systems, 19,* 9.

Demirkan, H., Kauffman, R. J., Vayghan, J. A., Fill, H. G., Karagiannis, D., & Maglio, P. P. (2008). Service-oriented technology and management: Perspectives on research and practice for the coming decade. *Perspectives on the Technology of Service Operations. Electronic Commerce Research and Applications,* (4): 356–376. doi:10.1016/j.elerap.2008.07.002

Department of Education and Early Childhood Development – Victoria. (2013). *S245-2013 DEECD and NEC sign agreement to facilitate continued provision of the Ultranet.* Author.

Department of Education and Early Childhood Development. (2010). *Ultranet - Students@Centre Trial.* Retrieved November 2010, from http://www.education.vic.gov.au/about/directions/ultranet/trial.htm

Department of Education and Early Childhood Development. (2011a). *CASES21.* Retrieved January 2012, from http://www.education.vic.gov.au/management/ictsupportservices/cases21/functionality.htm

Department of Education and Early Childhood Development. (2011b). *Ultranet.* Retrieved January 2012, from http://www.education.vic.gov.au/about/directions/ultranet/default.htm

DePhillips, H. A. (2007). Initiatives and Barriers to Adopting Health Information Technology: A US Perspective. *Disease Management & Health Outcomes, 15*(1). doi:10.2165/00115677-200715010-00001

DesRoches, C. M., Campbell, E. G., Rao, S. R., Donelan, K., Ferris, T. G., & Jha, A. et al. (2008). Electronic health records in ambulatory care–A national survey of physicians. *The New England Journal of Medicine, 359*(1), 50–60. doi:10.1056/NEJMsa0802005 PMID:18565855

Detmer, D., Bloomrosen, M., Raymond, B., & Tang, P. (2008). Integrated Personal Health Records: Transformative Tools for Consumer-Centric Care. *BMC Medical Informatics and Decision Making, 8*(1), 45. doi:10.1186/1472-6947-8-45 PMID:18837999

Diabcost Australia. (2002). *Assessing the Burden of Type 2 Diabetes in Australia.* Adelaide, Australia: DiabCost Australia.

Diabetes Australia. (2008). *Diabetes in Australia.* Author.

Die gesetzlichen Krankenkassen. (2007). *Das elektronische Rezept.* Retrieved August 26, 2011, from http://www.g-k-v.de/gkv/fileadmin/user_upload/Projekte/Telematik_im_Gesundheitswesen/2.1.7_das_elektronische_rezept.pdf

DiMicco, J. M., Geyer, W., Millen, D. R., & Dugan, C. (2009). People Sensemaking and Relationship Building on an Enterprise Social Network Site. In *Proceedings of Hawaii International Conference on System Sciences,* (pp. 1-10). IEEE.

Doe, J. (2003). *The story of Jane Doe.* Toronto, Canada: Vintage Canada.

Doe, J. (2012). Who benefits from the sexual assault evidence kit? In E. Sheehy (Ed.), *Sexual assault law in Canada: Law, legal practice and women's activism* (pp. 355–388). Ottawa, Canada: University of Ottawa Press.

DoHA. (2010). *Building a 21st Century - Primary Health Care System - Australia's First National Primary Health Care Strategy*. Australian Government Department of Health and Ageing. Retrieved from http://www.yourhealth. gov.au/internet/yourhealth/publishing.nsf/Content/Building-a-21st-Century-Primary-Health-Care-System-TOC

Dolwick, J. S. (2009). The social and beyond: Introducing actor-network theory. *Journal of Maritime Archaeology, 4*(1), 21–49. doi:10.1007/s11457-009-9044-3

Doolin, B., & Lowe, A. (2002). To reveal is to critique: Actor-network theory and critical information systems research. *Journal of Information Technology*, 69–78. doi:10.1080/02683960210145986

Du Mont, J., & Parnis, D. (2001). Constructing bodily evidence through sexual assault evidence kits. *Griffith Law Review, 10*, 63–76.

Du Mont, J., White, D., & McGregor, M. (2009). Investigating the medical forensic examination from the perspectives of sexually assaulted women. *Social Science & Medicine, 68*, 774–780. doi:10.1016/j.socscimed.2008.11.010 PMID:19095341

Dunning-Lewis, P., & Townson, C. (2004). *Using actor network theory ideas in information systems research: A case study of action research*. Lancaster University.

Dwyer, C., Hiltz, S. R., & Passerini, K. (2007). Trust and privacy concern within social networking sites: A comparison of Facebook and MySpace. In *Proceedings of the Thirteenth Americas Conference on Information Systems*. Keystone, CO: Academic Press.

Eisenhardt, K. M. (1989). Building theories from case study research. *Academy of Management Journal, 14*(4), 532–550.

Ekundayo, S., & Diaz Andrade, A. (2011). Mediated Action and Network of Actors: From Ladders, Stairs and Lifts to Escalators (and Travelators). *International Journal of Actor-Network Theory and Technological Innovation, 3*(3), 21–34. doi:10.4018/jantti.2011070102

Elbanna, A. (2009). Actor network theory and IS research. In Y. K. Dwivedi, B. Lal, M. D. Williams, S. L. Schneberger, & M. Wade (Eds.), *Handbook of research on contemporary theoretical models in information systems*. Hershey, PA: IGI Global. doi:10.4018/978-1-60566-659-4.ch023

Ellison, N. B., & Steinfield, C. (2007). The Benefits of Facebook Friends: Social Capital and College Students' Use of Online Social Network Sites. *Journal of Computer-Mediated Communication, 12*(4), 1143–1168. doi:10.1111/j.1083-6101.2007.00367.x

Elrod, J., & Androwich, I. M. (2009). Applying human factors analysis to the design of the electronic health record. *Studies in Health Technology and Informatics, 146*, 132–136. PMID:19592822

Emery, F. E., & Trist, E. L. (1960). Socio-Technical Systems. In *Management sciences, models and technique*. London: Pergamon.

Enders, A., Hungenberg, H., Denker, H. P., & Mauch, S. (2008). The long tail of social networking: Revenue models of social networking sites. *European Management Journal, 26*, 199–211. doi:10.1016/j.emj.2008.02.002

Enserink, B., Hermans, L., Kwakkel, J., Thissen, W., Koppenjan, J., & Bots, P. (2010). *Policy Analysis of Multi-Actor Systems*. The Hague, The Netherlands: Lemma.

European Commission. (2011). *The European Health Insurance Card*. Retrieved August 26, 2011, from http://ec.europa.eu/social/main.jsp?catId=559

Everitt-Deering, P. (2008). *The adoption of information and communication technologies by rural general practitioners: A socio-technical analysis*. (Ph.D dissertation). Faculty of Business and Law, Victoria University, Melbourne, Australia.

EVM. (2005). *Practice Standard for Earned Value Management* (2nd ed.). PMI.

Eysenbach, G. (2001). What is e-health? *Journal of Medical Internet Research, 3*(2). doi:10.2196/jmir.3.2.e20

Fang, L., & LeFevre, K. (2010). *Privacy Wizards for Social Networking Sites*. Paper presented at WWW 2010. Raleigh, NC.

Faraj, S., Kwon, D., & Watts, S. (2004). Contested artifact: technology sensemaking, actor networks, and the shaping of the Web browser. *Information Technology & People, 17*(2), 186–209. doi:10.1108/09593840410542501

Feagin, J., Orum, A., & Sjoberg, G. (Eds.). (1991). *A case for case study*. Chapel Hill, NC: University of North Carolina Press.

Feldman, M. S., Khademian, A. M., Ingram, H., & Schneider, A. S. (2006). Ways of knowing and inclusive management practices. *Public Administration Review, 66*(1), 88–89.

Feller, J., Finnegan, P., & Nilsson, O. (2008). We have everything to win: Collaboration and open innovation in public administration. In *Proceedings of ICIS*. Retrieved on 19/07/2011 from http://aisel.aisnet.org/icis2008/214

Fenwick, T. (2006). Toward enriched conceptions of work learning: Participation, expansion, and translation among individuals with/in activity. *Human Resource Development Review, 5*(3), 285–302. doi:10.1177/1534484306290105

Ferguson, R. W. (2004). A Project Risk Metric. *CrossTalk: The J. Defense Software Eng., 17*(4), 12–15.

Fisher, J. S. (1999). *Electronic Records in Clinical Practice*. Cinical Diabetes.

Fitzsimmons, E. G., & Rubin, B. M. (2008). *Social networking sites viewed by admission officers: Survey shows some use of Facebook, MySpace as another aspect to college application*. Retrieved from http://www.chicagotribune.com/news/local/chi-facebook

Fitzsimmons, J. A., & Fitzsimmons, M. J. (2004). *Service Management – Operations, Strategy, and Information Technology* (4th ed.). McGraw-Hill.

Fogel, J., & Nehmad, E. (2009). Internet social network communities: Risk taking, trust, and privacy concerns. *Computers in Human Behavior, 25*, 153–160. doi:10.1016/j.chb.2008.08.006

Ford, R. M., & Coulston. (2008). *Design for Electrical and Computer Engineers: Theory, Concepts, and Practice*. McGraw-Hill Science Engineering.

Frame, J., Watson, J., & Thomson, K. (2008). Deploying a culture change programme management approach in support of information and communication technology developments in Greater Glasgow NHS Board. *Health Informatics Journal, 14*(2), 125–139. doi:10.1177/1081180X08089320 PMID:18477599

Fraser, M., & Dutta, S. (2008). *Throwing Sheep in the Boardroom*. London: John Wiley & Sons.

Friedman, T. L. (2006). *The world is flat*. New York: Penguin Books.

Frost and Sullivan Country Industry Forecast – European Union Healthcare Industry. (2004, May 11). Retrieved from http://www.news-medical.net/print_article.asp?id=1405

Fullan, M. G., & Stiegelbauer, S. (1991). *The New Meaning of Educational Change* (2nd ed.). New York: Teachers College Press.

Furuta, R., & Marshall, C. C. (1996). Genre as Reflection of Technology in the World Wide Web. In *Hypermedia Design*. London: Springer-Verlag. doi:10.1007/978-1-4471-3082-6_19

Gad, C., & Jensen, C. B. (2010). On the Consequences of Post-ANT. *Science, Technology & Human Values, 35*(55). PMID:20526462

Gao, P. (2005). Using actor-network theory to analyse strategy formulation. *Information Systems Journal, 15*, 255–275. doi:10.1111/j.1365-2575.2005.00197.x

Geisler, E., & Wickramasinghe, N. (2009). *The Role and Use of Wireless Technology in the Management and Monitoring of Chronic Disease IBM Report*. Retrieved from www.businessofgovernment.org

gematik – Gesellschaft für Telematikanwendungen der Gesundheitskarte mbH. (2011a). *Versichertenstammdaten*. Retrieved August 26, 2011, from http://www.gematik.de/cms/de/egk_2/anwendungen/verfuegbare_anwendungen/versichertendaten/versichertendaten_1.jsp

gematik – Gesellschaft für Telematikanwendungen der Gesundheitskarte mbH. (2011b). *Anwendungen der eGK*. Retrieved August 26, 2011, from http://www.gematik.de/cms/de/egk_2/anwendungen/anwendungen_1.jsp

Gerlach, N. (2004). *The genetic imaginary: DNA in the Canadian criminal justice system*. Toronto, Canada: University of Toronto Press.

Germonprez, M., Hovorka, D., & Callopy, F. (2007). A Theory of Tailorable Technology Design. *Journal of the Association for Information Systems, 8*(6), 351–367.

Ghani, M., Bali, R., Naguib, R., Marshall, I., & Wickramasinghe, N. (2010). Critical issues for implementing a Lifetime Health Record in the Malaysian Public Health System.[IJHTM]. *International Journal of Healthcare Technology and Management, 11*(1/2), 113–130. doi:10.1504/IJHTM.2010.033279

Giddens, A. (1984). *The constitution of society: Outline of the theory of structuration*. Cambridge, MA: Polity Press.

Gillard, J. (2010). *My School website launched*. Retrieved May 2010, from http://www.deewr.gov.au/ministers/gillard/media/releases/pages/article_100128_102905.aspx

Giordano, R., Mysiak, J., Raziyeh, F., & Vurro, M. (2007). An integration between cognitive map and casual loop diagram for knowledge structuring in river basin management. In *Proceedings of International Conference on Adaptive & Integrated Water Management (CAIWA), Coping with Complexity and Uncertainty*. Basel, Switzerland: CAIWA.

GKV-Spitzenverband. (2011). *Alle gesetzlichen Krankenkassen*. Retrieved August 26, 2011, from http://www.gkv-spitzenverband.de/ITSGKrankenkassenListe.gkvnet

Goff, L. (1992). Patchwork of Laws Slows EC data Flow. *Computerworld, 26*(15), 80.

Goh, K., Heng, C., & Lin, Z. (2013). Social Media Brand Community and Consumer Behavior: Quantifying the Relative Impact of User- and Marketer-Generated Content. *Information Systems Research, 24*(1), 88–107. doi:10.1287/isre.1120.0469

Goldberg, S., et al. (2002a). Building the Evidence For A Standardized Mobile Internet (wireless) Environment In Ontario, Canada, January Update, internal INET documentation. Academic Press.

Goldberg, S. et al. (2002b). *HTA Presentational Selection and Aggregation Component Summary, internal documentation*. Academic Press.

Goldberg, S. et al. (2002c). *Wireless POC Device Component Summary, internal INET documentation*. Academic Press.

Goldberg, S. et al. (2002d). *HTA Presentation Rendering Component Summary, internal INET documentation*. Academic Press.

Goldberg, S. et al. (2002e). *HTA Quality Assurance Component Summary, internal INET documentation*. Academic Press.

Gonçalves, A., & Figueiredo. (2009). Organizing Competences: Actor-Network Theory in Virtual Settings. *International Journal of Networking and Virtual Organisations, 6*(1), 22–35. doi:10.1504/IJNVO.2009.022481

Goncalves, F. A., & Figueiredo, J. (2010). How to recognize an Immutable mobile when you find one: Translations on Innovation and design. *International Journal of Actor-Network Theory and Technological Innovation, 2*(2), 39–53. doi:10.4018/jantti.2010040103

Goodhue, D. L., & Thompson, R. L. (1995, June). Task-technology fit and individual performance. *Management Information Systems Quarterly*, 213–227. doi:10.2307/249689

Goodwin, P., Fildes, R., Lee, W. Y., Nikolopoulos, K., & Lawrence, M. (2005). *Unearthing some causes of BPR failure: An actor-network theory perspective*. Bath University. Retrieved from http://www.bath.ac.uk/management/research/pdf/2007-14.pdf

Goswami, D., & Ragahavendran, S. (2009). Mobile-banking: can elephants and hippos tango? *The Journal of Business Strategy, 30*(1), 14–20. doi:10.1108/02756660910926920

Greenhalgh, T., & Stones, R. (2010). Theorising big IT programmes in healthcare: Strong structuration theory meets actor-network theory. *Social Science & Medicine, 70*(9), 1285–1294. doi:10.1016/j.socscimed.2009.12.034 PMID:20185218

Gregor, S., & Jones, D. (2007). The Anatomy of a Design Theory. *Journal of the Association for Information Systems, 8*(5), 312–335.

Gregor, S., & Juhani, I. (2007). *Designing for mutability in information systems artifacts*. Canberra, Australia: ANU Press.

Guide, P. M. B. O. K. (2008). *A Guide to the Project Management Body of Knowledge* (4th ed.). PMI.

Gupta, U. (1992). Global Networks: Promises and Challenges. *Information Systems Management, 9*(4), 28–32. doi:10.1080/10580539208906896

Gustad, H. (2006). *Implications on system integration and standardisation within complex and heterogeneous organisational domains: Difficulties and critical success factors in open industry standards development*. (Master of Science thesis). Norwegian University of Science and Technology, Department of Computer and Information Science.

GVG – Gesellschaft für Versicherungswissenschaft und – gestaltung. (2004). *Managementpaper Elektronische Patientenakte*. Retrieved August 26, 2011, from http://ehealth.gvg-koeln.de//cms/medium/676/MP_ePa_050124.pdf

Haefliger, S., Monteiro, E., Foray, D., & Von Krogh, G. (2011). Social Software and Strategy. *Long Range Planning, 44*(5–6), 297–316. doi:10.1016/j.lrp.2011.08.001

Haghighat, M. (2008). *Determining the best date of cultivating rain fed wheat considering climatic parameters and soil characteristics in appropriate places for rain-fed agriculture in Fars Province*. (M.Sc. Thesis). Islamic Azad University, Science and Research branch.

Halamka, J. D., Mandl, K. D., & Tang, P. C. (2007). Early experiences with personal health records. *Journal of the American Medical Informatics Association, 15*(1), 1–7. doi:10.1197/jamia.M2562 PMID:17947615

Hall, E. (2005). The 'geneticisation' of heart disease: A network analysis of the production of new genetic knowledge. *Social Science & Medicine, 60*(12), 2673–2683. doi:10.1016/j.socscimed.2004.11.024 PMID:15820579

Hannemeyr, G. (2003). The Internet as hyperbole: A critical examination of adoption rates. *The Information Society, 19*, 111–121. doi:10.1080/01972240309459

Hansen, J. W., & Sivakumar, M. V. K. (2006). Advances in applying climate prediction to agriculture. *Climate Research, 33*, 1–2. doi:10.3354/cr033001

Hanseth, O., & Braa. (2000). Who's in Control: Designers, Managers, or Technology? Infrastructure at North Hydro. In *From Control to Drift, The Dynamics of Corporate Information Infrastructures*. Oxford University Press.

Hanseth, O., & Monteiro, E. (1998). *Understanding information infrastructure*. Retrieved on 19th July 2011 from http://heim.ifi.uio.no/oleha/Publications/bok.pdf

Hanseth, O. (2007). Integration-complexity-risk: The making of information systems out-of-control. In *Risk, Complexity and ICT*. Oslo: Edward Elgar. doi:10.4337/9781847207005.00005

Hanseth, O., & Aanestad, M. (2004). Actor-network theory and information systems: What's so special? *Information Technology & People, 17*(2), 116–123. doi:10.1108/09593840410542466

Hanseth, O., & Monteiro, E. (1996). Developing Information Infrastructure: The tension between standardization and flexibility. *Science, Technology & Human Values, 21*(4), 407–426. doi:10.1177/016224399602100402

Hanseth, O., & Monteiro, E. (1997). Inscribing behaviour in information infrastructure standards. *Accounting. Management and Information Technology, 7*(4), 183–211. doi:10.1016/S0959-8022(97)00008-8

Haraway, D. (1991). *Simians, cyborgs, and women: The reinvention of nature*. New York: Routledge.

Harding, S. (1991a). *Whose science? Whose knowledge: Thinking from women's lives*. New York: Cornell University.

Harding, S. (1991b). Rethinking standpoint epistemologies: What is strong objectivity? In L. Alcoff, & E. Potter (Eds.), *Feminist Epistemologies* (pp. 49–82). New York: Routledge.

Harding, S. (2008). *Sciences from below: Feminism, postcolonialities, and modernities*. Durham, NC: Duke University Press. doi:10.1215/9780822381181

Harper, F. M., Moy, D., & Konstan, J. A. (2009). *Facts or Friends? Distinguishing Informational and Conversational Questions in Social Q&A Sites*. Paper presented at CHI 2009. Boston, MA.

Harrison, M., & Williams, J. B. (2007). Communicating seasonal forecasts. In A. Troccoli et al. (Eds.), *Seasonal Climate: Forecasting and Managing Risk*. Springer.

Hartman, M. et al. (2009). National health spending in 2007: Slower drug spending contributes to lowest rate of overall growth since 1998. *Health Affairs*, 28(1), 246. doi:10.1377/hlthaff.28.1.246 PMID:19124877

Hartsock, N. (2004). The feminist standpoint: Developing a ground for a specifically feminist historical materialism. In S. Harding (Ed.), *The feminist standpoint reader: Intellectual and political controversies* (pp. 283–310). New York: Routledge. doi:10.1007/0-306-48017-4_15

Hashim, N. H., & Jones, M. L. (2007). Activity theory: A framework for qualitative analysis. In *Proceedings of 4th International Qualitative Research Convention (QRC)*. Malaysia: QRC.

Hassard, J., Law, J., & Lee, N. (1999). Actor network theory and managerialism. *Organization*, 6(3), 387–390. doi:10.1177/135050849963001

Häyrinen, K., Saranto, K., & Nykänen, P. (2008). Definition, structure, content, use and impacts of electronic health records: A review of the research literature. *International Journal of Medical Informatics*, 77(5), 291–304. doi:10.1016/j.ijmedinf.2007.09.001 PMID:17951106

Health Insurance Portability and Accountability Act (HIPPA). (2001, May). *Privacy Compliance Executive Summary*. Protegrity Inc.

Heeks, R., & Stanforth, C. (2007). Understanding e-Government project trajectories from an actor-network perspective. *European Journal of Information Systems*, 16, 165–177. doi:10.1057/palgrave.ejis.3000676

Hennington, A. H., & Janz, B. (2007). *Information systems and healthcare XVI: Physician adoption of electronic medical records: Applying the UTAUT model in a healthcare context*. Academic Press.

Hermans, L. M., & Muluk, Ç. B. (2002). Actor analysis for the implementation of the European Water Framework Directive in Turkey. In R.S.E.W. Leuven, A.G. van Os, & P. H. Nienhuis (Eds.), *Proceedings NCR-Days 2002 Current Themes in Dutch River Research*, (pp. 74-75). Academic Press.

Hermans, L. M., & Muluk, Ç. B. (2008). *Facilitating policy learning about the multi-actor dimension of water governance*. Paper for the XIII World Water Congress. Montpellier, France.

Hermans, L. M. (2005). *Actor analysis for water resources management: Putting the promise into practice*. Delft, The Netherlands: Eburon Publishers.

Hermans, L. M. (2008). Exploring the promise of actor analysis for environmental policy analysis: Lessons from four cases in water resources management. *Ecology and Society*, 13(1), 21.

Hermans, L. M., & Thissen, W. A. H. (2009). Actor analysis methods and their use for public policy analysts. *European Journal of Operational Research*, 196(2), 808–818. doi:10.1016/j.ejor.2008.03.040

Heslop, L. (2010). *Patient and health care delivery systems in the US, Canada and Australia: A critical ethnographic analysis*. LAP LAMBERT Academic Publishing.

Hevner, A. R., March, S. T., Park, J., & Ram, S. (2004). Design science in information systems research. *Management Information Systems Quarterly*, 28(1), 75–106.

HFMA. (2006). *Overcoming Barriers to Electronic health Record adoption: Results of Survey and Roundtable Discussions Conducted by the Healthcare Financial Management Association*. Retrieved from http://mhcc.maryland.gov/electronichealth/mhitr/EHR%20Links/overcoming_barriers_to_ehr_adoption.pdf

Hillestad, R. et al. (2005). Can electronic medical record systems transform health care? Potential health benefits, savings, and costs. *Health Affairs*, 24(5), 1103. doi:10.1377/hlthaff.24.5.1103 PMID:16162551

Hillson, D., & Murray-Webster, R. (2004). Understanding and managing risk attitude. In *Proceedings of 7th Annual Risk Conference*. London, UK: Academic Press.

Hinduja, S., & Patchin, J. W. (2008). Personal information of adolescents on the Internet: A quantitative content analysis of MySpace. *Journal of Adolescence*, 31(1), 125–146. doi:10.1016/j.adolescence.2007.05.004 PMID:17604833

HKEX. (2008). *Retail Investor Survey 2007*. Hong Kong Exchanges and Clearing Ltd.

Hoffinan, L. (2009). Implementing electronic medical records. *Communications of the ACM*. doi:10114515927611592770

Hoffinan, S., & Podguski, A. (2008). *Finding a cure: The case for regulation and oversight of electronic health record systems*. Academic Press.

Honeycutt, C., & Herring, S. C. (2009). Beyond Microblogging: Conversation and Collaboration via Twitter. In *Proceedings of the 42nd Hawaii International Conference on System Sciences*. IEEE.

Hoover, J. N. (2007, September 22). Social Networking: a time waste or the next big thing in collaboration?. *InformationWeek*.

Hordern, A., Georgiou, A., Whetton, S., & Prgomet, M. (2011). Consumer e-health: An overview of research evidence and implications for future policy. *The HIM Journal*, *40*(2), 6–14. PMID:21712556

Hovorka, D. S., & Germonprez, M. (2009). Reflecting, Tinkering, and Bricolage: Implications for Theories of Design. In *Proceedings of the Fifteenth Americas Conference on Information Systems*. San Francisco, CA: Academic Press.

Huang, Z., & Palvia, P. (2001). ERP implementation issues in advanced and developing countries. *Business Process Management Journal*, *7*(3), 276–284. doi:10.1108/14637150110392773

Huber, M. (1999). Health Expenditure Trends in OECD Countries 1970-1997. *Health Care Financing Review*, *21*(2), 99–117. PMID:11481789

Hu, Q., PytlikZillg, L. M., Lynne, G. D., Tomkins, A. J., Waltman, W. J., Hayes, M. J., ... Wilhite, D. A. (2006). Understanding farmers' forecast use from their beliefs, values, social norms, and perceived obstacles. *Journal of Applied Meteorology and Climatology*, *45*, 1190–1201. doi:10.1175/JAM2414.1

IfM and IBM. (2007). *IfM and IBM, Succeeding through Service Innovation: A Discussion Paper*. Cambridge, UK: University of Cambridge Institute for Manufacturing.

Iyamu, T. (2010). Theoretical Analysis of Strategic Implementation of Enterprise Architecture. *International Journal of Actor-Network Theory and Technological Innovation*, *2*(3), 17–32. doi:10.4018/jantti.2010070102

Iyamu, T. (2013). *Enterprise Architecture: From Concept to Practice*. Melbourne, Australia: Heidelberg Press.

Iyamu, T., & Roode, D. (2010). The use of Structuration and Actor Network Theory for analysis: A case study of a financial institution in South Africa. *International Journal of Actor-Network Theory and Technological Innovation*, *2*(1), 1–26. doi:10.4018/jantti.2010071601

Iyamu, T., & Tatnall, A. (2009). An Actor-Network analysis of a case of development and implementation of IT strategy. *International Journal of Actor-Network Theory and Technological Innovation*, *1*(4), 35–52. doi:10.4018/jantti.2009062303

Jackson, K. M., & Trochim, W. M. K. (2002). Concept mapping as an alternative approach for the analysis of open-ended survey responses. *Organizational Research Methods*, *5*, 307–336. doi:10.1177/109442802237114

Jamshidi, M. (2009). *System of Systems Engineering: Innovations for the 21st Century*. Hoboken, NJ: John Wiley & Sons.

Jansen, A. (2005). Assessing E-government progress – Why and what. In *Proceeding of Norsk konferanse for organisasjoners bruk av IT (Nokobit 2005)*. Bergen, Norway: Academic Press.

Jha, A. K., Doolan, D., Grandt, D., Scott, T., & Bates, D. W. (2008). The use of health information technology in seven nations. *International Journal of Medical Informatics*, *77*, 848–854. doi:10.1016/j.ijmedinf.2008.06.007 PMID:18657471

Johnson, A. N., & Paine, C. B. (2007). Self-Disclosure, privacy and the Internet. In A. N. Johnson, K. McKenna, & U. Reips (Eds.), *Oxford Handbook of Internet Psychology*. Oxford University Press.

Johnson, G., Gersten, R., & Carmine, D. (1998). Effects of Instructional Design Variables on Vocabulary acquisition of LD students: A Study of computer-assisted Instruction. *Journal of Learning Disabilities, 20*(4). PMID:3553400

Jones, D. A., Shipman, J. P., Plaut, D. A., & Selden, C. R. (2010). Characteristics of personal health records: Findings of the Medical Library Association/National Library of Medicine Joint Electronic Personal Health Record Task Force. *Journal of the Medical Library Association, 98*(3), 243–249. doi:10.3163/1536-5050.98.3.013 PMID:20648259

Kaghan, W., & Bowker, G. (2001). Out of machine age? Complexity, sociotechnical systems and actor network theory. *Journal of Engineering and Technology Management, 18*, 253–269. doi:10.1016/S0923-4748(01)00037-6

Kaplan, A. M., & Haenlein, M. (2010). Users of the world, unite! The challenges and opportunities of Social Media. *Business Horizons, 53*, 59–68. doi:10.1016/j.bushor.2009.09.003

Kaplan, B., & Harris-Salamone, K. D. (2009). Health IT success and failure: Recommendations from literature and an AMIA workshop. *Journal of the American Medical Informatics Association, 16*(3), 291–299. doi:10.1197/jamia.M2997 PMID:19261935

Kaptelinin, V., & Nardi, B. A. (2006). *Acting with Technology: Activity Theory and Interaction Design*. Cambridge, MA: MIT Press.

Karimi, M., Kheiralipour, K., Tabatabaeefar, A., Khoubakht, G. M., Naderi, M., & Heidarbeigi, K. (2009). The Effect of Moisture Content on Physical Properties of Wheat. *Pakistan Journal of Nutrition, 8*, 90–95. doi:10.3923/pjn.2009.90.95

Karsh, B.-T., Weinger, M. B., Abbott, P. A., & Wears, R. L. (2010). Health information technology: Fallacies and sober realities. *Journal of the American Medical Informatics Association, 17*(6), 617–623. doi:10.1136/jamia.2010.005637 PMID:20962121

Kavale, K. A., & Forness, S. R. (1995). *The Nature of Learning Disabilities: Critical Elements of Diagnosis and Classification*. Lawrence Erlbaum Associates, Inc.

Kavale, S. (1996). Interviews, an introduction to Qualitative Research. *Sage (Atlanta, Ga.)*.

Kennan, M. A., Cecez-Kecmanovic, D., & Underwood, J. (2010). Having a Say: Voices for all the Actors in ANT Research? *International Journal of Actor-Network Theory and Technological Innovation, 2*(2), 1–16. doi:10.4018/jantti.2010040101

Kennedy, B. L. (2011). *Exploring the Sustainment of Health Information Technology: Successful Practices for Addressing Human Factors*. Northcentral University.

Kietzmann, J. H., Hermkens, K., McCarthy, I. P., & Silvestre, B. S. (2011). Social Media? Get serious! Understanding functional building blocks of social media. *Business Horizons, 54*, 241–251. doi:10.1016/j.bushor.2011.01.005

Kimaro, H., & Nhampossa, J. (2007). *The challenges of sustainability of health information systems in developing countries: Comparative case studies of Mozambique and Tanzania*. Academic Press.

Kim, W., Jeong, O., & Lee, S. (2010). On Social networking sites. *Information Systems, 35*, 215–236. doi:10.1016/j.is.2009.08.003

Kingsford-Smith, D., & Williamson, K. (2004). How Do Online Investors Seek Information, And What Does This Mean for Regulation?. *The Journal of Information, Law and Technology, (2)*.

Klein, H. K., & Myers, M. D. (1999). A set of principles for conducting and evaluating interpretive field studies in information systems. *Management Information Systems Quarterly, 23*(1), 67–94. doi:10.2307/249410

Klinger, J. K. (1998). Outcomes for Students With and Without Learning Disabilities in Inclusive Classrooms. *Learning Disabilities Research & Practice, 13*(3), 153–161.

Kling, R. (1991). Computerization and social transformations. *Science, Technology & Human Values, 16*(3), 342–367. doi:10.1177/016224399101600304

Klopper, E., Vogel, C. H., & Landman, W. A. (2006). Seasonal climate forecasts- potential agricultural- risk management tool? *Climatic Change*, 76, 73–90. doi:10.1007/s10584-005-9019-9

Knights, D., Murray, F., & Willmott, H. (1997). Networking as knowledge work: A study of strategic interorganizational development in the financial service industry. In B. P. Bloomfield, R. Coombs, D. Knights, & D. Littler (Eds.), *Information technology and organizations: Strategies, networks, and integration.* Oxford, UK: Oxford University Press. doi:10.1093/acprof:oso/9780198289395.003.0007

Knox, H., O'Doherty, D., Vurdubakis, T., & Westrup, C. (2007). Transformative capacity, information technology, and the making of business 'experts'. *The Sociological Review*, 55(1), 22–41. doi:10.1111/j.1467-954X.2007.00679.x

Kolkman, M. J., Kok, M., & Van der Veen, A. (2005). Mental model mapping as a new tool to analyse the use of information in decision-making in integrated water management. *Physics and Chemistry of the Earth*, 30, 317–332. doi:10.1016/j.pce.2005.01.002

Korhan. (2010). Retrieved from http://ezinearticles.com/?Social-Media-Vocabulary---Language-and-Slang-Corruption&id=1807049

Kralewski, J., Dowd, B. E., Zink, T., & Gans, D. N. (2010). Preparing your practice for the adoption and implementation of electronic health records. *Physician Executive*, 36(2), 30–33. PMID:20411843

Krasnova, H., Spiekermann, S., Korolevu, K., & Hidebrand, T. (2010). Online social networks: Why we disclose. *Journal of Information Technology*, 25, 109–125. doi:10.1057/jit.2010.6

Krohn, R. (2007). The consumer-centric personal health record—It's time. *Journal of Healthcare Information Management*, 21(1), 21.

Kulkarni, R., & Nathanson, L. A. (2005). *Medical Informatics in medicine, E-Medicine.* Retrieved from http://www.emedicine.com/emerg/topic879.htm

Kutsch, E. (2008). The effect of intervening conditions on the management of project risk. *International Journal of Managing Projects in Business*, 1(4), 602–610. doi:10.1108/17538370810906282

Kuutti, K. (1996). Activity Theory as a Potential Framework for Human-Computer Interaction Research. In *Context and Consciousness: Activity Theory and Human-Computer Interaction.* Cambridge, MA: MIT Press.

Labuschagne, L., Marnewick, C., & Jakovljevic, M. (2008). IT project management maturity: A South African perspective. In *Proceedings on the PMSA Conference: From Strategy to Reality.* PMSA.

Lacroix, A. (1999). International concerted action on collaboration in telemedicine: G8sub-project 4, Sted. *Health Technol. Inform.*, 64, 12–19.

Lamb, R. (2006). Alternative paths toward a social actor concept. In *Proceedings of the Twelfth Americas Conference on Information Systems.* Acapulco, Mexico: Academic Press.

Lancaster University. (n.d.). *Centre for Science Studies.* Retrieved from http://www.lancs.ac.uk/fass/centres/css/ant/ant.htm

Latour, B. (1986). The Powers of Association. In *Power, Action and Belief: A New Sociology of Knowledge?* London: Routledge & Kegan Paul.

Latour, B. (1987). *Science in Action: How to Follow Engineers and Scientists Through Society.* Milton Keynes, UK: Open University Press.

Latour, B. (1987). *Science in action: How to follow scientists and engineers through society.* Cambridge, MA: Harvard University Press.

Latour, B. (1988). *The Pasteurization of France.* Cambridge, MA: Harvard University Press.

Latour, B. (1990). Drawing Things Together. In M. Lynch, & S. Woolgar (Eds.), *Representation in Scientific Practice.* Cambridge, MA: MIT Press.

Latour, B. (1991). Technology is society made durable. In J. Law (Ed.), *A sociology of monsters: Essays on power, technology and domination*. Routledge.

Latour, B. (1992). On Recalling ANT. In *Actor Network Theory and After*. Blackwell Publishers.

Latour, B. (1992). Where are the missing masses? The sociology of a few mundane artifacts. In W. E. Bijker, & J. Law (Eds.), *Shaping technology/building society: Studies in sociotechnical change*. Cambridge, MA: MIT Press.

Latour, B. (1993). *We have never been modern*. Brighton, UK: Harvester Wheatsheaf.

Latour, B. (1993). *We Have Never Been Modern*. Cambridge, MA: Harvester University Press.

Latour, B. (1994). On Technical Mediation. *Common Knowledge*, *3*(2), 29–64.

Latour, B. (1996). *Aramis or the Love of Technology*. Cambridge, MA: Harvard University Press.

Latour, B. (1996). *Aramis, or the love of technology*. Cambridge, MA: MIT Press.

Latour, B. (1999). On recalling ANT. In J. Law, & J. Hassard (Eds.), *Actor Network Theory and After*. Oxford, UK: Blackwell.

Latour, B. (2005). *Reassembling the social: An introduction to actor-network theory*. Oxford, UK: Oxford University Press.

Latour, B. (2010). *The making of law: An ethnography of the council d'etat*. Cambridge, MA: Polity Press.

Latour, B., Harbers, H., & Koenis, S. (1996). *On actor-network theory: A few clarifications*. Academic Press.

Latour, B., & Law, J. (1986). *The Powers of Association: Power, Action and Belief*. Academic Press.

Latour, B., Mauguin, P., & Teil, G. (1992). A note on socio-technical graphs. *Social Studies of Science*, *22*(1), 33. doi:10.1177/0306312792022001002

Law, J. (1992). *Notes on the Theory of the Actor Network: Ordering, Strategy, and Heterogeneity*. Retrieved August 26, 2011, from http://www.lancs.ac.uk/fass/sociology/papers/law-notes-on-ant.pdf

Law, J. (1997). *Heterogeneities, published by the Centre for Science Studies*. Lancaster University. Retrieved from http://www.biomedcentral.com/1472-6947/10/67/

Law, J. (2007). *Actor Network Theory and Material Semiotics*. Retrieved from http://www.heterogeneities.net/publications/Law-ANTandMaterialsSemiotics.pdf

Law, J., & Hassard, J. (1999). *Actor network theory and after*. Wiley-Blackwell.

Law, J., & Mol. (2000). *Situating Technoscience: An Inquiry into Spatialities*. Lancaster University.

Law, J. (1986). On methods of long-distance control: Vessels, navigation and the Portuguese route to India. In J. Law (Ed.), *Power, Action and Belief* (pp. 196–233). Routledge & Kegan Paul.

Law, J. (1986). The Heterogeneity of Texts. In *Mapping the Dynamics of Science and Technology*. Macmillan Press.

Law, J. (1987). Technology and Heterogeneous Engineering: The Case of Portuguese Expansion. In W. E. Bijker, T. P. Hughes, & T. J. Pinch (Eds.), *The Social Construction of Technological Systems: New Directions in the Sociology and History of Technology* (pp. 111–134). Cambridge, MA: MIT Press.

Law, J. (1987a). On the social explanation of technical change: The case of the Portuguese maritime expansion. *Technology and Culture*, *28*(2), 227–252. doi:10.2307/3105566

Law, J. (1990). Technology and heterogeneous engineering: The case of Portuguese expansion. In *The social construction of technological systems: New directions in the sociology and history of technology*. Cambridge, MA: MIT Press.

Law, J. (1991). *A Sociology of monsters: Essays on power, technology, and domination*. Routledge.

Law, J. (1992). Notes on the Theory of the Actor-Network: Ordering, Strategy and Heterogeneity. *Systems Practice*, *5*(4), 379–393. doi:10.1007/BF01059830

Law, J. (1999). After ANT: Topology, naming and complexity. In J. Law, & J. Hassard (Eds.), *Actor Network Theory and After*. Oxford, UK: Blackwell.

Law, J. (2004). *After method: Mess in social science research*. New York: Routledge.

Law, J. (2008). On Sociology and STS. *The Sociological Review*, *56*(4), 623–649. doi:10.1111/j.1467-954X.2008.00808.x

Law, J. (Ed.). (1991). *A Sociology of Monsters: Essays on power, technology and domination*. London: Routledge.

Law, J., & Callon, M. (1988). Engineering and Sociology in a Military Aircraft Project: A Network Analysis of Technological Change. *Social Problems*, *35*(3), 284–297. doi:10.2307/800623

Law, J., & Callon, M. (1992). The life and death of an aircraft: A network analysis of technical change. In *Shaping Technology/Building Society: Studies in Sociological Change*. Cambridge, MA: MIT Press.

Law, J., & Mol, A. (Eds.). (2002). *Complexities: Social studies of knowledge practices*. Durham, NC: Duke University Press. doi:10.1215/9780822383550

Learning Disabilities Association of Canada. (2002). *Official Definition of Learning Disabilities*. Retrieved May 2007, from http://www.ldac-taac.ca/Defined/defined_new-e.asp

Lee, G., & Xia, W. (2003). *An Empirical Study on the Relationships between the Flexibility, Complexity and Performance of Information Systems Development Projects*. University of Minnesota.

Leila. (2009). The Bug's Blog: Bruno Latour, Action-Network-Theory, and Dismantling Heirarchies: Actor-Network-Theory: Terms and Concepts. *The Bug's Blog*. Retrieved from http://latourbugblog.blogspot.com/2009/01/actor-network-theory-terms-and-concepts.html

Leslie, H. (2011). *Australia's PCEHR Challenge | Archetypical*. Retrieved from http://omowizard.wordpress.com/2011/08/30/australias-pcehr-challenge/

Lester, V. M. (1997). *A Program for Research on Management Information*. Academic Press.

Levy, M., & Powell, P. (2005). *Strategies for Growth in SMEs: The Role of Information and Information Systems*. Amsterdam: Elsevier Butterworth-Heinemann.

Li, C., & Bernoff, J. (2008). Harnessing the Power of the Oh-So-Social Web. *MIT Sloan Management Review*, *49*(3), 36–42.

Linden, L., & Saunders, C. (2009). Linux Kernel Developers Embracing Authors Embracing Licenses. *International Journal of Actor-Network Theory and Technological Innovation*, *1*(3), 15–35. doi:10.4018/jantti.2009070102

Linderoth, H. C. J., & Pellegrino, G. (2005). Frames and Inscriptions: Tracing a Way to Understand IT-Dependent Change Projects. *International Journal of Project Management*, *23*(5), 415–420. doi:10.1016/j.ijproman.2005.01.005

Lipke, W. (2003, March). Schedule is Different. *The Measurable News*.

Liu, L. S., Shih, P. C., & Hayes, G. R. (2011). Barriers to the adoption and use of personal health record systems. In *Proceedings of the 2011 iConference* (pp. 363–370). Academic Press.

Li, X., & Zhang, Y. (2009). Analysis of the 3G diffusion and adoption in China. *Journal of Academy of Business and Economics*, *9*(3), 175–187.

Lorenzi, N. et al. (2009). How to successfully select and implement electronic health records (EHR) in small ambulatory practice settings. *BMC Medical Informatics and Decision Making*, *9*, 15. doi:10.1186/1472-6947-9-15 PMID:19236705

Luoma-Aho, V., & Paloviita, A. (2010). Actor-networking stakeholder theory for today's corporate communications. *Corporate Communications: An International Journal*, *15*(1), 49–67. doi:10.1108/13563281011016831

Lybbert, T. J., Barrett, C., McPeak, J. G., & Luseno, W. K. (2007). Bayesian herders: Updating of rainfall beliefs in response to external forecasts. *World Development*, *35*(3), 480–497. doi:10.1016/j.worlddev.2006.04.004

Lynn, L. E. (2005). Public management: A concise history of the field. In E. Ferlie, L. E. Lynn, & C. Pollitt (Eds.), *The Oxford handbook of public management* (pp. 27–50). Oxford, UK: Oxford University Press.

MacLeod, M. A. (2001). *Actor network theory: Examining the role of influence on technology adoption.* University of Hawai'i at Manoa.

MacKinnon, C. (2005). *Women's lives, men's laws.* Cambridge, MA: Belknap Press of Harvard University Press.

MacMillan, D. L., & Hendrick, I. G. (1993). Evolution and Legacies. In *Integrating General and Special Education.* Academic Press.

Macome, E. (2008). On Implementation of an information system in the Mozambican context: The EDM case viewed through ANT lenses. *Information Technology for Development, 14*(2), 154–170. doi:10.1002/itdj.20063

MacPhee, D., Fritz, J., & Heyl, J. M. (1996). Ethnic Variations in Personal Social Networks and Parenting. *Child Development, 67*(6), 3278–3295. doi:10.2307/1131779

Madhaven, K. P. C., & Goasguen, S. (2007). Integrating Cutting-edge Research into Learning through Web 2.0 and Virtual Environments. *International Workshop on Grid Computing Environments.* Retrieved from http://www.collabogce.org/gce07/images/1/15/Madhavan.pdf

Maguire, C., Kazlauskas, E. J., & Weir, A. D. (1994). *Information Services for Innovative Organizations.* Sandiego, CA: Academic Press.

Maheu, M. M., Whitten, P., & Allen, A. (2001). *E-Health, Telehealth, and Telemedicine: A Guide to Start-Up and Success.* San Francisco, CA: Jossey-Bass.

Mahmoud-Jouini, S. B. (2000). Coment passer d'une demarche réactive à une démarche proactive fondée sur des stratégies d'offres inovantes? Le cas des grandes entreprises générales de Bâtiment françaises. In *De l'idée au marché.* Vuibert.

Mahring, M., Holmstrom, J., Keil, M., & Montealegre, R. (2004). Trojan actor-networks and swift translation: Bringing actor-network theory to IT project escalation studies. *Information Technology & People, 17*(2), 210–238. doi:10.1108/09593840410542510

March, S., & Smith, G. F. (1995). Design & natural science research on information technology. *Decision Support Systems, 15*(4), 251–266. doi:10.1016/0167-9236(94)00041-2

May, C., & Finch, T. (2009). Implementing, embedding, and integrating practices: An outline of normalization process theory. *Sociology, 43*(3), 535–554. doi:10.1177/0038038509103208

McBride, N. (2000). *Using actor-network theory to predict the organizational success of a communications network.* Department of Information Systems, De Montfort University. Retrieved from http://www.cse.dmu.ac.uk/~nkm/WTCPAP.html

McBride, N. (2011). Using Actor Network Theory to Predict the Organisational Success of a Communications Network. In *Proceedings of the World Telecommunications Congress,* (pp. 1 – 6). Academic Press.

McGrath, K. (2001). The golden circle: A case study of organizational change at the London Ambulance Service. *ECIS 2001 Proceedings.* Retrieved from http://aisel.aisnet.org/ecis2001/52

McIvor, R., McHugh, M., & Cadden, C. (2002). Internet technologies: supporting transparency in the public sector. *International Journal of Public Sector Management, 15*(3), 170–187. doi:10.1108/09513550210423352

McLean, C., & Hassard, J. (2004). Symmetrical Absence/Symmetrical Absurdity: Critical Notes on the Production of Actor-Network Accounts. *Journal of Management Studies, 41*(3), 493–519. doi:10.1111/j.1467-6486.2004.00442.x

McMaster, T., Vidgen, R. T., & Wastell, D. G. (1997). Towards an understanding of technology in transition: Two conflicting theories. In *Proceedings of Information Systems Research in Scandinavia, IRIS20 Conference.* Hanko, Norway: University of Oslo.

McReavy, D., Toth, L., Tremonti, C., & Yoder, C. (2009). *The CFO's role in implementing EHR systems: the experiences of a Florida health system point to 10 actions that can help CFOs manage the revenue risks and opportunities of implementing an electronic health record.* Retrieved December 16, 2011, from http://findarticles.com/p/articles/mi_m3257/is_6_63/ai_n35529545/

Metcalfe, J. (2009, March 29). The celebrity Twitter ecosystems. *The New York Times* mhplus Krankenkasse. (2011). *Infos zu Vorteilen bei der eGK*. Retrieved August 26, 2011, from http://www.mhplus-krankenkasse.de/1228.html

Miles, M. B., & Huberman, A. M. (1984). *Qualitative Data Analysis*. Newbury Park, CA: Sage.

Mitchell, T. (2002). *Rule of Experts: Egypt, Techno-politics Modernity*. Berkeley, CA: University of California Press.

Mitchell, V., & Nault, B. (2008). *The emergence of functional knowledge in sociotechnical systems*. Academic Press.

Mitev, N. (2008). In and out of actor-network theory: a necessary but insufficient journey. *Information Technology & People, 22*(1), 9–25. doi:10.1108/09593840910937463

Moameni, A. (1999). *Soil quality changes under longterm wheat cultivation in the Marvdasht plain, south-central Iran*. (Ph.D. Thesis). Ghent University, Belgium.

Mohammed, S., Klimoski, R., & Rentsch, J. R. (2000). The measurement of team mental models: We have no shared schema. *Organizational Research Methods, 3*(2), 123–165. doi:10.1177/109442810032001

Mohanty, C. T. (2003). *Feminism without borders: Decolonizing theory, practicing solidarity*. Durham, NC: Duke University Press. doi:10.1215/9780822384649

Mohd, H., & Mohamad, S. M. S. (2005). Acceptance model of electronic medical record. *Management, 2*(1), 75.

Mol, A., & Law, J. (1994). Regions, Networks and Fluids: Anaemia and Social Topology. *Social Studies of Science, 24*, 641–671. doi:10.1177/030631279402400402 PMID:11639423

Monteiro, E. (2000). Actor-Network Theory and Information Infrastructure. In *From Control to drift: the dynamics of corporate information infrastructures* (pp. 71–83). Oxford University Press.

Monteiro, E., & Hanseth, O. (1995). Social shaping of information infrastructure: On being specific about the technology. In W. Orlikowski, G. Walsham, M. R. Jones, & J. I. DeGross (Eds.), *Information technology and changes in organisational work* (pp. 325–343). Chapman & Hall.

Monteiro, E., & Hanseth, O. (1996). Social shaping of information infrastructure: On being specific about technology. In *Information Technology and Changes in Organisational Work*. Chapman and Hall.

Monteiro, E., & Hepsø, V. (2000). Infrastructure strategy formation: seize the day at Statoil. In C. Ciborra (Ed.), *From control to drift, The dynamics of corporate information infrastructure*. Oxford University Press.

Morel, R., Tatnall, A., Ketamo, H., Lainema, T., Koivisto, J., & Tatnall, B. (2005). *Mobility and Education. E-Training Practices for Professional Organizations*. Kluwer Academic Publishers.

Morf, M. E., & Weber, W. G. (2000). I/O Psychology and the Bridging Potential of A. N. Leont'ev's Activity Theory. *Canadian Psychology, 41*, 81–93. doi:10.1037/h0088234

Morgan, G. (1980). Paradigms, metaphors, and puzzle solving in organization theory. *Administrative Science Quarterly, 25*(4), 605–622. doi:10.2307/2392283

Morris, M. R., Teevan, J., & Panovich, K. (2010). *What Do People Ask Their Social Networks, and Why? A Survey Study of Status Message Q&A Behavior*. Paper presented at CHI 2010. Atlanta, GA.

Morris, P. W. G. (2004). *The irrelevance of project management as a professional discipline*. INDECO Management Solutions.

Mort, M., Finch, T., & May, C. (2007). Making and unmaking telepatients: identity and governance in new health technologies. *Science, Technology & Human Values, 34*(1), 9–33. doi:10.1177/0162243907311274

Moser, I., & Law, J. (2006). Fluids or flows? Information and qualculation in medical practice. *Information Technology & People, 19*, 55–73. doi:10.1108/09593840610649961

Muhr, T., & Friese, S. (2004). *User's Manual for ATLAS.ti 5.0* (2nd ed.). Scientific Software Development.

Mulla, S. (2008). There is no place like home: The body as the scene of the crime in sexual assault intervention. *Home Cultures, 5*(3), 301–325. doi:10.2752/174063108X368337

Mutch, A. (2002). Actors and networks or agents and structures: Towards a realist view of information systems. *Organization, 9*(3), 477–496. doi:10.1177/135050840293013

Naaman, M., Boase, J., & Lai, C. H. (2010). *Is it Really About Me? Message Content in Social Awareness Streams*. Paper presented at CSCW 2010. Savannah, GA.

Nagle, L. M. (2007). EHR versus EMR versus EPR. *Nursing Leadership, 20*(1), 30–32. doi:10.12927/cjnl.2007.18782 PMID:17472137

Naidoo, R., & Leonard, A. (2012). Observing the 'Fluid' Continuity of an IT Artefact. *International Journal of Actor-Network Theory and Technological Innovation, 4*(4), 23–46. doi:10.4018/jantti.2012100102

Naím, M. (2007). The YouTube Effect. *Foreign Policy, 158,* 103–104.

Najafi, B., & Bakhshoodeh, M. (2002). Effectiveness of government protective policies on rice production in Iran. In *Proceedings of the Xth EAAE Congress on 'Exploring Diversity in the European Agri -Food System.* Zaragoza, Spain: EAAE.

Nandhakumar, J., & Jones, M. (1997). Too close to comfort? Distance and engagement in interpretive information systems research. *Information Systems Journal, 7,* 109–131. doi:10.1046/j.1365-2575.1997.00013.x

Narayanan, V. K., & Armstrong, D. J. (2005). *Casual mapping for research in information technology.* Hershey, PA: Idea Group Publishing Inc.

Nardi, B. A., Schiano, D. J., & Gumbrecht, M. (2004). Blogging as social activity, or, would you let 900 million people read your diary. In *Proceedings of the 2004 ACM Conference on Computer Supported Cooperative Work.* Chicago: ACM.

Navarra, D. D., & Cornford, T. (2009). Globalization, networks, and governance: Researching global ICT programs. *Government Information Quarterly, 26*(1), 35–41. doi:10.1016/j.giq.2008.08.003

Nazemos'sadat, S. M. J., Kamgar-Haghighi, A. A., Sharifzadeh, M., & Ahmadvand, M. (2006). Adoption of Long – Term Rainfall Forecasting: A Case of Fars Province Wheat Farmers. *Iranian Agricultural Extension and Education Journal, 2*(2), 1–16.

NEHTA – National E-Health Transition Authority Limited. (2008). *Quantitative Survey Report.* Retrieved August 26, 2011, from http://www.nehta.gov.au/component/docman/doc_download/585-public-opinion-poll-iehr

NEHTA – National E-Health Transition Authority Limited. (2010). *Healthcare Identifiers Service: Implementation Approach.* Retrieved August 26, 2011, from http://www.nehta.gov.au/component/docman/doc_download/1139-healthcare-identifiers-implementation-approach

NEHTA – National E-Health Transition Authority Limited. (2011a). *Delivering E-Health's Foundations.* Retrieved August 26, 2011, from http://www.nehta.gov.au/media-centre/nehta-news/416-delivering-e-healths-foundations

NEHTA – National E-Health Transition Authority Limited. (2011b). *The National E-Health Transition Authority Strategic Plan (2009-2012).* Retrieved August 26, 2011, from http://www.nehta.gov.au/about-us/strategy

NEHTA – National E-Health Transition Authority Limited. (2011c). *What is a PCEHR?* Retrieved August 26, 2011, from http://www.nehta.gov.au/ehealth-implementation/what-is-a-pcher

NEHTA – National E-Health Transition Authority Limited. (2011d). *Benefits of a PCEHR.* Retrieved August 26, 2011, from http://www.nehta.gov.au/ehealth-implementation/benefits-of-a-pcehr

NEHTA – National E-Health Transition Authority Limited. (2011e). *e-Communications in Practice.* Retrieved August 26, 2011, from http://www.nehta.gov.au/e-communications-in-practice

NEHTA – National E-Health Transition Authority Limited. (2011f). *Electronic Transfer of Prescription (ETP) Release 1.1 draft.* Retrieved August 26, 2011, from http://www.nehta.gov.au/media-centre/nehta-news/704-etp-11

NEHTA & DoHA. (2011). *Concept of Operations: Relating to the introduction of a Personally Controlled Electronic Health Record System.* Retrieved from http://www.yourhealth.gov.au/internet/yourhealth/publishing.nsf/Content/CA2578620005CE1DCA2578F800194110/$File/PCEHR%20Concept%20of%20Operations.pdf

Neisser, F. (2013). Fostering Knowledge Transfer for Space Technology Utilization in Disaster Management – An Actor-Network Perspective. *International Journal of Actor-Network Theory and Technological Innovation, 5*(1). doi:10.4018/jantti.2013010101

Nelson, K. M., & Nelson, H. J. (2000). Revealed causal mapping as an evocative method for information systems research. In *Proceedings of the 33rd Hawaii International Conference on System Sciences.* IEEE.

Neumann, M., O'Murchu, I., Breslin, J., Decker, S., Hogan, D., & MacDonaill, C. (2005). Semantic social network portal for collaborative online communities. *Journal of European Industrial Training, 29*(6), 472–487. doi:10.1108/03090590510610263

Neves, J. et al. (2008). Electronic Health Records and Decision Support Local and Global Perspectives. *WSEAS Transactions on Biology and Biomedicine, 5*, 8.

Ngosi, T., Helfert, M., Carcary, M., & Whelan, E. (2011). *Design science and Actor-Network Theory Nexus: A perspective on content development of a critical process for enterprise architecture management.* Paper presented at the 13th International Conference on Enterprise Information Systems (ICES). Beijing, China.

NHHRC. (2009). *A Healthier Future for All Australians: National Health and Hospitals Reform Commission - Final Report June 2009.* Retrieved from http://www.health.gov.au/internet/main/publishing.nsf/Content/nhhrc-report

Nielsen, P., & Hanseth, O. (2007). *Fluid standards: A case study of the Norwegian standard for mobile content services.* Retrieved from http://heim.ifi.uio.no/~oleha/Publications/FluidStandardsNielsenHanseth.pdf

Normann, R. (2001). *Reframing business: When the map changes the landscape.* Chichester, UK: Wiley.

O'Toole, L. J. (1997a). Implementing public innovations in network settings. *Administration & Society, 29*(2), 115–138. doi:10.1177/009539979702900201

O'Toole, L. J. (1997b). The implications for democracy in a networked bureaucratic world. *Journal of Public Administration: Research and Theory, 7*(3), 443–459. doi:10.1093/oxfordjournals.jpart.a024358

O'Toole, L. J. (1997c). Treating networks seriously: Practical and research-based agendas in public administration. *Public Administration Review, 57*(1), 45–52. doi:10.2307/976691

Oberholzer-Gee, F., & Strumpf, K. (2007). The Effect of File Sharing on Record Sales: An Empirical Analysis. *The Journal of Political Economy, 115*(1), 1–42. doi:10.1086/511995

OECD. (2010). *OECD health data 2010.* Retrieved from http://stats.oecd.org/Index.aspx?DatasetCode=HEALTH World Health Organization (WHO). (2005). *What is e-health?* Retrieved from http://www.emro.who.int/his/ehealth/AboutEhealth.htm

OECD. (2010a). *Growing health spending puts pressure on government budgets.* Retrieved July 29, 2011, from http://www.oecd.org/document/11/0,3343,en_2649_34631_45549771_1_1_1_37407,00.html

OECD. (2010b). *OECD-Gesundheitsdaten 2010: Deutschland im Vergleich.* Retrieved January 31st, 2011 from http://www.oecd.org/dataoecd/15/1/39001235.pdf

OECD. (2010c). *OECD health data 2010.* Retrieved June 06, 2011, from http://stats.oecd.org/Index.aspx?DatasetCode=HEALTH

Orlikowski, W. J. (1992). The duality of technology: Rethinking the concept of technology in organizations. *Organization Science, 3*(3), 398–427. doi:10.1287/orsc.3.3.398

Orlikowski, W. J. (1996). Improvising Organizational Transformation over Time: A Situated Change Perspective. *Information Systems Research, 7*(1), 63–92. doi:10.1287/isre.7.1.63

Orlikowski, W. J., & Baroudi, J. (1991). Studying information technology in organizations: Research approaches and assumptions. *Information Systems Research, 2*(1), 1–28. doi:10.1287/isre.2.1.1

Orlikowski, W. J., & Iacono, C. S. (2001). Research commentary: Desperately seeking IT in IT research - A call to theorizing the IT artifact. *Information Systems Research, 12*, 121–134. doi:10.1287/isre.12.2.121.9700

Oxford. (1973). *The Shorter Oxford English Dictionary.* Oxford, UK: Clarendon Press.

Panagariya, A. (2000). E-commerce, WTO and developing countries. *World Economy, 23*(8), 959–978. doi:10.1111/1467-9701.00313

Pandya, A., & Dholakia, N. (2005). B2C Failures: Towards an Innovation Theory Framework. *Journal of Electronic Commerce in Organizations, 3*(2), 68–81. doi:10.4018/jeco.2005040105

Papacharissi, Z. (2009). The virtual geographies of social networks: A comparative analysis of Facebook, LinkedIn and ASmallWorld. *New Media & Society, 11*, 199. doi:10.1177/1461444808099577

Pappa, D., & Stergiouglas, L. K. (2008). The emerging role of corporate systems: An example from the era of business process-oriented learning. *International Journal of Business Science and Applied Management, 3*(2), 38–48.

Paré, G. (2004). Investigating information systems with positivist case study research. *Communications of the Association for Information Systems, 13*(1), 233–264.

Parnis, D., & Du Mont, J. (2006). Symbolic power and the institutional response to rape: Uncovering the cultural dynamics of a forensic technology. *The Canadian Review of Sociology and Anthropology. La Revue Canadienne de Sociologie et d'Anthropologie, 43*(1), 73–93. doi:10.1111/j.1755-618X.2006.tb00855.x

Patchin, J. W., & Hinduja, S. (2010). Changes in adolescent online social networking behaviours from 2006 to 2009. *Computers in Human Behavior, 26*, 1818–1821. doi:10.1016/j.chb.2010.07.009

Patton, M. Q. (2002). *Qualitative research and evaluation methods* (3rd ed.). Sage Publications, Inc.

Pawłowska, A. (2004). Failures in large systems projects in Poland: Mission impossible? *Information Polity, 9*(4), 167–180.

Pearce, C., & Haikerwal, M. C. (2010). E-health in Australia: time to plunge into the 21st Century. *The Medical Journal of Australia, 193*(7), 397–398. PMID:20919969

Pedersen, M. K., Fountain, J., & Loukis, E. (2006). Preface to the focus theme section: electronic markets and e-government. *Electronic Markets, 16*(4), 263–273. doi:10.1080/10196780600999601

Pempek, T. A., Yermolayeva, Y. A., & Calvert, S. L. (2009). College students' social networking experiences on Facebook. *Journal of Applied Developmental Psychology, 30*, 227–238. doi:10.1016/j.appdev.2008.12.010

Perez, C. (2002). *Technological revolutions and financial capital: The dynamics of bubbles and golden ages*. Cheltenham, UK: Edward Elgar Publishing. doi:10.4337/9781781005323

Perkins, M. (2010). *Teach for tests, teachers told.* Retrieved January 2012, from http://www.theage.com.au/national/education/teach-for-tests-teachers-told-20100204-ng6o.html

Pich, M. T., Loch, & De Meyer. (2002). On Uncertainty, Ambiguity, and Complexity in Project Management. *Management Science, 48*(8), 1008–1023. doi:10.1287/mnsc.48.8.1008.163

PMBOK. (2004). *A Guide to the Project Management Body of Knowledge (PMBOK® Guide)* (3rd ed.). Newtown Square, PA: Project Management Institute.

Pons, D. (2008). Project Management for New Product Development. *Project Management Journal, 39*(2), 82–97. doi:10.1002/pmj.20052

Pons, D. J., & Raine, J. (2005). Design Mechanisms and Constrains. *Research in Engineering Design, 16*, 73–85. doi:10.1007/s00163-005-0008-9

Ponte, D., & Camussone, P. F. (2013). Neither Heroes nor Chaos: The Victory of VHS Against Betamax. *International Journal of Actor-Network Theory and Technological Innovation, 5*(1). doi:10.4018/jantti.2013010103

Porter, M. (1990). *The Competitive advantage of Nations.* New York: Free Press.

Porter, M., & Tiesberg, E. (2006). *Re-defining health care delivery.* Harvard Business Press.

Posey, C., Lowry, P. B., Roberts, T. L., & Ellis, T. S. (2010). Proposing the online community self-disclosure model: The case of working professionals in France and the U. K. who use online communities. *European Journal of Information Systems, 19*, 181–195. doi:10.1057/ejis.2010.15

Postman, N. (1992). *Technopoly: The surrender of culture to technology.* New York: Vintage Press.

Pouloudi, A., & Whitley, E. A. (2000). Representing Human and Nonhuman Stakeholders: On Speaking with Authority. In R. Baskerville, J. Stage, & J. I. DeGross (Eds.), *Organizational and Social Perspectives on Information Technology.* Boston: Kluwer Academic Publishers. doi:10.1007/978-0-387-35505-4_20

Požgaj, Ž., & Sertić, H., & MarijaBoban, M. (2007). Effective information systems development as a key to successful enterprise. *Management, 12*(1), 65–86.

Preiss, B. (2013). Plan to Recoup Funds from Troubled Ultranet System. *Age*, 5.

PricewaterhouseCoopers Healthcare Practice. (n.d.). Retrieved from www.pwchealth.com

Protti, D., & Smit, C. (2006). The Netherlands: Another European Country Where GPs Have Been Using EMRs for Over Twenty Years. - Google Scholar. *Healthcare Information Management & Communications, 30*(3).

Quinlan, A. (2012). Imagining a Feminist Actor-Network Theory. *International Journal of Actor-Network Theory and Technological Innovation, 4*(2). doi:10.4018/jantti.2012040101

Quinlan, A., Fogel, C., & Quinlan, E. (2010). Unmasking Scientific Controversies: Forensic DNA Analysis in Canadian Legal Cases of Sexual Assault. *Canadian Women's Studies, 21*(1), 98–107.

Rachlis, M. (2006). *Key to sustainable healthcare system.* Retrieved from http:www.improveingchroniccare.org

RahimiBadr. B., & Moghaddasi, R. (2009). Iran wheat price forecasting performance evaluation of economic. In *Proceedings of 4th International congress on Aspects and Visions of Applied Economics and Informatics (AVA).* Debrecen, Hungary: AVA.

Ramiller, N. C., & Wagner, E. L. (2009). The element of surprise: Appreciating the unexpected in (and through) actor-networks. *Information Technology & People, 22*(1), 36–50. doi:10.1108/09593840910937481

Randerson, J. (n.d.). *Social networking sites don't deepen friendships.* Retrieved from http://www.guardian.co.uk/science/2007/sep/10/socialwork

Rauber, A., & Kogler, A. (2001). Integrating automatic genre analysis into digital libraries. In *Proceedings of the First ACM/IEEE-CS Joint Conference on Digital Libraries.* Roanoke, VA: ACM Press.

Reid, S. (2013). Distance, climate, demographics and the development of online courses in Newfoundland and Labrador. *International Journal of Actor-Network Theory and Technological Innovation, 5*(2). doi:10.4018/jantti.2013040102

Reite, I. C. (2013). Between Blackboxing and Unfolding: Professional Learning Networks of Pastors. *International Journal of Actor-Network Theory and Technological Innovation, 5*(4). doi:10.4018/ijantti.2013100104

Rethemeyer, R. K. (2005). Conceptualizing and measuring collaborative networks. *Public Administration Review, 65*(1), 117–121. doi:10.1111/j.1540-6210.2005.00436.x

Rethemeyer, R. K. (2009). Making sense of collaboration and governance issues and challenges. *Public Performance & Management Review, 32*(4), 565–573. doi:10.2753/PMR1530-9576320405

Rhodes, J. (2009). Using Actor-Network Theory to Trace an ICT (Telecenter) Implementation Trajectory in an African Women's Micro-Enterprise Development Organization. *Information Technologies & International Development, 5*(3), 1–20.

Richetin, J., Perugini, M., Adjali, I., & Hurling, R. (2008). Comparing leading theoretical models of behavioral predictions and post-behavior evaluations. *Psychology and Marketing, 25*(12), 1131–1150. doi:10.1002/mar.20257

Riddell, S., & Watson, N. (2003). *Disability, Culture and Identity*. Harlow, UK: Pearson Prentice Hall.

Rimpiläinen, S. (2011). Knowledge in Networks – Knowing in Transactions? *International Journal of Actor-Network Theory and Technological Innovation, 3*(2), 46–56. doi:10.4018/jantti.2011040104

Rivalland, J. (2000). Definitions & identification: Who are the children with learning difficulties? *Australian Journal of Learning Difficulties, 5*(2), 12–16. doi:10.1080/19404150009546621

Rodrigues, A. (2010). Effective Measurement of Time Performance using Earned Value Management – A proposed modified version for SPIU tested across various industries and project types. *PM World Today, 12*(10).

Rogers, E. (2003). *Diffusion of innovation*. New York: Free Press.

Rogers, E. M. (1962). *Diffusion of Innovations*. New York: The Free Press.

Rogers, E. M. (1976). New product adoption and diffusion. *The Journal of Consumer Research*, 290–301. doi:10.1086/208642

Rogers, E. M. (1995). *Diffusion of innovations (4ᵗʰedn.)*. New York: Free Press.

Rogers, E. M. (2010). *Diffusion of innovations*. New York: Simon and Schuster.

Rogers, E. M., & Shoemaker, F. F. (1971). *Communication of Innovations: A Cross-Cultural Approach*. Academic Press.

Rosebaugh, D. (2004). Getting ready for the software in your future. *Home Health Care Management Practice*. doi: 1011771084822303259877

Rosen, P., & Shermon, P. (2006). Hedonic Information Systems: Acceptance of social networking websites. In *Proceedings of AMCIS*. Acapulco, Mexico: AMCIS.

Rosenbloom, S. T. et al. (2006). Implementing Pediatric Growth Charts into an Electronic Health Record System. *Journal of the American Medical Informatics Association, 13*, 302–308. doi:10.1197/jamia.M1944 PMID:16501182

Ross, S. E., Schilling, L. M., Fernald, D. H., Davidson, A. J., & West, D. R. (2010). Health information exchange in small-to-medium sized family medicine practices: Motivators, barriers, and potential facilitators of adoption. *International Journal of Medical Informatics, 79*(2), 123–129. doi:10.1016/j.ijmedinf.2009.12.001 PMID:20061182

Rowan, L., & Bigum, C. (2009). What's your problem? ANT reflections on a research project studying girls enrolment in information technology subjects in post-compulsory education. *International Journal of Actor-Network Theory and Technological Innovation, 1*(4), 1–20. doi:10.4018/jantti.2009062301

Royal Australian College of General Practitioners. (2010). *Model Healthcare Community*. Retrieved August 26, 2011, from http://www.racgp.org.au/scriptcontent/ehealth/201002MHC_Handout.pdf

Rubas, D. J., Hill, H. S. J., & Mjelde, J. W. (2006). Economics and climate applications: Exploring the frontier. *Climate Research, 33*, 43–54. doi:10.3354/cr033043

Ruddin, L. (2006). You can generalize stupid! Social scientists, Bent Flyvbjerg, and case study methodology. *Qualitative Inquiry*, (4), 797–812. doi:10.1177/1077800406288622

Rusaw, C. A. (2007). Changing public organizations: Four approaches. *International Journal of Public Administration, 30*, 347–361. doi:10.1080/01900690601117853

Rydin, Y. (2010). Actor-network theory and planning theory: A response to Boelens. *Planning Theory, 9*, 265–268.

Samiee, S. (1998). The Internet and International Marketing: Is There a Fit? *Journal of Interactive Marketing, 12*(4), 5–21. doi:10.1002/(SICI)1520-6653(199823)12:4<5::AID-DIR2>3.0.CO;2-5

Sarker, S., Sarker, S., & Sidorova, A. (2006). Understanding Business Process Change Failure: An Actor-Network Perspective. *Journal of Management Information Systems, 23*(1), 51–86. doi:10.2753/MIS0742-1222230102

Scheer, A.-W. (2009). *BITCOM-Pressekonferenz E-Health*. Retrieved August 26, 2011, from http://www.bitkom.org/files/documents/BITKOM_Praesentation_E-Health_PK_04_03_2009.pdf

Schultze, U., & Orlikowski, W. (2004). A practice perspective on technology-mediated network relations: The use of Internet-based self-serve technologies. *Information Systems Research*, *15*(1), 87–106. doi:10.1287/isre.1030.0016

Schweitzer, S. (2005). When students open up – a little too much. *The Boston Globe*.

Sena, J. A. (2009). The Impact of Web 2.0 on Technology. *IJCSNS*, *9*(2), 378–385.

Shakespeare, T. (1999). Disability, Genetics and Global Justice. *Social Policy and Society*, *4*(1), 87–95. doi:10.1017/S1474746404002210

Sharifzadeh, M., Zamani, G. H., Karami, E., Khalili, D., & Tatnall, A. (2012). The Iranian wheat growers' climate information use: An Actor-Network Theory perspective. *International Journal of Actor-Network Theory and Technological Innovation*, *4*(4), 1–22. doi:10.4018/jantti.2012100101

Sharifzadeh, M., Zamani, G. H., Khalili, D., & Karami, E. (2012). Agricultural climate information use: An application of the planned behaviour theory. *Journal of Agricultural Science and Technology*, *14*, 479–492.

Shaw, S., Grimes, D., & Bulman, J. (2005). Educating Slow Learners: Are Charter Schools the Last, Best Hope for Their Educational Success?. *The Charter Schools Resource Journal*, *1*(1).

Sheehy, E. (Ed.). (2012). *Sexual assault in Canada: Law, legal practice, and women's activism*. Ottawa, Canada: University of Ottawa Press.

Sherrick, B. J., Sonka, S. T., Lamb, P. J., & Mazzocco, M. A. (2000). Decision-maker expectations and the value of climate prediction information: Conceptual considerations and preliminary evidence. *Meteorological Applications*, *7*(4), 377–386. doi:10.1017/S1350482700001584

Shoib, G., Nandhakumar, J., & Jones, M. (2006). Using social theory in information systems research: A reflexive account. In A. Ruth (Ed.), *Quality and Impact of Qualitative Research, Proceedings of the 3rd International Conference on Qualitative Research in IT & IT in Qualitative Research*. Brisbane, Australia: Institute for Integrated and Intelligent Systems, Griffith University.

Showell, C.M. (2011). Citizens, patients and policy: A challenge for Australia's national electronic health record. *Health Information Management Journal OnLine*, *40*(2).

Simon, H. A. (1996). *The Sciences of the Artificial* (3rd ed.). MIT Press.

Singh, M., & Waddell, D. (2006). Innovation and Entrepreneurial Opportunities with Internet Ventures. In *Proceedings of the European Conference on Entrepreneurship and Innovation*, (pp. 247 – 254). Academic Press.

Singh, M., Davison, C., & Wickramasinghe, N. (2010). Organisational Use of Web 2.0 Technologies: An Australian Perspective. In *Proceedings of AMCIS*. AMCIS. Retrieved from http://aisel.aisnet.org/amcis2010/

Singleton, V., & Michael, M. (1993). Actor-Networks and Ambivalence: General Practitioners in the UK Cervical Screening Programme. *Social Studies of Science*, *23*, 227–264. doi:10.1177/030631293023002001

Sismondo, S. (2010). *Introduction to science & technology studies*. Oxford, UK: Wiley-Blackwell.

Skeels, M. M., & Grudin, J. (2009). *When Social Networks Cross Boundaries: A Case Study of Workplace Use of Facebook and LinkedIn*. Paper presented at GROUP'09. Sanibel Island, FL.

Smart, C. (1989). *Feminism and the power of law*. New York: Routledge. doi:10.4324/9780203206164

Smith, D. (2005). *Institutional ethnography: A sociology for people*. New York: Altamira Press.

Solove, D. J. (2007). *The future of reputation: Gossip, Rumor, and Privacy on the Internet*. New Haven, CT: Yale University Press.

Sonnemans, P. J. M., Geudens, W. H. J., & Brombacher, A. C. (2003). Organizing product releases in time-driven development processes: A probabilistic analysis. *IMA Journal of Management Mathematics*, *14*, 337–356. doi:10.1093/imaman/14.4.337

Sophonthummapharn, K. (2008). *A comprehensive framework for the adoption of techno-relationship innovations: Empirical evidence from eCRM in manufacturing SMEs*. (University Dissertation). Handelshögskolan vid Umeåuniversitet.

Spertus, E., Sahami, M., & Buyukkokten, O. (2005). Evaluating similarity measures: A large-scale study in the Orkut social network. In *Proc. SIGKDD'05* (pp. 678-684). ACM.

Spohrer, J., & Maglio, P. P. (2008). The emergence of service science: Toward systematic service innovations to accelerate co-creation of value. *Production and Operations Management, 17*(3), 238–246. doi:10.3401/poms.1080.0027

Spohrer, J., Maglio, P. P., Bailey, J., & Gruhl, D. (2007). Steps Toward a Science of Service Systems. *IEEE Computer, 40*(1), 71–77. doi:10.1109/MC.2007.33

Spöhrer, M. (2013). The (re-)socialization of technical objects in patient networks: The case of the cochlear implant. *International Journal of Actor-Network Theory and Technological Innovation, 5*(3). doi:10.4018/jantti.2013070103

Star, S. L. (1991). Power, technologies and the phenomenology of conventions: On being allergic to onions. In J. Law (Ed.), *A sociology of monsters? Essays on power, technology and domination.* London: Routledge.

Star, S. L., & Griesemer, J. R. (1989). Institutional ecology, 'translations' and boundary objects: Amateurs and professionals in Berkeley's Museum of Vertebrate Zoology, 1907-39. *Social Studies of Science, 19*(3), 387–420. doi:10.1177/030631289019003001

Star, S. L., & Ruhleder, K. (1996). Steps Toward an Ecology of Infrastructure: Design and Access for Large Information Spaces. *Information Systems Research, 7*, 111–133. doi:10.1287/isre.7.1.111

Steinfield, C., DiMicco, J. M., Ellison, N. B., & Lampe, C. (2009). *Bowling Online: Social Networking and Social Capital within the Organization.* Paper presented at C&T'09. University Park, PA.

Stone, B. (2008, July 2). Our practical attitudes toward privacy. *The New York Times.*

Stross, R. (2009, March 8). When everyone's a friend, is anything private? *The New York Times.*

Subbiah, A. R., Kalsi, S. R., & Yap, K. (2004). *Climate information application forehancing: Resilience to climate risks.* Paper presented at the the International Committee of the Third International Workshop on Monsoons (IWM-III). Hangzhou, China.

Subrahmanyam, K., Reich, S. M., Waechter, N., & Espinoza, G. (2008). Online and offline social networks: Use of social networking sites by emerging adults. *Journal of Applied Developmental Psychology, 29*, 420–433. doi:10.1016/j.appdev.2008.07.003

Subramani, M. R., & Rajagopalan, B. (2003). Knowledge-Sharing and Influence in Online Social Networks via Viral Marketing. *Communications of the ACM, 46*(12). doi:10.1145/953460.953514

Suchman, L. A. (1987). *Plans and situated actions: The problem of human-machine communications.* Cambridge, UK: Cambridge University Press.

Swanson, E. B., & Ramiller, N. C. (2004). Innovating mindfully with information technology. *Management Information Systems Quarterly, 28*(4), 553–583.

Tang, P. C., Ash, J. S., Bates, D. W., Overhage, J. M., & Sands, D. Z. (2006). Personal health records: Definitions, benefits, and strategies for overcoming barriers to adoption. *Journal of the American Medical Informatics Association, 13*(2), 121–126. doi:10.1197/jamia.M2025 PMID:16357345

Tapscott, D., & Caston, A. (1993). *Paradigm shift: The new promise of information technology.* New York: McGraw-Hill, Inc.

Tatnall, A. (2000). *Innovation and change in the information systems curriculum of an Australian University: A socio-technical perspective.* (Dissertation). Central Queensland University.

Tatnall, A. (20002). *Innovation and Change in the Information Systems Curriculum of an Australian University: A Socio-Technical Perspective.* (PhD thesis). Central Queensland University, Rockhampton, Australia.

Tatnall, A. (2007a). *Business Culture and the Death of a Portal*. Paper presented at the 20th Bled e-Conference - eMergence: Merging and Emerging Technologies, Processes and Institutions. Bled, Slovenia.

Tatnall, A., & Gilding, A. (1999). Actor-Network Theory and Information Systems Research. In *Proceedings of the 10th Australasian Conference on Information Systems*. Wellington, New Zealand: Academic Press.

Tatnall, A., Michael, I., & Dakich, E. (2011). *The Victorian Schools Ultranet - An Australian eGovernment Initiative*. Paper presented at International Information Systems Conference - iiSC 2011. Muscat, Oman.

Tatnall, A. (1995). Information Technology and the Management of Victorian Schools - Providing Flexibility or Enabling Better Central Control? In *Information Technology in Educational Management*. London: Chapman & Hall. doi:10.1007/978-0-387-34839-1_13

Tatnall, A. (2002). Socio-Technical Processes in the Development of Electronic Commerce Curricula. In A. Wenn (Ed.), *Skilling the e-Business Professional* (pp. 9–21). Melbourne, Australia: Heidelberg Press.

Tatnall, A. (2005). Actor-Network Theory in Information Systems. In *Encyclopaedia of Information Science and Technology*. Hershey, PA: Idea Group Reference.

Tatnall, A. (2005). Technological Change in Small Organisations: An Innovation Translation Perspective. *International Journal of Knowledge. Culture and Change Management, 4*(1), 755–761.

Tatnall, A. (2007b). Innovation, Lifelong Learning and the ICT Professional. In *Education, Training and Lifelong Learning*. Laxenburg, Austria: IFIP.

Tatnall, A. (2009). Actor-Network Theory Applied to Information Systems Research. In M. Khosrow-Pour (Ed.), *Encyclopedia of Information Science and Technology* (2nd ed., Vol. 1, pp. 20–24). Hershey, PA: Information Science Reference.

Tatnall, A. (2009). Web Portal Research Issues. In *Encyclopedia of Information Science and Technology* (2nd ed.). Hershey, PA: Idea Group Reference.

Tatnall, A. (2011a). *Information Systems Research, Technological Innovation and Actor-Network Theory*. Melbourne, Australia: Heidelberg Press.

Tatnall, A. (2011b). Innovation Translation, Innovation Diffusion and the Technology Acceptance Model: Comparing Three Different Approaches to Theorising Technological Innovation. In *Actor-Network Theory and Technology Innovation: Advancements and New Concepts*. Hershey, PA: IGI Global.

Tatnall, A., & Burgess, S. (2006). Innovation Translation and E-Commerce in SMEs. In *Encyclopaedia of Information Science and Technology*. Hershey, PA: Idea Group Reference.

Tatnall, A., & Dakich, E. (2011). *Informing Parents with the Victorian Education Ultranet*. Novi Sad, Serbia. *Informing Science*.

Tatnall, A., Dakich, E., & Davey, W. (2011). The Ultranet as a Future Social Network: An Actor-Network Analysis. In *Proceedings of 24th Bled eConference*. University of Maribor.

Tatnall, A., & Davey, B. (2001). How Visual Basic Entered the Curriculum at an Australian University: An Account Informed by Innovation Translation. In *Challenges to Informing Clients: A Transdisciplinary Approach*. Krakow, Poland: Academic Press.

Tatnall, A., & Davey, B. (2002). Understanding the Process of Information Systems and ICT Curriculum Development: Three Models. In *Human Choice and Computers: Issues of Choice and Quality of Life in the Information Society*. Kluwer Academic Publishers. doi:10.1007/978-0-387-35609-9_23

Tatnall, A., & Davey, B. (2003b). Modelling the Adoption of Web-Based Mobile Learning - An Innovation Translation Approach. In *Advances in Web-Based Learning*. Springer Verlag. doi:10.1007/978-3-540-45200-3_40

Tatnall, A., & Davey, W. (2007). *Researching the Portal*. Vancouver, Canada: Information Management Resource Association.

Tatnall, A., & Davey, W. (2013). The Ultranet and School Management: Creating a new management paradigm for education. In *Next Generation of Information Technology and Educational Management*. Heidleberg, Germany: Springer. doi:10.1007/978-3-642-38411-0_15

Tatnall, A., & Lepa, J. (2003). The Internet, e-commerce and older people: an actor-network approach to researching reasons for adoption and use. *Logistics Information Management*, *16*(1), 56–63. doi:10.1108/09576050310453741

Tatnall, C., & Tatnall, A. (2007). Using Educational Management Systems to Enhance Teaching and Learning in the Classroom: An Investigative Study. In *Knowledge Management for Educational Innovation*. New York: Springer. doi:10.1007/978-0-387-69312-5_10

Taylor, H. (2006). Risk Management and Problem Resolution Strategies for IT Projects. *Project Management Journal*, *37*(5), 49–63.

Tesch, D., Kloppenborg, T., & Frolick, M. N. (2007). IT project risk factors: The project management professional's perspective. *Journal of Computer Information Systems*, *47*(4), 61–69.

Tetteh, G. K., & Snaith, J. (2006). Information system strategy: Applying galliers and Sutherland's stages of growth model in a developing country. *The Consortium Journal*, *11*(1), 5–16.

Thompson, P., & Alvesson, M. (2005). Bureaucracy at Work: Misunderstandings and Mixed Blessings. In P. Du Gay (Ed.), *The Values of Bureaucracy*. Oxford University Press.

Thompson, T. G., & Brailer, D. J. (2004). *The decade of health information technology: Delivering consumer-centric and information-rich health care. Framework for Strategic Action, Office for the National Coordinator for Health Information Technology (ONCHIT)*. Department of Health and Human Services, and the United States Federal Government.

Thweatt, E. G., & Kleiner, B. H. (2007). New Developments in Health Care Organisational Management. *Journal of Health Management*, *9*(3), 433–441. doi:10.1177/097206340700900308

Timpka, T. et al. (2007). Information infrastructure for inter-organizational mental health services: An actor network theory analysis of psychiatric rehabilitation. *Journal of Biomedical Informatics*, *40*(4), 429–437. doi:10.1016/j.jbi.2006.11.001 PMID:17182285

Tobler, N. (2008). *Technology, organizational change, and the nonhuman agent: Exploratory analysis of electronic health record implementation in a small practice ambulatory care*. The University of Utah.

Toffler, A. (1970). *Future shock*. New York: Bantam Books.

Topi, H., Valacich, J. S., Wright, R. T., Kaiser, K. M., Nunamaker, J. F., Sipior, J. C., & Vreede, G. J. (2012). *IS 2010–Curriculum Guidelines for Undergraduate Degree Programs in Information Systems. Association for Information Systems*. AIS.

Torda, P., Han, E. S., & Scholle, S. H. (2010). Easing the adoption and use of electronic health records in small practices. *Health Affairs (Project Hope)*, *29*(4), 668–675. doi:10.1377/hlthaff.2010.0188 PMID:20368597

Trudel, M. C. (2010). *Challenges to personal information sharing in interorganizational settings: Learning from the Quebec Health Smart Card project*. The University of Western Ontario.

Tuman, J. (1993). *Project management decision-making and risk management in a changing corporate environment*. Paper presented at the Project Management Institute 24th Annual Seminar/Symposium. Vancouver, Canada.

Tummons, J. (2009). Higher Education in Further Education in England: An Actor-Network Ethnography. *International Journal of Actor-Network Theory and Technological Innovation*, *1*(3), 55–69. doi:10.4018/jantti.2009070104

Turban, E., Bolloju, N., & Liang, T. (2011). Enterprise Social Networking: Opportunities, Adoption, and Risk Mitigation. *Journal of Organizational Computing and Electronic Commerce*, *21*(3), 202–220. doi:10.1080/10919392.2011.590109

Turban, E., Lee, J. K., King, D., Liang, T. P., & Turban, D. (2010). *Electronic Commerce: A managerial perspective* (6th ed.). Prentice Hall.

Turner, J. H. (1986). The Theory of Structuration. *American Journal of Sociology*, *91*(4), 969–977. doi:10.1086/228358

Uden, L., & Francis, J. (2009). Actor-Network Theory for service innovation. *International Journal of Actor-Network Theory and Technological Innovation*, *1*(1), 23–44. doi:10.4018/jantti.2009010102

United Nations Conference on trade and Development. (2002). Retrieved from http://r0.unctad.org/ecommerce/ecommerce_en

Urry, J. (2003). *Global Complexities*. Polity Press.

Valenzuela, S., Park, N., & Kee, K. F. (2009). Is There Social Capital in a Social Network Site? Facebook Use and College Students' Life Satisfaction, Trust, and Participation. *Journal of Computer-Mediated Communication*, *14*(4), 875–901. doi:10.1111/j.1083-6101.2009.01474.x

Van den Hooff, B., & de Winter, M. (2011). Us and them: A social capital perspective on the relationship between business and IT departments. *European Journal of Information Systems*, *20*(3), 255–266. doi:10.1057/ejis.2011.4

Van House, N. A. (2009). Collocated photo sharing, storytelling, and the performance of self. *International Journal of Human-Computer Studies archive*, *67*(12), 1073-1086.

Vargo, S. L., & Lusch, R. F. (2008). From Goods to Service(s), Divergences and Convergences of Logics. *Industrial Marketing Management*, *37*, 254–259. doi:10.1016/j.indmarman.2007.07.004

vdek – Verband der Ersatzkassen e. V. (2009). *Fragen und Antworten zur elektronischen Gesundheitskarte (eGK)*. Retrieved August 26, 2011, from http://www.vdek.com/presse/Fragen_und_Antworten/egk.pdf

Venkatesh, V., Morris, M. G., Davis, G. B., & Davis, F. D. (2003). User Acceptance of Information Technology: Toward a Unified View. *Management Information Systems Quarterly*, *27*(3), 425–478.

Venkatesh, V., & Thong, J., & Xu. (2012). Consumer acceptance and use of information technology: Extending the unified theory of acceptance and use of technology. *Management Information Systems Quarterly*, *36*(1), 157–178.

Verband der privaten Krankenversicherung e. V. (2011). *Die elektronische Gesundheitskarte in der privaten Krankenversicherung*. Retrieved August 26, 2011, from http://www.pkv.de/zahlen/

Verenikina, I. (2001). Cultural-historical psychology and activity theory in everyday practice. In information systems and activity theory: Volume 2 theory and practice. Wollongong, Australia: University of Wollongong Press.

Vitacca, M., Mazzù, M., & Scalvini, S. (2009). Socio-technical and organizational challenges to wider e-Health implementation. *Chronic Respiratory Disease*, *6*(2), 91–97. doi:10.1177/1479972309102805 PMID:19411570

Von Hippel, E. (1988). *The sources of innovation*. New York: Oxford University Press.

von Lubitz, D. K. J. E., & Wickramasinghe, N. (2006). Healthcare and technology: The doctrine of network centric healthcare. *International Journal of Electronic Healthcare*, *2*(4), 322–344. PMID:18048253

Wajcman, J. (2000). Reflections on gender and technology studies: In what state is the art? *Social Studies of Science*, *30*(3), 447–464. doi:10.1177/030631200030003005

Walker, S. K., & Riley, D. A. (2001). Involvement of the Personal Social Network as a Factor in Parent Education Effectiveness. *Family Relations*, *50*(2), 186–193. doi:10.1111/j.1741-3729.2001.00186.x

Walsham, G. (1995). Interpretive case studies in IS research. *European Journal of Information Systems*, *4*, 74–81. doi:10.1057/ejis.1995.9

Walsham, G. (1997). Actor-network theory and IS research: Current status and future prospects. In A. S. Lee, J. Liebenau, & J. I. DeGross (Eds.), *Information Systems and Qualitative Research*. Chapman and Hall. doi:10.1007/978-0-387-35309-8_23

Walsham, G., & Sahay, S. (1999). GIS for district-level administration in India: Problems and opportunities. *Management Information Systems Quarterly*, *23*(1), 39–66. doi:10.2307/249409

Ward, J. C., & Ostrom, A. L. (2006). Complaining to the masses: The role of protest framing in customer-created complaint web sites. *The Journal of Consumer Research*, *33*(2), 220–230. doi:10.1086/506303

Weber, M. (1919). Bureaucracy. In *From Max Weber: Essays in sociology*. London: Routledge.

Weill, P., Subramani, M., & Broadbent, M. (2002). Building IT infrastructure for strategic agility. *Sloan Management Review, 44*(1), 57–65.

Weimar, C. (2009). Electronic health care advances, physician frustration grows. *Physician Executive, 35*(2), 8–15. PMID:19452840

Wernick, P., Hall, T., & Nehaniv, C. L. (2008). Software evolutionary dynamics modelled as the activity of an actor-network. *The Institution of Engineering and Technology, 2*(4), 321–336.

Westbrook, J. I., & Braithwaite, J. (2010). Will Information and Communication Technology Disrupt the Health System and Deliver on Its Promise? *The Medical Journal of Australia, 193*(7), 399–400. PMID:20919970

Whitley, E. A., & Darking, M. (2006). Object lessons and invisible technologies. *Journal of Information Technology, 21*, 176–184. doi:10.1057/palgrave.jit.2000065

Whittaker. (2008). Retrieved from http://www.webuser.co.uk/news/top-stories/452168/teens-create-secret-social-network-language

Whitworth, B., & de Moor, A. (2004). Legitimate by Design: Towards Trusted Virtual Community Environments. In *Proceedings of 35th Hawaii International Conference on System Sciences*. IEEE.

WHO – World Health Organization. (2005). *What is e-health?* Retrieved August 26, 2011, from http://www.emro.who.int/his/ehealth/AboutEhealth.htm

Wickramasinghe, N. (2010). *Christmas Carol for IBM*. Paper presented at IBM Healthcare Executives Dinner. New York, NY.

Wickramasinghe, N., & Goldberg, S. (2003). The Wireless Panacea for Healthcare. In *Proceedings of the 36th Hawaii International Conference on System Sciences (HICSS-35)*. IEEE.

Wickramasinghe, N., & Schaffer, J. (2010). *Realizing Value Driven e-Health Solutions*. IBM centre for the Business of government.

Wickramasinghe, N., Fadllala, A. M. A., Geisler, E., & Schaffer, J. L. (2005). A framework for assessing e-health preparedness. *International Journal e-Health, 1*(3), 316–334.

Wickramasinghe, N., Tatnall, A., & Goldberg, S. (2011). The Advantages Of Mobile Solutions For Chronic Disease Management. *PACIS 2011 Proceedings*. Retrieved from http://aisel.aisnet.org/pacis2011/214.

Wickramasinghe, N. (2000). IS/IT as a Tool to Achieve Goal Alignment: A theoretical framework. *International Journal of Healthcare Technology and Management, 2*(1-4), 163–180. doi:10.1504/IJHTM.2000.001089

Wickramasinghe, N. (2007). Fostering knowledge assets in healthcare with the KMI model. *International Journal of Management and Enterprise Development, 4*(1), 52–65. doi:10.1504/IJMED.2007.011455

Wickramasinghe, N., & Bali, R. K. (2009). The S'ANT Imperative for Realizing the Vision of Healthcare Network Centric Operations. *International Journal of Actor-Network Theory and Technological Innovation, 1*(1), 45–58. doi:10.4018/jantti.2009010103

Wickramasinghe, N., Bali, R. K., & Goldberg, S. (2009). The S'ANT approach to facilitate a superior chronic disease self-management model. *International Journal of Actor-Network Theory and Technological Innovation, 1*(4), 21–34. doi:10.4018/jantti.2009062302

Wickramasinghe, N., Bali, R. K., & Lehaney, B. (2009). *Knowledge Management Primer*. Taylor & Francis.

Wickramasinghe, N., Bali, R. K., & Tatnall, A. (2007). Using actor network theory to understand network centric healthcare operations. *International Journal of Electronic Healthcare, 3*(3), 317–328. doi:10.1504/IJEH.2007.014551 PMID:18048305

Wickramasinghe, N., Bali, R., Kirn, S., & Sumoi, R. (2012). *Critical Issues for the Development of Sustainable E-health Solutions*. New York: Springer. doi:10.1007/978-1-4614-1536-7

Wickramasinghe, N., & Goldberg, S. (2004). How M=EC2 in Healthcare. *International Journal of Mobile Communications, 2*(2), 140–156. doi:10.1504/IJMC.2004.004664

Wickramasinghe, N., & Goldberg, S. (2007a). Adaptive mapping to realisation methodology (AMR) to facilitate mobile initiatives in healthcare. *International Journal of Mobile Communications*, *5*(3), 300–318. doi:10.1504/IJMC.2007.012396

Wickramasinghe, N., Goldberg, S., & Bali, R. (2008). Enabling superior m-health project success: A tri-country validation. *International Journal of Services and Standards*, *4*(1), 97–117. doi:10.1504/IJSS.2008.016087

Wickramasinghe, N., & Mills, G. (2001). MARS: The Electronic Medical Record System The Core of the Kaiser Galaxy. *International Journal of Healthcare Technology and Management*, *3*(5/6), 406–423. doi:10.1504/IJHTM.2001.001119

Wickramasinghe, N., & Misra, S. (2004). A Wireless Trust Model for Healthcare. *International Journal of Electronic Healthcare*, *1*(1), 60–77. doi:10.1504/IJEH.2004.004658 PMID:18048204

Wickramasinghe, N., & Schaffer, J. (2006). Creating Knowledge Driven Healthcare Processes With The Intelligence *Continuum. International Journal of Electronic Healthcare*, *2*(2), 164–174. PMID:18048242

Wickramasinghe, N., & Schaffer, J. L. (2010). *Realizing Value Driven e-Health Solutions*. Washington, DC: IBM Center for the Business of Government, Improving Healthcare Series.

Wickramasinghe, N., Schaffer, J., & Geisler, E. (2005). Assessing e-health. In T. Spil, & R. Schuring (Eds.), *E-Health Systems Diffusion and Use: The Innovation, The User and the User IT Model*. Hershey, PA: Idea Group Publishing. doi:10.4018/978-1-59140-423-1.ch017

Wickramasinghe, N., & Silvers, J. B. (2003). IS/IT The Prescription To Enable Medical Group Practices To Manage Managed Care. *Health Care Management Science*, *6*, 75–86. doi:10.1023/A:1023376801767 PMID:12733611

Wickramasinghe, N., Tumu, S., Bali, R. K., & Tatnall, A. (2007). Using actor-network theory (ANT) as an analytic tool in order to effect superior PACS implementation. *International Journal of Networking and Virtual Organisations*, *4*(3), 257–279. doi:10.1504/IJNVO.2007.015164

Wikipedia. (2010). *Wikipedia About*. Retrieved from http://en.wikipedia.org/wiki/Help:About

Wild, S., Roglic, G., Green, A., Sicree, R., & King, H. (2004). Global prevalence of diabetes: Estimates for the year 2000 and projections for 2030. *Diabetes Care*, *27*(5), 1047–1053. doi:10.2337/diacare.27.5.1047 PMID:15111519

Willcocks, L. (1994). Managing information system in the UK public administration: issues and prospects. *Public Administration*, *72*, 13–32. doi:10.1111/j.1467-9299.1994.tb00997.x

Williams, R. (2007). Managing complex adaptive networks. In *Proceedings of the 4th International Conference on Intellectual Capital, Knowledge Management & Organizational Learning* (pp. 441–452). Academic Press.

Winner, L. (1977). *Autonomous Technology*. Cambridge, MA: MIT Press.

Winner, L. (1993). Upon opening the black box and finding it empty: Social constructivism and the philosophy of technology. *Science, Technology & Human Values*, *18*(3), 362–378. doi:10.1177/016224399301800306

Woolgar, S. (1991). The turn to technology in social studies of science. *Science, Technology & Human Values*, *16*(1), 20–50. doi:10.1177/016224399101600102

World Health Organization (WHO). (2003). Retrieved February 23, 2010 from http://www.emro.who.int/ehealth/

Wortham, J. (2009, April 17). With Oprah onboard, Twitter grows. *The New York Times*.

Wu, L. (2013). Social Network Effects on Productivity and Job Security: Evidence from the Adoption of a Social Networking Tool. *Information Systems Research*, *24*(1), 30–51. doi:10.1287/isre.1120.0465

Yin, R. (1994). *Case Study Research: Design and Methods* (2nd ed.). Sage Publications.

Yin, R. K. (1999a). Case study research, design and methods. *Sage (Atlanta, Ga.)*.

Yin, R. K. (1999b). Enhancing the quality of case studies in health services research. *Health Services Research*, *34*(1), 1209–1224. PMID:10591280

Yin, R. K. (2002). *Case Study Research*. Design and Methods.

Yin, R. K. (2009). *Case study research: Design and methods* (4th ed.). Thousand Oaks, CA: Sage.

Yusof, M., Stergioulas, L., & Zugic, J. (2007). *Health information systems adoption: Findings from a systematic review*. Academic Press.

Zachry, M. (1999). Constructing usable documentation: A study of communicative practices and the early uses of mainframe computing in industry. In *Proceedings of the 17th Annual International Conference on Computer Documentation* (pp. 22-25). New Orleans, LA: ACM Press.

Zhang, W., Johnson, T. J., Seltzer, T., & Bichard, S. L. (2010). The Revolution Will be Networked: The Influence of Social Networking Sites on Political Attitudes and Behavior. *Social Science Computer Review*, 28, 75. doi:10.1177/0894439309335162

Ziervogel, G. (2004). Targeting seasonal climate forecasts for integration into household level decisions: The case of smallholder farmers in Lesotho. *The Geographical Journal*, *170*(1), 6–21. doi:10.1111/j.0016-7398.2004.05002.x

Ziervogel, G., Bithell, M., Washington, R., & Downing, T. (2005). Agent-based social simulation: A method for assessing the impact of seasonal climate forecast applications among smallholder farmers. *Agricultural Systems*, *83*(1), 1–26. doi:10.1016/j.agsy.2004.02.009

Ziervogel, G., & Downing, T. E. (2004). Stakeholder networks: Improving seasonal climate forecasts. *Climatic Change*, *65*, 73–101. doi:10.1023/B:CLIM.0000037492.18679.9e

Zigmond, N., Jenkins, J., Fuchs, L. S., Deno, S., Fuchs, D., & Baker, J. N. et al. (1995). Special education in restructured schools: Findings from three multi-year studies. *Phi Delta Kappan*, *76*, 531–540.

Zimmer, B. (2009). *The Language of Social Media: Unlike Any Other*. Retrieved from http://www.visualthesaurus.com/cm/wordroutes/1845/

ZTG – Zentrum für Telematik im Gesundheitswesen GmbH. (2010). *Notfalldaten*. Retrieved August 26, 2011, from http://www.egesundheit.nrw.de/content/elektronische_gesundheitskarte/e3338/e3380/index_ger.html

About the Contributors

Arthur Tatnall is an Associate Professor in Information Systems at Victoria University in Melbourne. His PhD used actor-network theory to investigate adoption of Visual Basic in the curriculum of an Australian university. His research interests include technological innovation, history of technology, project management, information systems curriculum, and IT in educational management.

* * *

Tas Adam has recently retired from the position of Senior Lecturer in the School of Management and Information Systems at Victoria University, Melbourne. His PhD is titled "Determining an e-learning Model for Students with Learning Disabilities: An Analysis of Web-Based Technologies and Curriculum," and he has worked extensively in this area.

Arthur Adamopoulos is a Lecturer at RMIT University, Melbourne, Australia in the school of Business Information Technology and Logistics. He is currently researching technology adoption and usage, with a specific focus on online investing.

Noel Carroll completed his PhD with the Department of Computer Science and Information Systems at the University of Limerick, Ireland. His research focuses on Service Science initiatives and examines methods to investigate the value of IT on service networks.

Bill Davey is a Senior Lecturer in Business Information Technology at RMIT University, Melbourne, Australia. He holds degrees in Science, Education, Computing and a Master of Business. His research interests include methodologies for systems analysis and development, Visual Basic programming, information systems curriculum, and information technology in educational management.

Martin Dick is a Senior Lecturer at RMIT University, Melbourne, Australia in the school of Business Information Technology and Logistics.

José Figueiredo has a PhD in Industrial Engineering and Management with a thesis "Sociotechnical Approaches to Inter-Institutional Information Systems Design," MBA in Information Management, Engineer Degree in Electronics and Digital Systems. He is a Professor at the Engineering and Management Department (DEG) of IST, Technical University of Lisbon, currently teaching Project and Knowledge Management.

Steve Goldberg founded INET International Inc. in 1998. As INET's President, he leads the firm in data collection for international research studies, wireless healthcare programs, and INET mini-conferences. He is on the Editorial Board of the *International Journal of Networking and Virtual Organizations* and has many peer-reviewed papers.

Fernando Abreu Gonçalves has a PhD in Industrial Engineering and Management with a thesis on "Actor-Network Theory in Management and Engineering Design," MBA in Information Management, Engineer Degree in Chemical Engineer. He is Emeritus Professor from ISCAL Accounting and Administration Institute, Polytechnic Institute of Lisbon.

Tiko Iyamu holds a PhD in Information Systems and is currently Professor of Informatics and Chair of Health Informatics at the Namibia University of Science and Technology, Windhoek, Namibia. Prior to his fulltime appointment in academic in 2009, Tiko held several positions in both Public and Private Institutions in South Africa.

Ezatollah Karami is Professor of Agricultural Development and Extension at College of Agriculture, Shiraz University, Iran. He has conducted research and published widely on issues of agricultural extension and sustainable agriculture.

Davar Khalili is Associate Professor in the Department of Water Engineering, College of Agriculture, Shiraz University. His research interests include system engineering, hydrologic systems, developing efficient and realistic operating policies for multi-purpose reservoir systems, decision methodology for the resource utilization of rangeland watersheds, and stochastic dynamic games in solving river basin disputes.

Jayan Kurian is a lecturer at RMIT University, Vietnam.

Awie Leonard joined the University of Pretoria in 1992 as senior lecturer after spending several years in the private sector as programmer/systems analyst, project manager, systems development, and maintenance manager. Apart from lecturing a variety of Information Technology subjects, his research focus is in the broad field of IT Service Management.

Imran Muhammad is a PhD student at RMIT University, Australia. His primary research interests are in the area of business information systems especially in the area of healthcare information systems management and electronic health records.

Rennie Naidoo is a Senior Lecturer in Information Systems at the University of the Pretoria. His research interest is in applying a practice lens using social theories such as actor-network theory and structuration to better understand general IS management, IS design, IS implementation, IS decision-making, and IS Human Resources challenges.

Andrea Quinlan is a PhD Candidate at York University, Canada. Her research examines interconnections between law, science, technology, and sexual violence. Her PhD draws on Actor-Network Theory and feminist theory to examine the Sexual Assault Evidence Kit, a tool used to collect forensic evidence in legal cases of sexual assault.

Petronnell Sehlola is an IT practitioner who started her career as a software developer and currently provides consulting services to Government Administration in South Africa. Her focus includes IT Projects, Business Analysis, and Risk Management. She is a Business Information System postgraduate student at Tshwane University of Technology, South Africa.

Juergen Seitz received his PhD from Viadrina European University, Frankfurt (Oder), Germany. He is Professor for Business Information Systems and Finance, and Chair of the Business Information Systems Department at Baden-Wuerttemberg Cooperative State University Heidenheim, Germany. Dr. Seitz is also editor, associate editor, and editorial board member of several international journals.

Tefo Sekgweleo is from the Department of Informatics, Tshwane University of Technology, Pretoria, South Africa.

Maryam Sharifzadeh is an Assistant Professor in the Rural Management Department, Yasouj University. Her research interests and experience have been concentrated on human and non-human actors of agricultural climate networks, network analysis tools, and information systems, especially in agricultural development processes in Iran.

Mohini Singh is Professor of Information Systems at RMIT University in Australia. She has published well over 100 scholarly papers in the areas of e-business, e-government, and new technology and innovation management. Her current research focus is on social media and Websites.

Nilmini Wickramasinghe received her PhD from Case Western Reserve University, USA and currently is the Epworth Chair in Health Information Management and a Professor at RMIT University, Australia. She researches within the IS domain with a special focus on IS/IT solutions to effect superior, patient-centric healthcare delivery.

Gholam H. Zamani is a Professor in the Agricultural Extension and Education Department at the Shiraz University. His research emphasizes the importance of educating human resource in development process.

Manuel Zwicker is a PhD student at RMIT University, Australia. His research is in business information systems, especially in e-health. He received his MBA from ESB Business School, Reutlingen, Germany and his diploma degree from University of Cooperative Education Heidenheim, Germany (nowadays Baden-Wuerttemberg Cooperative State University Heidenheim, Germany).

Index